Basic Electrical Installation Work

To Joyce, Samantha and Victoria

Basic Electrical Installation Work

FIFTH EDITION

TREVOR LINSLEY

Senior Lecturer
Blackpool and The Fylde College

AMSTERDAM • BOSTON • HEIDELBERG • LONDON • NEW YORK • OXFORD
PARIS • SAN DIEGO • SAN FRANCISCO • SINGAPORE • SYDNEY • TOKYO
Newnes is an imprint of Elsevier

Newnes
An imprint of Elsevier
Linacre House, Jordan Hill, Oxford OX2 8DP
30 Corporate Drive, Burlington, MA 01803

First published by Arnold 1998
Reprinted by Butterworth-Heinemann 2001, 2002, 2003 (twice), 2004 (twice)
Fourth edition 2005
Fifth edition 2008

Written to meet the requirements of the 2330 Level 2 Certificate in Electrotechnical Technology – Installation route
(buildings and structures), and the Level 2 NVQ in Installing Electrotechnical Systems (2356) from City & Guids.

British Library Cataloguing in Publication Data
A catalogue record for this book is available from the British Library

For information on all Newnes publications
visit our website at www.newnespress.com

ISBN 978-0-7506-8751-5

Typeset by Charon Tec Ltd., A Macmillan Company (www.macmillansolutions.com)

Printed and bound in Slovenia

Working together to grow
libraries in developing countries

www.elsevier.com | www.bookaid.org | www.sabre.org

ELSEVIER   BOOK AID International   Sabre Foundation

Contents

Preface

The 5th Edition of *Basic Electrical Installation Work* has been completely rewritten in 14 Chapters to closely match the 14 Outcomes of the City and Guilds qualification. The technical content has been revised and updated to the requirements of the new 17th Edition of the IEE Regulations BS 7671: 2008. Improved page design with new coloured illustrations give greater clarity to each topic.

This book of electrical installation theory and practice will be of value to the electrical trainee working towards:

- The City and Guilds 2330 Level 2 Certificate in Electrical Technology, Installation Route.
- The City and Guilds 2356 Level 2 NVQ in Installing Electrotechnical Systems.
- The SCOTVEC and BTEC Electrical Utilisation Units at Levels I and II.
- Those taking Engineering and modern Apprenticeship Courses.

Basic Electrical Installation Work provides a sound basic knowledge of electrical practice which other trades in the construction industry will find of value, particularly those involved in multi-skilling activities.

The book incorporates the requirements of the latest Regulations, particularly:

- 17th Edition IEE Wiring Regulations.
- British Standards BS 7671: 2008.
- Part P of the Building Regulation, Electrical Safety in Dwellings: 2006.
- Hazardous Waste Regulations: 2005.
- Work at Height Regulations: 2005.

Trevor Linsley
2008

Acknowledgements

I would like to acknowledge the assistance given by the following manufacturers and professional organizations in the preparation of this book:

- The Institution of Engineering and Technology for permission to reproduce Regulations and Tables from the 17th Edition IEE Regulations.

- The British Standards Institution for permission to reproduce material from BS 7671: 2008.

- Crabtree Electrical Industries for technical information and data.

- RS Components Limited for technical information and photographs.

- Stocksigns Limited for technical information and photographs.

- Wylex Electrical Components for technical information and photographs.

- Jason Vann Smith MIET MIEEE MBCS BOOKS for the photograph used in the page design.

I would like to thank the many College Lecturers who responded to the questionnaire from Elsevier the publishers, regarding the proposed new edition of this book. Their recommendations have been taken into account in producing this improved 5th Edition.

I would also like to thank the editorial and production staff at Elsevier the publishers for their enthusiasm and support. They were able to publish this 5th Edition within the very short timescale created by the publication of the 17th Edition of the IEE Regulations.

Finally, I would like to thank Joyce, Samantha and Victoria for their support and encouragement.

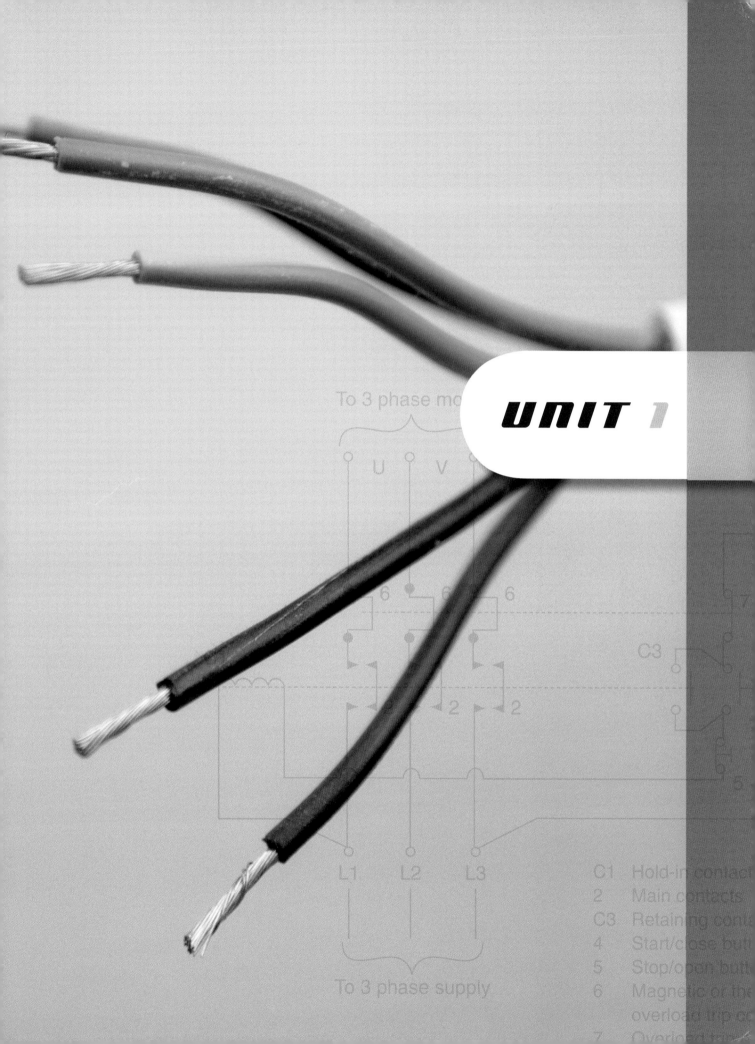

UNIT 1

To 3 phase mo

U   V

6   6   6

C3

2   2

L1   L2   L3

C1   Hold-in contac
2   Main contacts
C3   Retaining cont
4   Start/close but
5   Stop/open butt
6   Magnetic or t
     overload trip
7   Over

To 3 phase supply

The legal responsibilities of both employers and employees

Unit 1 – Working effectively and safely in the electrotechnical environment – Outcome 1

Underpinning knowledge: when you have completed this chapter you should be able to:

- state the legal responsibilities under the Health and Safety at Work Act
- identify statutory and non-statutory regulations
- identify PPE for the task being carried out
- state the need for electrical isolation
- identify the types and meanings of safety signs
- identify the types of fire extinguisher
- state the actions to be taken following an electric shock

Safety regulations and laws

At the beginning of the nineteenth century children formed a large part of the working population of Great Britain. They started work early in their lives and they worked long hours for unscrupulous employers or masters.

The Health and Morals of Apprentices Act of 1802 was introduced by Robert Peel in an attempt at reducing apprentice working hours to 12 h per day and improving the conditions of their employment. The Factories Act of 1833 restricted the working week for children aged 13–18 years to 69 h in any working week.

With the introduction of the Factories Act of 1833, the first four full time Factory Inspectors were appointed. They were allowed to employ a small number of assistants and were given the responsibility of inspecting factories throughout England, Scotland, Ireland and Wales. This small overworked band of men were the forerunners of the modern HSE Inspectorate, enforcing the safety laws passed by Parliament. As the years progressed, new Acts of Parliament increased the powers of the Inspectorate and the growing strength of the trade unions meant that employers were increasingly being pressed to improve health, safety and welfare at work.

The most important recent piece of health and safety law was passed by Parliament in 1974 called the Health and Safety at Work Act. This Act gave added powers to the Inspectorate and is the basis of all modern statutory health and safety laws. This Law not only increased the employer's liability for safety measures, but also put the responsibility for safety on employees too.

Health, safety and welfare legislation has increased the awareness of everyone to the risks involved in the workplace. All statutes within the Acts of Parliament must be obeyed and, therefore, we all need an understanding of the laws as they apply to our electrotechnical industry.

Statutory laws

Acts of Parliament are made up of Statutes. **Statutory Regulations** have been passed by Parliament and have, therefore, become laws. Non-compliance with the laws of this land may lead to prosecution by the Courts and possible imprisonment for offenders.

We shall now look at eight Statutory Regulations as they apply to the electrotechnical industry.

Definition

Statutory Regulations have been passed by Parliament and have, therefore, become laws.

The Health and Safety at Work Act 1974

Many governments have passed laws aimed at improving safety at work, but the most important recent legislation has been the Health and Safety at Work Act 1974. The purpose of the Act is to provide the legal framework for stimulating and encouraging high standards of health and safety at work; the Act puts the responsibility for safety at work on both workers and managers.

4

The employer has a duty to care for the health and safety of employees (Section 2 of the Act). To do this he must ensure that:

- the working conditions and standard of hygiene are appropriate;

- the plant, tools and equipment are properly maintained;

- the necessary safety equipment – such as personal protective equipment (PPE), dust and fume extractors and machine guards – is available and properly used;

- the workers are trained to use equipment and plant safely.

Employees have a duty to care for their own health and safety and that of others who may be affected by their actions (Section 7 of the Act). To do this they must:

- take reasonable care to avoid injury to themselves or others as a result of their work activity;

- co-operate with their employer, helping him or her to comply with the requirements of the Act;

- not interfere with or misuse anything provided to protect their health and safety.

Failure to comply with the Health and Safety at Work Act is a criminal offence and any infringement of the law can result in heavy fines, a prison sentence or both.

Enforcement

Laws and rules must be enforced if they are to be effective. The system of control under the Health and Safety at Work Act comes from the Health and Safety Executive (HSE) which is charged with enforcing the law. The HSE is divided into a number of specialist inspectorates or sections which operate from local offices throughout the United Kingdom. From the local offices the inspectors visit individual places of work.

The HSE inspectors have been given wide-ranging powers to assist them in the enforcement of the law. They can:

1. enter premises unannounced and carry out investigations, take measurements or photographs;

2. take statements from individuals;

3. check the records and documents required by legislation;

4. give information and advice to an employee or employer about safety in the workplace;

5. demand the dismantling or destruction of any equipment, material or substance likely to cause immediate serious injury;

6. issue an improvement notice which will require an employer to put right, within a specified period of time, a minor infringement of the legislation;

7. issue a prohibition notice which will require an employer to stop immediately any activity likely to result in serious injury, and which will be enforced until the situation is corrected;

8. prosecute all persons who fail to comply with their safety duties, including employers, employees, designers, manufacturers, suppliers and the self-employed.

Safety documentation

Under the Health and Safety at Work Act, the employer is responsible for ensuring that adequate instruction and information is given to employees to make them safety conscious. Part 1, Section 3 of the Act instructs all employers to prepare a written health and safety policy statement and to bring this to the notice of all employees. Figure 1.1 shows a typical Health and Safety Policy Statement of the type which will be available within your Company. Your employer must let you know who your safety representatives are and the new Health and Safety poster shown in Fig. 1.2 has a blank section into which the names and contact information of your specific representatives can be added. This is a large laminated poster, $595 \times 415\,\text{mm}$ suitable for wall or notice board display.

All workplaces employing five or more people must display the type of poster shown in Fig. 1.2 after 30th June 2000.

To promote adequate health and safety measures the employer must consult with the employees' safety representatives. In companies which employ more than 20 people this is normally undertaken by forming a safety committee which is made up of a safety officer and employee representatives, usually nominated by a trade union. The safety officer is usually employed full-time in that role. Small companies might employ a safety supervisor, who will have other duties within the company, or alternatively they could join a 'safety group'. The safety group then shares the cost of employing a safety adviser or safety officer, who visits each company in rotation. An employee who identifies a dangerous situation should initially report to his site safety representative. The safety representative should then bring the dangerous situation to the notice of the safety committee for action which will remove the danger. This may mean changing company policy or procedures or making modifications to equipment. All actions of the safety committee should be documented and recorded as evidence that the company takes seriously its health and safety policy.

The Electricity Safety, Quality and Continuity Regulations 2002 (formerly Electricity Supply Regulations 1989)

The Electricity Safety, Quality and Continuity Regulations 2002 are issued by the Department of Trade and Industry. They are statutory regulations which are enforceable by the laws of the land. They are designed to ensure a proper and safe supply of electrical energy up to the consumer's terminals.

FLASH-BANG ELECTRICAL

Statement of Health and Safety at Work Policy in accordance with the Health and Safety at Work Act 1974

Company objective

The promotion of health and safety measures is a mutual objective for the Company and for its employees at all levels. It is the intention that all the Company's affairs will be conducted in a manner which will not cause risk to the health and safety of its members, employees or the general public. For this purpose it is the Company policy that the responsibility for health and safety at work will be divided between all the employees and the Company in the manner outlined below.

Company's responsibilities

The Company will, as a responsible employer, make every endeavour to meet its legal obligations under the Health and Safety at Work Act to ensure the health and safety of its employees and the general public. Particular attention will be paid to the provision of the following:

1 Plant equipment and systems of work that are safe.
2 Safe arrangements for the use, handling, storage and transport of articles, materials and substances.
3 Sufficient information, instruction, training and supervision to enable all employees to contribute positively to their own safety and health at work and to avoid hazards.
4 A safe place of work, and safe access to it.
5 A healthy working environment.
6 Adequate welfare services.

Note: Reference should be made to the appropriate safety etc. manuals.

Employees' responsibilities

Each employee is responsible for ensuring that the work which he/she undertakes is conducted in a manner which is safe to himself or herself, other members of the general public, and for obeying the advice and instructions on safety and health matters issued by his/her superior. If any employee considers that a hazard to health and safety exists it is his/her responsibility to report the matter to his/her supervisor or through his/her Union Representative or such other person as may be subsequently defined.

Management and Supervisors' responsibilities

Management and supervisors at all levels are expected to set an example in safe behaviour and maintain a constant and continuing interest in employee safety, in particular by:

1 acquiring the knowledge of health and safety regulations and codes of practice necessary to ensure the safety of employees in the workplace,
2 acquainting employees with these regulations on codes of practice and giving guidance on safety matters,
3 ensuring that employees act on instructions and advice given.

General Managers are ultimately responsible to the Company for the rectification or reporting of any safety hazard which is brought to their attention.

Joint consultations

Joint consultation on health and safety matters is important. The Company will agree with its staff, or their representatives, adequate arrangements for joint consultation on measures for promoting safety and health at work, and make and maintain satisfactory arrangements for the participation of their employees in the development and supervision of such measures. Trade Union representatives will initially be regarded as undertaking the role of Safety Representatives envisaged in the Health and Safety at Work Act. These representatives share a responsibility with management to ensure the health and safety of their members and are responsible for drawing the attention of management to any shortcomings in the Company's health and safety arrangements. The Company will in so far as is reasonably practicable provide representatives with facilities and training in order that they may carry out this task.

Review

A review, addition or modification of this statement may be made at any time and may be supplemented as appropriate by further statements relating to the work of particular departments and in accordance with any new regulations or codes of practice.
    This policy statement will be brought to the attention of all employees.

FIGURE 1.1
Typical Health and Safety Policy Statement.

FIGURE 1.2

New health and safety law poster. *Source*: HSE © Crown copyright material is reproduced with the permission of the Controller of HMSO and Her Majesty's Stationery Office, Norwich.

These regulations impose requirements upon the regional electricity companies regarding the installation and use of electric lines and equipment. The regulations are administered by the Engineering Inspectorate of the Electricity Division of the Department of Energy and will not normally concern the electrical contractor except that it is these regulations which lay down the earthing requirement of the electrical supply at the meter position.

The regional electricity companies must declare the supply voltage and maintain its value between prescribed limits or tolerances.

The government agreed on 1 January 1995 that the electricity supplies in the United Kingdom would be harmonized with those of the rest of Europe. Thus the voltages used previously in low-voltage supply systems of 415V and 240V have become 400V for three-phase supplies and 230V for single-phase supplies. The permitted tolerances to the nominal voltage have also been changed from ±6% to +10% and −6%. This gives a voltage range of 216–253V for a nominal voltage of 230V and 376–440V for a nominal supply voltage of 400V.

The next proposed change is for the tolerance levels to be adjusted to ±10% of the declared nominal voltage (IEE Regulation, Appendix 2:14).

The frequency is maintained at an average value of 50Hz over 24h so that electric clocks remain accurate.

Regulation 29 gives the area boards the power to refuse to connect a supply to an installation which in their opinion is not constructed, installed and protected to an appropriately high standard. This regulation would only be enforced if the installation did not meet the requirements of the IEE Regulations for Electrical Installations.

The Electricity at Work Regulations 1989 (EWR)

This legislation came into force in 1990 and replaced earlier regulations such as the Electricity (Factories Act) Special Regulations 1944. The Regulations are made under the Health and Safety at Work Act 1974, and enforced by the Health and Safety Executive. The purpose of the Regulations is to 'require precautions to be taken against the risk of death or personal injury from electricity in work activities'.

Section 4 of the EWR tells us that 'all systems must be constructed so as to prevent danger ..., and be properly maintained.... Every work activity shall be carried out in a manner which does not give rise to danger.... In the case of work of an electrical nature, it is preferable that the conductors be made dead before work commences'.

The EWR do not tell us specifically how to carry out our work activities and ensure compliance, but if proceedings were brought against an individual for breaking the EWR, the only acceptable defence would be 'to prove that all reasonable steps were taken and all diligence exercised to avoid the offence' (Regulation 29).

An electrical contractor could reasonably be expected to have 'exercised all diligence' if the installation was wired according to the IEE Wiring Regulations (see below). However, electrical contractors must become more 'legally aware' following the conviction of an electrician for manslaughter at Maidstone Crown Court in 1989. The Court accepted that an electrician had caused the death of another man as a result of his shoddy work in wiring up a central heating system. He received a 9-month suspended prison sentence. This case has set an important legal precedent, and in future any tradesman or professional who causes death through negligence or poor workmanship risks prosecution and possible imprisonment.

The Management of Health and Safety at Work Regulations 1999

The Health and Safety at Work Act 1974 places responsibilities on employers to have robust health and safety systems and procedures in the workplace. Directors and managers of any company who employ more than five employees can be held personally responsible for failures to control health and safety.

The Management of Health and Safety at Work Regulations 1999 tell us that employers must systematically examine the workplace, the work activity and the management of safety in the establishment through a process of 'risk assessments'. A record of all significant risk assessment findings must be kept in a safe place and be available to an HSE inspector if required. Information based on these findings must be communicated to relevant staff and if changes in work behaviour patterns are recommended in the interests of safety, then they must be put in place. The process of risk assessment is considered in detail in Chapter 3 of this book.

Risks, which may require a formal assessment in the electrotechnical industry, might be:

- working at heights;
- using electrical power tools;
- falling objects;
- working in confined places;
- electrocution and personal injury;
- working with 'live' equipment;
- using hire equipment;
- manual handling – pushing – pulling – lifting;
- site conditions – falling objects – dust – weather – water – accidents and injuries.

And any other risks which are particular to a specific type of workplace or work activity.

The Control of Substances Hazardous to Health Regulations 2002 (COSHH)

The original COSHH Regulations were published in 1988 and came into force in October 1989. They were re-enacted in 1994 with modifications and improvements, and the latest modifications and additions came into force in 2002.

The COSHH Regulations control people's exposure to hazardous substances in the workplace. Regulation 6 requires employers to assess the risks to health from working with hazardous substances, to train employees in techniques which will reduce the risk and provide personal protective equipment (PPE) so that employees will not endanger themselves or others through exposure to hazardous substances. Employees should also know what cleaning, storage and disposal procedures are required and what emergency procedures to follow. The necessary information must be available to anyone using hazardous substances as well as to visiting HSE inspectors.

Hazardous substances include:

1. any substance which gives off fumes causing headaches or respiratory irritation;

2. man-made fibres which might cause skin or eye irritation (e.g. loft insulation);

3. acids causing skin burns and breathing irritation (e.g. car batteries, which contain dilute sulphuric acid);

4. solvents causing skin and respiratory irritation (strong solvents are used to cement together PVC conduit fittings and tube);

5. fumes and gases causing asphyxiation (burning PVC gives off toxic fumes);

6. cement and wood dust causing breathing problems and eye irritation;

7. exposure to asbestos – although the supply and use of the most hazardous asbestos material is now prohibited, huge amounts were installed between 1950 and 1980 in the construction industry and much of it is still in place today. In their latest amendments, the COSHH Regulations focus on giving advice and guidance to builders and contractors on the safe use and control of asbestos products. These can be found in Guidance Notes EH 71.

Where PPE is provided by an employer, employees have a duty to use it to safeguard themselves.

Provision and Use of Work Equipment Regulations 1998

These regulations tidy up a number of existing requirements already in place under other regulations such as the Health and Safety at Work Act 1974, the Factories Act 1961 and the Offices, Shops and Railway Premises Act 1963.

The Provision and Use of Work Equipment Regulations 1998 places a general duty on employers to ensure minimum requirements of plant and equipment. If an employer has purchased good quality plant and equipment, which is well maintained, there is little else to do. Some older equipment may require modifications to bring it in line with modern standards of dust extraction, fume extraction or noise, but no assessments are required by the regulations other than those generally required by the Management Regulations 1999 discussed previously.

The Construction (Health, Safety and Welfare) Regulations 1996

An electrical contractor is a part of the construction team, usually as a subcontractor, and therefore the regulations particularly aimed at the construction industry also influence the daily work procedures and environment of an electrician. The most important recent piece of legislation is the Construction Regulations.

The temporary nature of construction sites makes them one of the most dangerous places to work. These regulations are made under the Health and Safety at Work Act 1974 and are designed specifically to promote safety at work in the construction industry. Construction work is defined as any building or civil engineering work, including construction, assembly, alterations, conversions, repairs, upkeep, maintenance or dismantling of a structure.

The general provision sets out minimum standards to promote a good level of safety on site. Schedules specify the requirements for guardrails, working platforms, ladders, emergency procedures, lighting and welfare facilities. Welfare facilities set out minimum provisions for site accommodation: washing facilities, sanitary conveniences and protective clothing. There is now a duty for all those working on construction sites to wear head protection, and this includes electricians working on site as subcontractors.

Personal Protective Equipment (PPE) at Work Regulations 1998

PPE is defined as all equipment designed to be worn, or held, to protect against a risk to health and safety. This includes most types of protective clothing, and equipment such as eye, foot and head protection, safety harnesses, life jackets and high visibility clothing.

Under the Health and Safety at Work Act, employers must provide free of charge any PPE and employees must make full and proper use of it. Safety signs such as those shown at Fig. 1.3 are useful reminders of the type of

Definition

PPE is defined as all equipment designed to be worn, or held, to protect against a risk to health and safety.

(a)   (b)

(c)   (d)

(e)   (f)

(g)   (h)

FIGURE 1.3
Safety signs showing type of PPE to be worn.

PPE to be used in a particular area. The vulnerable parts of the body which may need protection are the head, eyes, ears, lungs, torso, hands and feet and, additionally, protection from falls may need to be considered. Objects falling from a height present the major hazard against which head protection is provided. Other hazards include striking the head against projections and hair becoming entangled in machinery. Typical methods of protection include helmets, light duty scalp protectors called 'bump caps' and hairnets.

The eyes are very vulnerable to liquid splashes, flying particles and light emissions such as ultraviolet light, electric arcs and lasers. Types of eye protectors include safety spectacles, safety goggles and face shields. Screen-based workstations are being used increasingly in industrial and commercial locations by all types of personnel. Working with VDUs (visual display units) can cause eye strain and fatigue and, therefore, this hazard is the subject of a separate section at the beginning of Chapter 3 headed 'VDU operation hazards'.

Noise is accepted as a problem in most industries and surprisingly there has been very little control legislation. The Health and Safety Executive have published a 'Code of Practice' and 'Guidance Notes' HSG 56 for reducing the exposure of employed persons to noise. A continuous exposure limit of below 85 dB for an 8-hour working day is recommended by the code.

Noise may be defined as any disagreeable or undesirable sound or sounds, generally of a random nature, which do not have clearly defined frequencies. The usual basis for measuring noise or sound level is the decibel scale. Whether noise of a particular level is harmful or not also depends on the length of exposure to it. This is the basis of the widely accepted limit of 85 dB of continuous exposure to noise for 8 hours per day.

A peak sound pressure of above 200 pascals or about 120 dB is considered unacceptable and 130 dB is the threshold of pain for humans. If a person has to shout to be understood at 2 m, the background noise is about 85 dB. If the distance is only 1 m, the noise level is about 90 dB. Continuous noise at work causes deafness, makes people irritable, affects concentration, causes fatigue and accident proneness and may mask sounds which need to be heard in order to work efficiently and safely.

It may be possible to engineer out some of the noise, for example by placing a generator in a separate sound-proofed building. Alternatively, it may be possible to provide job rotation, to rearrange work locations or provide acoustic refuges.

Where individuals must be subjected to some noise at work, it may be reduced by ear protectors. These may be disposable ear plugs, reusable ear plugs or ear muffs. The chosen ear protector must be suited to the user and suitable for the type of noise and individual personnel should be trained in its correct use.

Breathing reasonably clean air is the right of every individual, particularly at work. Some industrial processes produce dust which may present

Safety First

PPE

Always wear or use the PPE (personal protective equipment) provided by your employer for your safety.

a potentially serious hazard. The lung disease asbestosis is caused by the inhalation of asbestos dust or particles and the coal dust disease pneumoconiosis, suffered by many coal miners, has made people aware of the dangers of breathing in contaminated air.

Some people may prove to be allergic to quite innocent products such as flour dust in the food industry or wood dust in the construction industry. The main effect of inhaling dust is a measurable impairment of lung function. This can be avoided by wearing an appropriate mask, respirator or breathing apparatus as recommended by the company's health and safety policy and indicated by local safety signs.

A worker's body may need protection against heat or cold, bad weather, chemical or metal splash, impact or penetration and contaminated dust. Alternatively, there may be a risk of the worker's own clothes causing contamination of the product, as in the food industry. Appropriate clothing will be recommended in the company's health and safety policy. Ordinary working clothes and clothing provided for food hygiene purposes are not included in the Personal Protective Equipment at Work Regulations.

Hands and feet may need protection from abrasion, temperature extremes, cuts and punctures, impact or skin infection. Gloves or gauntlets provide protection from most industrial processes, but should not be worn when operating machinery because they may become entangled in it. Care in selecting the appropriate protective device is required; for example, barrier creams provide only a limited protection against infection.

Try This

PPE
- Make a list of any PPE which you have used at work.
- What was this PPE protecting you from?

Boots or shoes with in-built toe caps can give protection against impact or falling objects and, when fitted with a mild steel sole plate, can also provide protection from sharp objects penetrating through the sole. Special slip resistant soles can also be provided for employees working in wet areas.

Whatever the hazard to health and safety at work, the employer must be able to demonstrate that he or she has carried out a risk analysis, made recommendations which will reduce that risk and communicated these recommendations to the workforce. Where there is a need for PPE to protect against personal injury and to create a safe working environment, the employer must provide that equipment and any necessary training which might be required and the employee must make full and proper use of such equipment and training.

Non-statutory regulations

Statutory laws and regulations are written in a legal framework, some don't actually tell us how to comply with the laws at an everyday level.

Safety First

Safety Signs
- Always follow the instructions given in the safety signs where you are working.
- It will help to keep you safe.

Definition

Statutory laws and regulations are written in a legal framework.

14

Definition

Non-statutory regulations and codes of practice interpret the statutory regulations telling us how we can comply with the law.

Non-statutory regulations and codes of practice interpret the statutory regulations telling us how we can comply with the law.

They have been written for every specific section of industry, commerce and situation, to enable everyone to comply with, or obey the written laws.

When the Electricity at Work Regulations (EWR) tell us to 'ensure that all systems are constructed so as to prevent danger' they do not tell us how to actually do this in a specific situation. However, the IEE Regulations tell us precisely how to carry out our electrotechnical work safely in order to meet the statutory requirements of the EWR. In Part 1 of the IEE Regulations, at 114, it states 'the Regulations are non-statutory. They may, however, be used in a court of law in evidence to claim compliance with a statutory requirement'. If your electrotechnical work meets the requirements of the IEE Regulations, you will also meet the requirements of EWR.

Over the years, non-statutory regulations and codes of practice have built upon previous good practice and responded to changes by bringing out new editions of the various regulations and codes of practice to meet the changing needs of industry and commerce.

We will now look at one non-statutory regulation, what is sometimes called 'the electrician's bible', the most important set of regulations for any one working in the electrotechnical industry, the BS 7671: 2008 Requirements for Electrical Installations, IEE Wiring Regulations 17th Edition.

The IEE Wiring Regulations 17th edition requirements for electrical installations to BS 7671: 2008

The Institution of Electrical Engineers Requirements for Electrical Installations (the IEE Regulations) are non-statutory regulations. They relate principally to the design, selection, erection, inspection and testing of electrical installations, whether permanent or temporary, in and about buildings generally and to agricultural and horticultural premises, construction sites and caravans and their sites. Paragraph 7 of the introduction to the EWR says: 'the IEE Wiring Regulations is a code of practice which is widely recognized and accepted in the United Kingdom and compliance with them is likely to achieve compliance with all relevant aspects of the Electricity at Work Regulations'. The IEE Wiring Regulations are the National Standard in the United Kingdom and apply to installations operating at a voltage up to 1000V a.c. They do not apply to electrical installations in mines and quarries, where special regulations apply because of the adverse conditions experienced there.

The current edition of the IEE Wiring Regulations is the 17th edition 2008. The main reason for incorporating the IEE Wiring Regulations into British Standard BS 7671: 2008 was to create harmonization with European Standards.

The IEE Regulations take account of the technical intent of the CENELEC European Standards, which in turn are based on the IEC International Standards.

The purpose in harmonizing British and European Standards is to help develop a single European market economy so that there are no trade barriers to electrical goods and services across the European Economic Area.

To assist Electricians, etc.

To assist electricians in their understanding of the Regulations a number of guidance notes have been published. The guidance notes which I will frequently make reference to in this book are those contained in the *On Site Guide*. Eight other guidance notes booklets are also currently available. These are:

- *Selection and Erection*
- *Isolation and Switching*
- *Inspection and Testing*
- *Protection against Fire*
- *Protection against Electric Shock*
- *Protection against Overcurrent*
- *Special Locations*
- *Earthing and Bonding*

These guidance notes are intended to be read in conjunction with the Regulations.

The IEE Wiring Regulations are the electrician's bible and provide the authoritative framework of information for anyone working in the electrotechnical industry.

Health and safety responsibilities

We have now looked at statutory and non-statutory regulations which influence working conditions in the electrotechnical industry today. So, who has *responsibility* for these workplace Health and Safety Regulations?

In 1970, a Royal Commission was set up to look at the health and safety of employees at work. The findings concluded that the main cause of accidents at work was apathy on the part of *both* employers and employees.

The Health and Safety at Work Act 1974 was passed as a result of recommendations made by the Royal Commission and, therefore, the Act puts legal responsibility for safety at work on *both* the employer and employee.

In general terms, the employer must put adequate health and safety systems in place at work and the employee must use all safety systems and procedures responsibly.

In specific terms the employer must:

- provide a Health and Safety Policy Statement if there are five or more employees such as that shown in Fig. 1.1;

- display a current employers liability insurance certificate as required by the Employers Liability (Compulsory Insurance) Act 1969;

- report certain injuries, diseases and dangerous occurrences to the enforcing authority (HSE area office – see Appendix for address);

- provide adequate first aid facilities (see Tables 1.1 and 1.2);

- provide PPE;

- provide information, training and supervision to ensure staffs' health and safety;

- provide adequate welfare facilities;

- put in place adequate precautions against fire, provide a means of escape and means of fighting fire;

- ensure plant and machinery are safe and that safe systems of operation are in place;

- ensure articles and substances are moved, stored and used safely;

Table 1.1 Suggested numbers of first aid personnel

Category of risk	Numbers employed at any location	Suggested number of first aid personnel
Lower risk e.g. shops and offices, libraries	Fewer than 50 50–100 More than 100	At least one appointed person At least one first aider One additional first aider for every 100 employed
Medium risk e.g. light engineering and assembly work, food processing, warehousing	Fewer than 20 20–100 More than 100	At least one appointed person At least one first aider for every 50 employed (or part thereof) One additional first aider for every 100 employed
Higher risk e.g. most construction, slaughterhouses, chemical manufacture, extensive work with dangerous machinery or sharp instruments	Fewer than five 5–50 More than 50	At least one appointed person At least one first aider One additional first aider for every 50 employed

Table 1.2 Contents of first aid boxes

Item	Number of employees				
	1–5	6–10	11–50	51–100	101–150
Guidance card on general first aid	1	1	1	1	1
Individually wrapped sterile adhesive dressings	10	20	40	40	40
Sterile eye pads, with attachment (Standard Dressing No. 16 BPC)	1	2	4	6	8
Triangular bandages	1	2	4	6	8
Sterile covering for serious wounds (where applicable)	1	2	4	6	8
Safety pins	6	6	12	12	12
Medium sized sterile unmedicated dressings (Standard Dressings No. 9 and No. 14 and the Ambulance Dressing No. 1)	3	6	8	10	12
Large sterile unmedicated dressings (Standard Dressings No. 9 and No. 14 and the Ambulance Dressing No. 1)	1	2	4	6	10
Extra large sterile unmedicated dressings (Ambulance Dressing No. 3)	1	2	4	6	8

Where tap water is not available, sterile water or sterile normal saline in disposable containers (each holding a minimum of 300 ml) must be kept near the first aid box. The following minimum quantities should be kept:

Number of employees	1–10	11–50	51–100	101–150
Quantity of sterile water	1 × 300 ml	3 × 300 ml	6 × 300 ml	6 × 300 ml

- make the workplace safe and without risk to health by keeping dust, fumes and noise under control.

In specific terms the employee must:

- take reasonable care of his/her own health and safety and that of others who may be affected by what they do;

- co-operate with his/her employer on health and safety issues by not interfering or misusing anything provided for health, safety and welfare in the working environment;

- report any health and safety problem in the workplace to, in the first place, a supervisor, manager or employer.

Employment – rights and responsibilities

As a trainee in the electrotechnical industry you will be employed by a member company and receive a weekly or monthly wage, which will be dependent upon your age and grade as agreed by the appropriate trade union, probably Amicus.

We have seen in the beginning of this chapter that there are many rules and regulations which your employer must comply with in order to make your workplace healthy and safe. There are also responsibilities that apply to you, as an employee (or worker) in the electrotechnical industry, in order to assist your employer to obey the law.

As an Employee you must:

- obey all lawful and reasonable requests;
- behave in a sensible and responsible way at work;
- work with care and reasonable skill.

Your Employer must:

- take care for your safety;
- not ask you to do anything unlawful or unreasonable;
- pay agreed wages;
- not change your contract of employment without your agreement.

Most of the other things that can be expected of you are things like honesty, punctuality, reliability and hard work. Really, just common sense things like politeness will help you to get on at work.

If you have problems relating to your employment rights you should talk it through with your supervisor or trade union representative at work.

Wages and tax

When you start work you will be paid either weekly or monthly. It is quite common to work a week in hand if you are paid weekly, which means that you will be paid for the first week's work at the end of the second week. When you leave that employment, if you have worked a week in hand, you will have a week's wage to come. Money that you have worked for belongs to you and cannot be kept by your employer if you leave without giving notice.

Every employee is entitled to a payslip along with their wages, which should show how much you have earned (gross), how much has been taken off for tax and national insurance and what your take home pay (net) is.

If you are not given a payslip, ask for one, it is your legal right and you may be required to show payslips as proof of income. Always keep your payslips in a safe place.

We all pay tax on the money we earn (income tax). The Government uses tax to pay for services such as health, education, defence, social security and pensions.

We are all allowed to earn a small amount of money tax free each year and this is called the personal allowance. The personal allowance for the tax year in which I am writing this book 2007/2008 is £5225. So every pound that we earn above £5225 is taxed. The tax year starts on the 6th of April each year and finishes on the 5th of April the following year. Your personal tax code enables the personal allowance to be spread out throughout the year and you pay tax on each of your wages on a system called PAYE, pay as you earn.

At the end of the tax year your employers will give you a form called a P.60 which shows your tax code, how much you have earned and how much tax you have paid during a particular tax year. It is important to keep your P.60 somewhere safe along with your payslips. If at some time you want to buy a house a building society will want proof of your earnings, which these documents show.

When leaving a particular employment you must obtain from your employer a form P.45. On starting new employment this form will be required by your new employer and will ensure that you do not initially pay too much tax.

Working hours

Employees cannot be forced to work more than 48h each week on average, and 40h for 16–18 year old trainees. Trainees must also have 12h uninterrupted rest from work each day. Older workers, required to work for more than 6h continuously, are entitled to a 20-min rest break, to be taken within the 6h, and must have 11h uninterrupted rest from work each day. If you think you are not getting the correct number of breaks, talk to your supervisor or trade union representative.

Sickness

If you are sick and unable to go to work, you should contact your employer or supervisor as soon as you can on the first day of illness. When you go back to work, if you have been sick for up to 7 days, you will have to fill in a self-certification form. After 7 days you will need a medical certificate from your doctor and you must send it to work as soon as you can. If you are sick for 4 days or more your employer must pay you statutory sick pay (SSP), which can be paid for up to 28 weeks. If you are sick after 28 weeks you can claim incapacity benefit. To claim this you will need a form from your employer or Social Security Office. If you have a sickness problem, talk to your supervisor or someone at work who you trust, or telephone the local Social Security Office.

Accidents

It is the employer's duty to protect the health and safety and welfare of its employees, so if you do have an accident at work, however small, inform your supervisor, safety officer or first aid person. Make sure that the details are recorded in the accident/first aid book. Failure to do so may affect compensation if the accident proves to be more serious than you first thought.

Always be careful, use common sense and follow instructions. If in doubt, ask someone. A simple accident might prevent you playing your favourite sport for a considerable period of time.

Holidays

Most employees are entitled to at least 4 weeks paid holiday each year. Your entitlement to paid holidays builds up each month, so a month after you start work you are entitled to one-twelfth of the total holiday entitlement for the year. After 2 months it becomes two-twelfths and so on. Ask your supervisor or the kind lady in the office who makes up the wages to explain your holiday entitlement to you.

Problems at work

It is not unusual to find it hard to fit in when you start a new job. Give it a chance, give it time and things are likely to settle down. As a new person you might seem to get all the rotten jobs, but sometimes, being new, these are the only jobs that you can do for now.

In some companies there can be a culture of 'teasing', which may be OK, if everyone is treated the same, but not so good if you are always the one being teased. If this happens, see if it stops after a while, if not, talk to someone about it. Don't give up your job without trying to get the problem sorted out.

If you feel that you are being discriminated against or harassed because of your race, sex or disability, then talk to your supervisor, trainer or someone you trust at work. There are laws about discrimination that are discussed in Advanced Electrical Installation Work.

You can join a trade union when you are 16 years of age or over. Trade unions work toward fair deals for their members. If you join a trade union there will be subscriptions (subs) to pay. These are often reduced or suspended during the training period. As a member of a Trade Union you can get advice and support from them. If there is a problem of any kind at work, you can ask for union's support. However, you cannot get this support unless you are a member.

Resignation/dismissal

Most employers like to have your resignation or 'Notice' in writing. Your Contract of Employment will tell you how much Notice is expected. The minimum Notice you should give is 1 week if you have been employed for 1 month or more by that employer. However, if your Contract states a longer period, then that is what is expected.

If you have worked for 1 month or more, but less than 2 years, you are entitled to 1 week's Notice. If you have worked for 2 years you are entitled to 2 week's Notice and a further week's Notice for every additional continuous year of employment (with the same employer) up to 12 weeks for 12 years service.

If you are dismissed or 'sacked' you are entitled to the same periods of Notice. However, if you do something very serious, like stealing or hitting someone, your employer can dismiss you without Notice.

You can also be dismissed if you are often late or your behaviour is inappropriate to the type of work being done. You should have verbal or written warnings before you are dismissed.

If there are 20 or more employees at your place of employment then there should be a disciplinary procedure written down, which must be followed. If you do get a warning, then you might like to see this as a second chance to start again.

If you have been working for the same employer for 1 year or more, you can complain to an Employment Tribunal if you think you have been unfairly dismissed. If you haven't worked for the same employer for this length of time, then you should talk to your training officer or trade union.

I do not want to finish this section in a negative way, talking about problems at work, so let me finally say that each year over 8000 young people are in apprenticeships in the electrical contracting industry and very few of them have problems. The small problems that may arise, because moving into full-time work is very different to school, can usually be resolved by your training officer or supervisor. Most of the trainees go on to qualify as craftsmen and enjoy a well-paid and fulfilling career in the electrotechnical industry.

Safety signs

The rules and regulations of the working environment are communicated to employees by written instructions, signs and symbols. All signs in the working environment are intended to inform. They should give warning of possible dangers and must be obeyed. At first there were many different safety signs, but British Standard BS 5499 Part 1 and the Health and Safety (Signs and Signals) Regulations 1996 have introduced a standard system which gives health and safety information with the minimum use of words. The purpose of the regulations is to establish an internationally understood system of safety signs and colours which draw attention to equipment and situations that do, or could, affect health and safety. Text-only safety signs became illegal from 24th December 1998. From that date, all safety signs have had to contain a pictogram or symbol such as those shown in Fig. 1.4. Signs fall into four categories: prohibited activities; warnings; mandatory instructions and safe conditions.

PROHIBITION SIGNS

These are *must not do* signs. These are circular white signs with a red border and red cross-bar, and are given in Fig. 1.5. They indicate an activity *which must* not be done.

FIGURE 1.4
Text only safety signs do not comply.

WARNING SIGNS

These give safety information. These are triangular yellow signs with a black border and symbol, and are given in Fig. 1.6. They *give warning* of a hazard or danger.

MANDATORY SIGNS

These are *must do* signs. These are circular blue signs with a white symbol, and are given in Fig. 1.7. They *give instructions* which must be obeyed.

ADVISORY OR SAFE CONDITION SIGN

These give safety information. These are square or rectangular green signs with a white symbol, and are given in Fig. 1.8. They *give information* about safety provision.

FIGURE 1.5

Prohibition signs. These are *must not* do signs.

FIGURE 1.6

Warning signs. These give safety information.

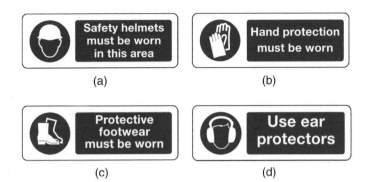

FIGURE 1.7
Mandatory signs. These are *must do* signs.

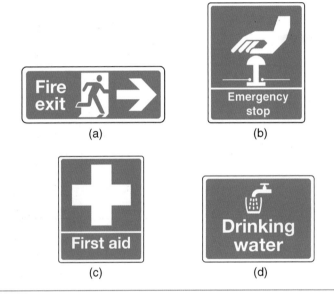

FIGURE 1.8
Advisory or Safe condition signs. These also give safety information.

Accidents at work

Definition

An *accident* may be defined as an uncontrolled event causing injury or damage to an individual or property.

Despite new legislation, improved information, education and training, accidents at work do still happen. An accident may be defined as an uncontrolled event causing injury or damage to an individual or property. An accident can nearly always be avoided if correct procedures and methods of working are followed. Any accident which results in an absence from work for more than 3 days, causes a major injury or death, is notifiable to the HSE. There are more than 40,000 accidents reported to the HSE each year which occur as a result of some building-related activity. To avoid having an accident you should:

1. follow all safety procedures (e.g. fit safety signs when isolating supplies and screen off work areas from the general public);

2. not misuse or interfere with equipment provided for health and safety;

3. dress appropriately and use PPE when appropriate;

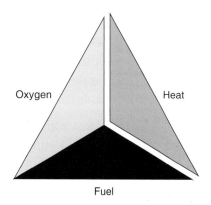

FIGURE 1.9
The fire triangle.

4. behave appropriately and with care;

5. avoid over-enthusiasm and foolishness;

6. stay alert and avoid fatigue;

7. not use alcohol or drugs at work;

8. work within your level of competence;

9. attend safety courses and read safety literature;

10. take a positive decision to act and work safely.

If you observe a hazardous situation at work, first make the hazard safe, using an appropriate method, or screen it off, but only if you can do so without putting yourself or others at risk, then report the situation to your safety representative or supervisor.

Fire control

Definition

Fire is a chemical reaction which will continue if fuel, oxygen and heat are present.

Fire is a chemical reaction which will continue if fuel, oxygen and heat are present. To eliminate a fire *one* of these components must be removed. This is often expressed by means of the fire triangle shown in Fig. 1.9; all three corners of the triangle must be present for a fire to burn.

Fuel

Fuel is found in the construction industry in many forms: petrol and paraffin for portable generators and heaters; bottled gas for heating and soldering. Most solvents are flammable. Rubbish also represents a source of fuel: off-cuts of wood, roofing felt, rags, empty solvent cans and discarded packaging will all provide fuel for a fire.

To eliminate fuel as a source of fire, all flammable liquids and gases should be stored correctly, usually in an outside locked store. The working environment should be kept clean by placing rags in a metal bin with a lid. Combustible waste material should be removed from the work site or burned outside under controlled conditions by a competent person.

Safety First

Fire

If you discover a fire:

- raise the alarm;
- attack small fires with an extinguisher;
- BUT only if you can do so without risk to your own safety.

Oxygen

Oxygen is all around us in the air we breathe, but can be eliminated from a small fire by smothering with a fire blanket, sand or foam. Closing doors and windows, but not locking them will limit the amount of oxygen available to a fire in a building and help to prevent it spreading.

Most substances will burn if they are at a high enough temperature and have a supply of oxygen. The minimum temperature at which a substance will burn is called the 'minimum ignition temperature' and for most materials this is considerably higher than the surrounding temperature. However, a danger does exist from portable heaters, blow torches and hot air guns which provide heat and can cause a fire by raising the temperature of materials placed in their path above the minimum ignition temperature. A safe distance must be maintained between heat sources and all flammable materials.

Heat

Heat can be removed from a fire by dousing with water, but water must not be used on burning liquids since the water will spread the liquid and the fire. Some fire extinguishers have a cooling action which removes heat from the fire.

Fires in industry damage property and materials, injure people and sometimes cause loss of life. Everyone should make an effort to prevent fires, but those which do break out should be extinguished as quickly as possible.

In the event of fire you should:

- raise the alarm;
- turn off machinery, gas and electricity supplies in the area of the fire;
- close doors and windows but without locking or bolting them;
- remove combustible materials and fuels away from the path of the fire, if the fire is small, and if this can be done safely;
- attack small fires with the correct extinguisher.

Only attack the fire if you can do so without endangering your own safety in any way. Always leave your own exit from the danger zone clear. Those not involved in fighting the fire should walk to a safe area or assembly point.

Fires are divided into four classes or categories:

- Class A are wood, paper and textile fires.
- Class B are liquid fires such as paint, petrol and oil.
- Class C are fires involving gas or spilled liquefied gas.
- Class D are very special types of fire involving burning metal.

Electrical fires do not have a special category because, once started, they can be identified as one of the four above types.

Fire extinguishers are for dealing with small fires, and different types of fire must be attacked with a different type of extinguisher. Using the wrong type of extinguisher could make matters worse. For example, water must not be used on a liquid or electrical fire. The normal procedure when dealing with electrical fires is to cut off the electrical supply and use an extinguisher which is appropriate to whatever is burning. Figure 1.10 shows the correct type of extinguisher to be used on the various categories of fire. The colour coding shown is in accordance with BS EN3: 1996.

Electrical safety and isolation

Electrical supplies at voltages above extra low voltages (ELV) – that is, above 50V a.c. – can kill human beings and livestock and should therefore be treated with the greatest respect. As an electrician working on electrical installations and equipment, you should always make sure that the supply

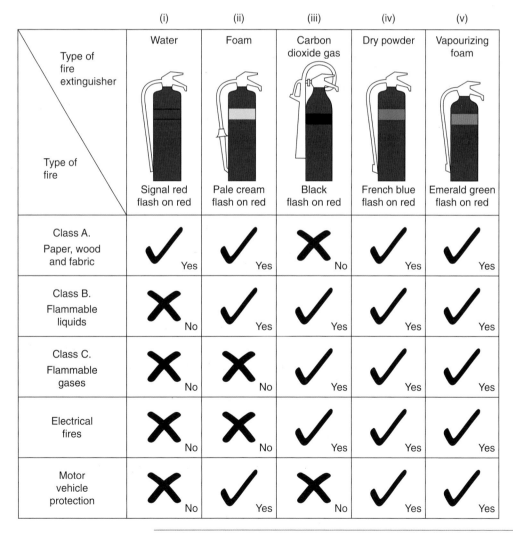

	(i)	(ii)	(iii)	(iv)	(v)
Type of fire extinguisher / Type of fire	Water	Foam	Carbon dioxide gas	Dry powder	Vapourizing foam
	Signal red flash on red	Pale cream flash on red	Black flash on red	French blue flash on red	Emerald green flash on red
Class A. Paper, wood and fabric	✓ Yes	✓ Yes	✗ No	✓ Yes	✓ Yes
Class B. Flammable liquids	✗ No	✓ Yes	✓ Yes	✓ Yes	✓ Yes
Class C. Flammable gases	✗ No	✗ No	✓ Yes	✓ Yes	✓ Yes
Electrical fires	✗ No	✗ No	✓ Yes	✓ Yes	✓ Yes
Motor vehicle protection	✗ No	✓ Yes	✗ No	✓ Yes	✓ Yes

FIGURE 1.10

Fire extinguishers and their applications (colour codes to BS EN3: 1996). The base colour of all fire extinguishers is red, with a different coloured flash to indicate the type.

is first switched off. Every circuit must be provided with a means of isolation (Regulation 132.10) and you should isolate and lock off before work begins. In order to deter anyone from reconnecting the supply, a 'Danger Electrician at Work' sign should be displayed on the isolation switch. Where a test instrument or voltage indicator such as that shown in Fig. 1.11 is used to prove conductors dead, Regulation 4(3) of EWR 1989 recommends that the following procedure should be adopted so that the device itself is 'proved':

Safety First

Working 'Live'

- NEVER work 'live'
- ALWAYS isolate first
- SECURE the isolation
- PROVE the supply 'dead' before starting work.

1. Connect the test device to the supply which is to be isolated; this should indicate mains voltage.
2. Isolate the supply and observe that the test device now reads 0V.
3. Connect the test device to another source of supply to 'prove' that the device is still working correctly.
4. Lock off the supply and place warning notices. Only then should work commence on the 'dead' supply.

The test device must incorporate fused test leads to comply with HSE Guidance Note GS 38, *Electrical Test Equipment Used by Electricians*. Electrical isolation of supplies is further discussed in Chapter 3 of this book.

Temporary electrical supplies on construction sites can save many person-hours of labour by providing energy for fixed and portable tools and

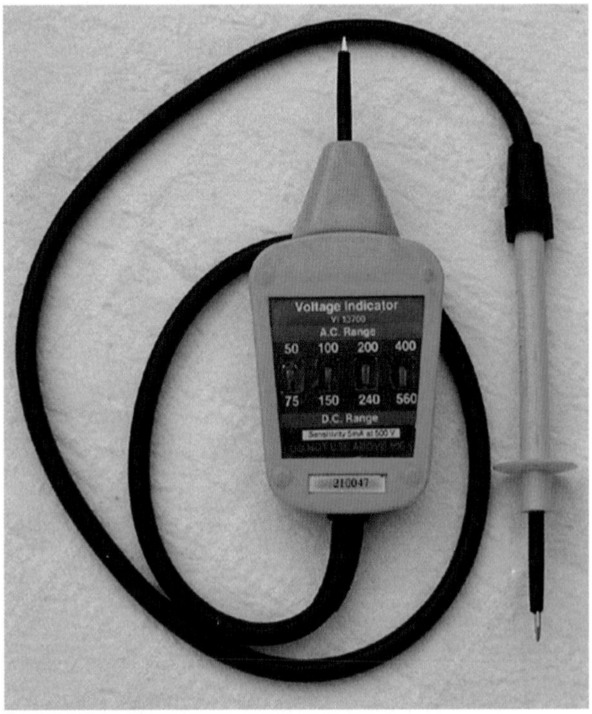

FIGURE 1.11
Typical voltage indicator.

lighting. However, as stated previously in this chapter, construction sites are dangerous places and the temporary electrical supplies must be safe. IEE Regulation 110.1 tells us that *ALL* the regulations apply to temporary electrical installations such as construction sites. The frequency of inspection of construction sites is increased to every 3 months because of the inherent dangers. Regulation 704.313.4 recommends the following voltages for distributing to plant and equipment on construction sites:

400V – fixed plant such as cranes

230V – site offices and fixed floodlighting robustly installed

110V – portable tools and hand lamps

SELV – portable lamps used in damp or confined places.

Portable tools must be fed from a 110V socket outlet unit (see Fig. 1.12(a)) incorporating splash-proof sockets and plugs with a keyway which prevents a tool from one voltage being connected to the socket outlet of a different voltage.

Socket outlet and plugs are also colour-coded for voltage identification: 25V violet, 50V white, 110V yellow, 230V blue and 400V red, as shown in Fig. 1.12(b).

29

Electric shock

Electric shock occurs when a person becomes part of the electrical circuit, as shown in Fig. 1.13. The level or intensity of the shock will depend upon many factors, such as age, fitness and the circumstances in which the shock is received. The lethal level is approximately 50 mA, above which muscles contract, the heart flutters and breathing stops. A shock above the 50 mA level is therefore fatal unless the person is quickly separated from the supply. Below 50 mA only an unpleasant tingling sensation may be experienced or you may be thrown across a room, roof or ladder, but the resulting fall may lead to serious injury.

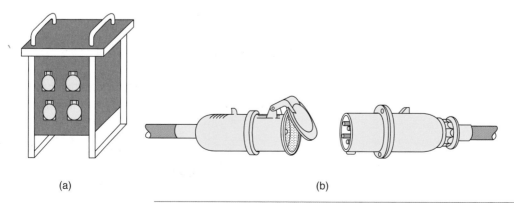

(a)　　　　　　　　　　　　　　　(b)

FIGURE 1.12

110 volts distribution unit and cable connector suitable for construction site electrical supplies: (a) reduced-voltage distribution unit incorporating industrial sockets to BS 4343; (b) industrial plug and connector.

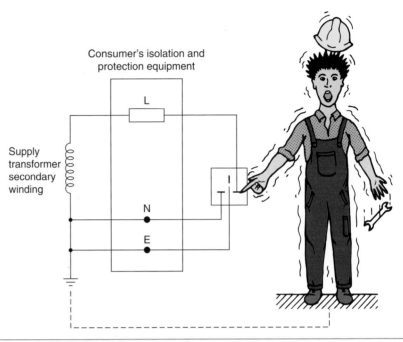

Consumer's isolation and
protection equipment

L

Supply
transformer
secondary
winding

N

E

FIGURE 1.13

Touching live and earth or live and neutral makes a person part of the electrical circuit and can lead to an electric shock.

To prevent people receiving an electric shock accidentally, all circuits must contain protective devices. All exposed metal must be earthed; fuses and miniature circuit breakers (MCBs) are designed to trip under fault conditions and residual current devices (RCDs) are designed to trip below the fatal level as described in Chapter 4.

Construction workers and particularly electricians do receive electric shocks, usually as a result of carelessness or unforeseen circumstances. When this happens it is necessary to act quickly to prevent the electric shock becoming fatal. Actions to be taken upon finding a workmate receiving an electric shock are as follows:

- Switch off the supply if possible.

- Alternatively, remove the person from the supply *without touching him*, e.g. push him off with a piece of wood, pull him off with a scarf, dry towel or coat.

- If breathing or heart has stopped, immediately call professional help by dialling 999 or 112 and asking for the ambulance service. Give precise directions to the scene of the accident. The casualty stands the best chance of survival if the emergency services can get a rapid response paramedic team quickly to the scene. They have extensive training and will have specialist equipment with them.

- Only then should you apply resuscitation or cardiac massage until the patient recovers, or help arrives.

- Treat for shock.

First aid

Despite all the safety precautions taken on construction sites to prevent injury to the workforce, accidents do happen and *you* may be the only other person able to take action to assist a workmate. If you are not a qualified first aider limit your help to obvious common sense assistance and call for help, *but* do remember that if a workmate's heart or breathing has stopped as a result of an accident he has only minutes to live unless you act quickly. The Health and Safety (First Aid) Regulations 1981 and relevant approved codes of practice and guidance notes place a duty of care on all employers to provide *adequate* first aid facilities appropriate to the type of work being undertaken. *Adequate* facilities will relate to a number of factors such as:

- How many employees are employed?

- What type of work is being carried out?

- Are there any special or unusual hazards?

- Are employees working in scattered and/or isolated locations?

- Is there shift work or 'out of hours' work being undertaken?

- Is the workplace remote from emergency medical services?

- Are there inexperienced workers on site?

- What were the risks of injury and ill health identified by the company's Hazard Risk Assessment?

The regulations state that:

Employers are under a duty to provide such numbers of suitable persons as is *adequate and appropriate in the circumstances* for rendering first aid to his employees if they are injured or become ill at work. For this purpose a person shall not be suitable unless he or she has undergone such training and has such qualifications as the Health and Safety Executive may approve.

This is typical of the way in which the health and safety regulations are written. The regulations and codes of practice do not specify numbers, but set out guidelines in respect of the number of first aiders needed, dependent upon the type of company, the hazards present and the number of people employed.

Let us now consider the questions 'what is first aid'? and 'who might become a first aider?' The regulations give the following definitions of first aid. '*First aid* is the treatment of minor injuries which would otherwise receive no treatment or do not need treatment by a doctor or nurse' *or* 'In cases where a person will require help from a doctor or nurse, first aid is treatment for the purpose of preserving life and minimizing the

Definition

First aid is the initial assistance or treatment given to a casualty for any injury or sudden illness before the arrival of an ambulance, doctor or other medically qualified person.

Definition

A *first aider* is someone who has undergone a training course to administer first aid at work and holds a current first aid certificate.

Definition

An *appointed person* is someone who is nominated to take charge when someone is injured or becomes ill, including calling an ambulance if required. The appointed person will also look after the first aid equipment, including re-stocking the first aid box.

consequences of an injury or illness until such help is obtained'. A more generally accepted definition of first aid might be as follows: **first aid** is the initial assistance or treatment given to a casualty for any injury or sudden illness before the arrival of an ambulance, doctor or other medically qualified person.

Now having defined first aid, who might become a first aider? A **first aider** is someone who has undergone a training course to administer first aid at work and holds a current first aid certificate. The training course and certification must be approved by the HSE. The aims of a first aider are to preserve life, to limit the worsening of the injury or illness and to promote recovery.

A first aider may also undertake the duties of an **appointed person**. An appointed person is someone who is nominated to take charge when someone is injured or becomes ill, including calling an ambulance if required. The appointed person will also look after the first aid equipment, including re-stocking the first aid box.

Appointed persons should not attempt to give first aid for which they have not been trained, but should limit their help to obvious common sense assistance and summon professional assistance as required. Suggested numbers of first aid personnel are given in Table 1.1. The actual number of first aid personnel must take into account any special circumstances such as remoteness from medical services, the use of several separate buildings and the company's hazard risk assessment. First aid personnel must be available at all times when people are at work, taking into account shift working patterns and providing cover for sickness absences.

Every company must have at least one first aid kit under the regulations. The size and contents of the kit will depend upon the nature of the risks involved in the particular working environment and the number of employees. Table 1.2 gives a list of the contents of any first aid box to comply with the HSE Regulations.

There now follows a description of some first aid procedures which should be practised under expert guidance before they are required in an emergency.

Bleeding

If the wound is dirty, rinse it under clean running water. Clean the skin around the wound and apply a plaster, pulling the skin together.

If the bleeding is severe apply direct pressure to reduce the bleeding and raise the limb if possible. Apply a sterile dressing or pad and bandage firmly before obtaining professional advice.

To avoid possible contact with hepatitis or the AIDS virus, when dealing with open wounds, first aiders should avoid contact with fresh blood by wearing plastic or rubber protective gloves, or by allowing the casualty to apply pressure to the bleeding wound.

Burns

Remove heat from the burn to relieve the pain by placing the injured part under clean cold water. Do not remove burnt clothing sticking to the skin. Do not apply lotions or ointments. Do not break blisters or attempt to remove loose skin. Cover the injured area with a clean dry dressing.

Broken bones

Make the casualty as comfortable as possible by supporting the broken limb either by hand or with padding. Do not move the casualty unless by remaining in that position he is likely to suffer further injury. Obtain professional help as soon as possible.

Contact with chemicals

Wash the affected area very thoroughly with clean cold water. Remove any contaminated clothing. Cover the affected area with a clean sterile dressing and seek expert advice. It is a wise precaution to treat all chemical substances as possibly harmful; even commonly used substances can be dangerous if contamination is from concentrated solutions. When handling dangerous substances, it is also good practice to have a neutralizing agent to hand.

Disposal of dangerous substances must not be into the main drains since this can give rise to an environmental hazard, but should be undertaken in accordance with local authority regulations.

Exposure to toxic fumes

Get the casualty into fresh air quickly and encourage deep breathing if conscious. Resuscitate if breathing has stopped. Obtain expert medical advice as fumes may cause irritation of the lungs.

Sprains and bruising

A cold compress can help to relieve swelling and pain. Soak a towel or cloth in cold water, squeeze it out and place it on the injured part. Renew the compress every few minutes.

Breathing stopped

Remove any restrictions from the face and any vomit, loose or false teeth from the mouth. Loosen tight clothing around the neck, chest and waist. To ensure a good airway, lay the casualty on his back and support the shoulders on some padding. Tilt the head backwards and open the mouth. If the casualty is faintly breathing, lifting the tongue, clearing of the airway may be all that is necessary to restore normal breathing. However, if the casualty does not begin to breathe, open your mouth wide and take a deep breath, close the casualty's nose by pinching with your fingers, and, sealing your lips around his mouth, blow into his lungs until the chest rises. Remove your mouth and watch the casualty's chest fall. Continue this procedure at your natural breathing rate. If the mouth is damaged or you have difficulty making a seal around the casualty's mouth, close his mouth and inflate the lungs through his nostrils. Give artificial respiration until natural breathing is restored or until professional help arrives.

Heart stopped beating

This sometimes happens following a severe electric shock. If the casualty's lips are blue, the pupils of his eyes widely dilated and the pulse in his neck cannot be felt, then he may have gone into cardiac arrest. Act quickly and lay the casualty on his back. Kneel down beside him and place the heel of one hand in the centre of his chest. Cover this hand with your other hand and interlace the fingers. Straighten your arms and press down on his chest sharply with the heel of your hands and then release the pressure. Continue to do this 15 times at the rate of one push per second. Check the casualty's pulse. If none is felt, give two breaths of artificial respiration and then a further 15 chest compressions. Continue this procedure until the heartbeat is restored and the artificial respiration until normal breathing returns. Pay close attention to the condition of the casualty while giving heart massage. When a pulse is restored the blueness around the mouth will quickly go away and you should stop the heart massage. Look carefully at the rate of breathing. When this is also normal, stop giving artificial respiration. Treat the casualty for shock, place him in the recovery position and obtain professional help.

Shock

Everyone suffers from shock following an accident. The severity of the shock depends upon the nature and extent of the injury. In cases of severe shock the casualty will become pale and his skin become clammy from sweating. He may feel faint, have blurred vision, feel sick and complain of thirst. Reassure the casualty that everything that needs to be done is being done. Loosen tight clothing and keep him warm and dry until help arrives. *Do not* move him unnecessarily or give him anything to drink.

Accident reports

Every accident must be reported to an employer and the details of the accident and treatment given are suitably documented. A first aid Log book or accident book such as that shown in Fig. 1.14 containing first aid treatment

FIGURE 1.14
First aid log book/accident book with data protection compliant removable sheets.

record sheets could be used to effectively document accidents which occur in the workplace and the treatment given. Failure to do so may influence the payment of compensation at a later date if an injury leads to permanent disability. To comply with the Data Protection Regulations, from the 31 December 2003 all First Aid Treatment Log books or Accident Report books must contain perforated sheets which can be removed after completion and filed away for personal security.

Check your Understanding

When you have completed the questions, check out the answers at the back of the book.
Note: more than one multiple choice answer may be correct.

1. For any fire to continue to burn, three components must be present. These are:
    a. fuel, wood and cardboard
    b. petrol, oxygen and bottled gas
    c. flames, fuel and heat
    d. fuel, oxygen and heat.

2. A water fire extinguisher is suitable for use on small fires of burning
    a. wood, paper and fabric
    b. flammable liquids
    c. flammable gas
    d. all of the above.

3. A foam fire extinguisher is suitable for use on small fires of burning
    a. wood, paper and fabric
    b. flammable liquids
    c. flammable gas
    d. all of the above.

4. A carbon dioxide gas fire extinguisher is suitable for use on small fires of burning
    a. wood, paper and fabric
    b. flammable liquids
    c. flammable gas
    d. all of the above.

5. A dry powder fire extinguisher is suitable for use on small fires of burning
    a. wood, paper and fabric
    b. flammable liquids
    c. flammable gas
    d. all of the above.

6. A vapourizing foam fire extinguisher is suitable for use on small fires of burning
    a. wood, paper and fabric
    b. flammable liquids
    c. flammable gas
    d. all of the above.

7. You should only attack a fire with a fire extinguisher if
   a. it is burning brightly
   b. you can save someone's property
   c. you can save someone's life
   d. you can do so without putting yourself at risk.

8. A fire extinguisher should only be used to fight
   a. car fires
   b. electrical fires
   c. small fires
   d. big fires.

9. A Statutory Regulation
   a. is the law of the land
   b. must be obeyed
   c. tells us how to comply with the law
   d. is a code of practice.

10. A Non-statutory Regulation
    a. is the law of the land
    b. must be obeyed
    c. tells us how to comply with the law
    d. is a code of practice.

11. Under the Health and Safety at Work Act an employer is responsible for
    a. maintaining plant and equipment
    b. providing PPE
    c. wearing PPE
    d. taking reasonable care to avoid injury.

12. Under the Health and Safety at Work Act an employee is responsible for
    a. maintaining plant and equipment
    b. providing PPE
    c. wearing PPE
    d. taking reasonable care to avoid injury.

13. The IEE Wiring Regulations
    a. are Statutory regulations
    b. are Non-statutory regulations
    c. are codes of good practice
    d. must always be complied with.

14. Before beginning work on a 'live' circuit or piece of equipment you should
    a. only work 'live' if your supervisor is with you
    b. only work 'live' if you feel that you are 'competent' to do so
    c. isolate the circuit or equipment before work commences
    d. secure the isolation before work commences.

15. The initial assistance or treatment given to a casualty for any injury or sudden illness before the arrival of an ambulance or medically qualified person is one definition of
    a. an appointed person
    b. a first aider
    c. first aid
    d. an adequate first aid facility.

16. Someone who has undergone a training course to administer medical aid at work and holds a current qualification is one definition of
    a. a doctor
    b. a nurse
    c. a first aider
    d. a supervisor.

17. Use bullet points to describe a safe isolation procedure of a 'live' electrical circuit.

18. How does the law enforce the regulations of the Health and Safety at Work Act?

19. List the responsibilities under the Health and Safety at Work Act of:
    (a) an employer to his employees
    (b) an employee to his employer.

20. Safety signs are used in the working environment to give information and warning. Sketch and colour, one sign from each of the four categories of signs and state the message given by that sign.

21. State the name of two important Statutory Regulations and one Non-statutory Regulation relevant to the electrotechnical industry.

22. Define what is meant by PPE.

23. State five pieces of PPE which a trainee could be expected to wear at work and the protection given by each piece.

24. Describe the action to be taken upon finding a workmate apparently dead on the floor and connected to an electrical supply.

25. State how the Data Protection Act has changed the way in which we record accident and first aid information at work.

The occupational specialisms and individual roles within the electrotechnical industry

Unit 1 – Working effectively and safely in the electrotechnical environment – Outcome 2

Underpinning knowledge: when you have completed this chapter, you should be able to:

- identify the different organizations which have electrotechnical activities
- describe the services provided by the electrotechnical industry
- state the role of the people working in the electrotechnical industry
- identify the professional bodies supporting electrotechnical organizations

Organizations having electrotechnical activities

When we talk about the Electrotechnical Industry we are referring to all those different organizations or companies which provide an electrical service of some kind.

Electrical contractors install equipment and systems in new buildings. Once a building is fully operational, the electrical contractor may provide a maintenance service to that client or customer or alternatively the client may employ an 'in-house' electrician to maintain the installed electrical equipment. It all depends on the amount of work to be done and the complexity of the customers' systems.

The City and Guilds Syllabus directs us to look at *twelve* different organizations having electrotechnical activities and *ten* services provided by the electrotechnical industry, so here goes.

1. Electrical contractors

   - electrical contractors provide a design and installation service for all types of buildings and construction projects
   - the focus of this type of organization is on all types of electrotechnical activities in and around buildings
   - they install electrical equipment
   - they install electrical wiring systems
   - they carry out their installation work in domestic, commercial, industrial, agricultural and horticultural buildings

2. Factories

   - factories contain lots of electrical plant and equipment
   - the wheels of all types of industry are driven by electromechanical devices and electrotechnical activities; such as motors and control systems

3. Process plants

   - whether they process food or nuclear fuels, the prime mover for all processes is electrical plant, control and instrumentation equipment and machine drives

4. Local councils

   - they are responsible for many different types of community buildings from town halls to swimming pools
   - the buildings all have electrical systems which require installation, maintenance and repair

5. Commercial buildings and complexes

   - the 'office type' activities carried out in these buildings require that electrical communication and data transmission systems are installed, maintained and repaired

6. Leisure centres

   - these type of buildings contain lots of equipment driven by human sweat but which is also controlled and monitored by electrical and electronic systems

   - Leisure centres might contain a swimming pool or 'hot-air' sauna. Both types of electrical installation are considered 'Special Installations' by the IEE Regulations BS: 7671

7. Panel builders

   - build specialist control, protection and isolation main switch-gear systems for commerce and industry

   - the panel incorporates the isolation and protection systems required by the electrical installation

8. Motor re-wind and repair

   - electrical motors and their drives usually form an integral part of the industrial system or process

   - electrical motors and transformers sometimes break down or burn out

   - an exact new replacement can often be quickly installed

   - alternatively the existing motor can be re-wound and reconditioned by a specialist company if time permits

9. Railways

   - the prime mover for a modern inter-city type electric train is an electric motor

   - electric trains require an infrastructure of electrical transmission lines throughout the network

   - all rail movements require signal and control systems

   - railway station buildings contain electrical and electronic installations

10. The armed forces

   - operate in harsh, hostile and unpredictable environments

   - they need to adapt, modify and repair electrical and electronic systems in a war situation away from their home base and a comfortable well equipped workshop

   - a modern warship can contain as many people as an English village. They need electrotechnical systems to support them and to keep them safe 24 h per day, 7 days per week

11. Hospitals

   - contain a great deal of high technology equipment

   - this equipment requires power and electronic systems

- life monitoring equipment must continue to operate in a power failure

- standby electrical supplies are, therefore, often an important part of a hospital's electrical installations

12. Equipment and machine manufacturers

- white goods, brown goods, computer hardware, motors and transformers are manufactured to meet the increasing demands of the domestic, commercial and industrial markets

- they are manufactured to very high standards and often contain very sophisticated electrical and electronic circuits and systems

- they manufacture to British and European Standards

Try This

Information

Which type of electrotechnical organization do you belong to?

What types of work have you carried out so far?

Services provided by the electrotechnical industry

1. Lighting and power installations ensure that the building in which they are installed

- is illuminated to an appropriate level

- is heated to a comfortable level

- has the power circuits to drive the electrical and electronic equipment required by those who will use the buildings

2. Emergency lighting and security systems

- These ensure that the building is safe to use in unforeseen or adverse situations

- is secure from unwanted intruders

3. Building Management and control systems

- These systems provide a controlled environment for the people who use commercial buildings

- They provide a pleasant environment so that people can work effectively and efficiently

4. Instrumentation

- Electrical instrumentation allows us to monitor industrial processes and systems often at a safe distance

5. Electrical maintenance

  - A programme of planned maintenance allows us to maintain the efficiency of all installed systems

6. Live cable jointing

  - Making connections to 'live' cables provides a means of connecting new installations and services to existing live supply cables without inconvenience to existing supplies caused by electrical shutdown. This work requires special training

7. Highway electrical systems

  - Illuminated motorways, roads and traffic control systems make our roads and pavements safe for vehicles and pedestrians

8. Electrical panel building

  - Main electrical panels provide a means of electrical isolation and protection

  - They also provide a means of monitoring and measuring electrical systems in our commercial and industrial buildings

9. Electrical machine drive installations

  - Electrical machine drives drive everything that makes our modern life comfortable from

  - trains and trams to

  - lifts and air conditioning units

  - refrigerators, freezers and all types of domestic appliances

10. Consumer and commercial electronics

  - These give us data processing and number crunching

  - electronic mail and access to information on the world wide web

  - access to high quality audio and video systems

Try This

Information

Which services does the company you work for provide within the Electrotechnical Industry?

Roles and responsibilities of workers in the electrotechnical industry

Any electrotechnical organization is made up of a group of individuals with various duties, all working together for their own good, the good of their employer and their customers.

There is often no clear distinction between the duties of the individual employees, each do some of the others work activities.

Responsibilities vary, even by people holding the same job title and some individuals hold more than one job title. However, let us look at some of the roles and responsibilities of those working in the electrotechnical industry.

1. Design engineer

   - The design engineer will normally meet with clients and other trade professionals to interpret the customers' requirements

   - He or she will produce the design specification which enables the cost of the project to be estimated

2. Estimator/cost engineer

   - This person measures the quantities of labour and material necessary to complete the electrical project using the plans and specifications for the project

   - From these calculations and the company's fixed costs, a project cost can be agreed

3. Contracts manager

   - may oversee a number of electrical contracts on different sites

   - will monitor progress in consultation with the project manager on behalf of the electrical companies

   - cost out variations to the initial contract

   - may have health and safety responsibilities because he or she has an overview of all company employees and contracts in progress

4. Project manager

   - is responsible for the day-to-day management of one specific contract

   - will have overall responsibility on that site for the whole electrical installation

   - attends site meetings with other trades as the representative of the electrical contractor

5. Service manager

   - monitors the quality of the service delivered under the terms of the contractor

   - checks that the contract targets are being met

   - checks that the customer is satisfied with all aspects of the project

   - the service manager's focus is customer specific while the Project Manager's focus is job specific

6. Technician

- will be more office based than site based

- will carry out surveys of electrical systems

- update electrical drawings

- obtain quotations from suppliers

- maintain records such as ISO 9000 quality systems

- carry out testing inspections and commissioning of electrical installations

- trouble shoot

7. Supervisor/Foreman

- This person will probably be a mature electrician

- have responsibility for small contracts

- have responsibility for a small part of a large contract

- be the leader of a small team (e.g. electrician and trainee) installing electrical systems

8. Operative or skilled operative

- This person will carry out the electrical work under the direction and guidance of a supervisor

- will demonstrate a high degree of skill and competence in electrical work

- will have, or be working towards, a recognized electrical qualification and status as an electrician, approved electrician or electrical technician.

9. Mechanic/Fitter

- An operative who usually has a 'core skill' or 'basic skill' and qualification in mechanical rather than electrical engineering

- in production or process work, he or she would have responsibility for the engineering and fitting aspects of the contract, while the electrician and instrumentation technician would take care of the electrical and instrumentation aspects

- all three operatives must work closely in production and process work

- 'additional skilling' or 'multi-skilling' training produces a more flexible operative for production and process plant operations

10. Maintenance manager/engineer

- is responsible for keeping the installed electrotechnical plant and equipment working efficiently

- takes over from the builders and contractors the responsibility of maintaining all plant equipment and systems under his or her control

- might be responsible for a hospital or a commercial building, a university or college complex

- will set up routine and preventative maintenance programmes to reduce possible future breakdowns

- when faults or breakdowns do occur he or she will be responsible for the repair using the company's maintenance staff

Try This

Information

Where do you fit into your company's workforce?

What is your job title – apprentice/trainee?

What is your supervisor's job title?

What is the name of the foreman?

The electrical team

An electrical contractor is responsible for the installation of electrical equipment within a building. An electrical contracting firm is made up of a group of individuals with varying duties and responsibilities (see Fig. 2.1). There is often no clear distinction between the duties of the individuals, and the responsibilities carried by an employee will vary from one employer to another employer. If the firm is to be successful, the individuals must work together to meet the requirements of their customers. Good customer relationships are important for the success of the firm and the continuing employment of the employee.

The customer or his or her representatives will probably see more of the electrician and the electrical trainee than the managing director of the firm and, therefore, the image presented by them is very important. They should always be polite and be seen to be capable and in command of the situation. This gives a customer confidence in the firm's ability to meet his or her needs. The electrician and his or her trainee should be appropriately dressed for the job in hand, which probably means an overall of some kind. Footwear is also important, but sometimes a difficult consideration for an electrician. For example, if working in a factory, the safety regulations insist that protective footwear be worn, but rubber boots may be most appropriate for a building site. However, neither of these would be the most suitable footwear for an electrician fixing a new light fitting in the home of the managing director!

The electrical installation in a building is often carried out alongside other trades. It makes sound sense to help other trades where possible and to develop good working relationships with other employees.

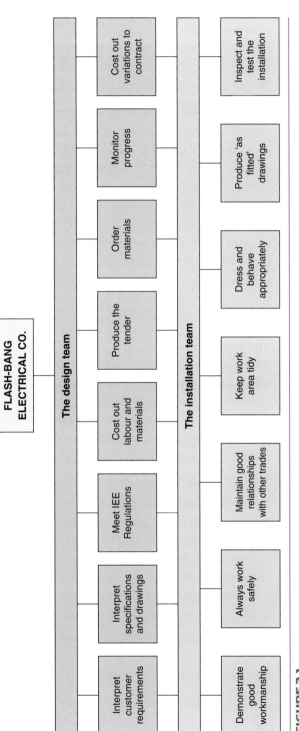

FLASH-BANG ELECTRICAL CO.

The design team

| Interpret customer requirements | Interpret specifications and drawings | Meet IEE Regulations | Cost out labour and materials | Produce the tender | Order materials | Monitor progress | Cost out variations to contract |

The installation team

| Demonstrate good workmanship | Always work safely | Maintain good relationships with other trades | Keep work area tidy | Dress and behave appropriately | Produce 'as fitted' drawings | Inspect and test the installation |

FIGURE 2.1

The electrical team.

The employer has the responsibility of finding sufficient work for his or her employees, paying government taxes and meeting the requirements of the Health and Safety at Work Act described earlier. The rates of pay and conditions for electricians and trainees are determined by negotiation between the Joint Industry Board and the Union, Amicus, which will also represent their members in any disputes. Electricians are usually paid at a rate agreed for their grade as an electrician, approved electrician or technician electrician; movements through the grades are determined by a combination of academic achievement and practical experience.

One of the installation team will have special responsibility for the specific contract being carried out. He or she might be called the project manager or supervisor and will be responsible to his or her electrical company to see that the design specification is carried out and will have overall responsibility on that site for the electrical installation. He or she will attend site meetings as the representative of the electrical contractor, supported by other members of the team, who will demonstrate a range of skills and responsibilities. The supervisor himself will probably be a mature electrician of 'technician' status. The trainee electrician will initially work alongside an electrician or 'approved electrician', who might have been given responsibility for a small part of a large installation by the supervisor on site.

The project manager or site supervisor will be supported by the design team. The design team might be made up of a contracts manager, who will oversee a number of individual electrical contracts at different sites, monitoring progress and costing out variations to the initial contractual agreement. He or she might also have responsibility for health and safety because he or she attends all sites, therefore, has an overview of all company employees and projects that are being carried out.

The contracts manager will also be supported by the design engineer. The design engineer will meet with clients, architects and other trade professionals, to interpret the customer's requirements. He or she will produce the design specifications, which will set out the detailed design of the electrical installation and provide sufficient information to enable a competent person to carry out such installation. The design specifications will also enable a cost for the project to be estimated and included in the legal contracts between the client or main contractor and the electrical contractor.

Electrotechnical industry

The electrical team discussed above are working for an electrical contracting company, which I have called the Flash-Bang Electrical Company (as a joke between you and me). Any electrical contractor is part of the electrotechnical industry. The work of an electrical contractor is one of *installing* electrical equipment and systems, but a very similar role is also carried out by electrical teams working for local councils, the railways, the armed forces and hospitals.

White goods and electrical control panels are *manufactured* and assembled to meet specific specifications by electrical teams working in the manufacturing sector of the electrotechnical industry.

Whatever section of the electrotechnical industry *you* work for, the organizational structure of your company will be similar to the one described above for the electrical contractor. The electrical team in any section of the electrotechnical industry is made up of a dedicated team of 'electrical craftsmen' or 'operatives', carrying out their work to a high standard of competence and skill while complying with the requirements of the relevant regulations. The craftsmen are supported by a supervisor or a foreman, who pulls together the various parts of that specific job or product, thereby meeting the requirements of the client or customer. The supervisor is supported by the manager, who is responsible for designing the electrotechnical product within the requirements of all relevant regulations and specifications.

The craftsman, supervisor, foreman or manager, might be called something different in your electrotechnical company's organization, but there will be a team of people *installing* equipment and systems or *maintaining* equipment and systems or *manufacturing* panels, equipment and machines or *re-winding* electrical machines and transformers. Each individual has a specific role to play within the teams discussed earlier for 'the electrical team'. Each individual is important to the success of the team and the success of the company.

The **electrotechnical industry** is made up of a variety of individual companies, all providing a service within their own specialism to a customer, client or user

The electrical contracting industry provides lighting and power installations so that buildings and systems may be illuminated to an appropriate level, heated to a comfortable level and have the power circuits to drive electrical and electronic equipment. Emergency lighting and security systems are installed so that buildings are safe in unforeseen and adverse situations.

Building management and control systems provide a controlled environment for the people who use commercial buildings.

Instrumentation allows us to monitor industrial processes and systems.

Electrical maintenance allows us to maintain the efficiency of all installed systems.

Computer installations, fibre optic cables and data cabling provide speedy data processing and communications.

High voltage/low voltage (HV/LV) jointing provides a means of connecting new installations and services to live cables without the need to inconvenience existing supplies by electrical shutdown.

Highway electrical systems make our roads, pavements and alleyways safer for vehicle users and pedestrians.

Electrical panels provide electrical protection, isolation and monitoring for the electrical systems in commercial and industrial buildings.

Definition

The electrotechnical industry is made up of a variety of individual companies, all providing a service within their own specialism to a customer, client or user.

51

Electrical machine drive installations drive everything that makes our modern life comfortable, from trains and trams to lifts and air conditioning units.

Finally, *consumer commercial electronics* allows us to live our modern life of rapid personal communication systems while listening to popular or classical music and watching wide screen television.

Designing an electrical installation

The designer of an electrical installation must ensure that the design meets the requirements of the IEE Wiring Regulations for electrical installations and any other regulations which may be relevant to a particular installation. The designer may be a professional technician or engineer whose only job is to design electrical installations for a large contracting firm. In a smaller firm, the designer may also be the electrician who will carry out the installation to the customer's requirements. The designer of any electrical installation is the person who interprets the electrical requirements of the customer within the regulations, identifies the appropriate types of installation, the most suitable methods of protection and control and the size of cables to be used.

A large electrical installation may require many meetings with the customer and his or her professional representatives in order to identify a specification of what is required. The designer can then identify the general characteristics of the electrical installation and its compatibility with other services and equipment, as indicated in Part 3 of the Regulations. The protection and safety of the installation, and of those who will use it, must be considered, with due regard to Part 4 of the Regulations. An assessment of the frequency and quality of the maintenance to be expected (Chapter 34 of the IEE Regulations) will give an indication of the type of installation which is most appropriate.

The size and quantity of all the materials, cables, control equipment and accessories can then be determined. This is called 'bill of quantities'.

It is common practice to ask a number of electrical contractors to tender or submit a price for work specified by the bill of quantities. The contractor must cost all the materials, assess the labour cost required to install the materials and add on profit and overhead costs in order to arrive at a final estimate for the work. The contractor tendering the lowest cost is usually, but not always, awarded the contract.

To complete the contract in the specified time the electrical contractor must use the management skills required by any business to ensure that men and materials are on site as and when they are required. If alterations or modifications are made to the electrical installation as the work proceeds which are outside the original specification, then a variation order must be issued so that the electrical contractor can be paid for the additional work.

The specification for the chosen wiring system will be largely determined by the building construction and the activities to be carried out in the completed building.

An industrial building, for example, will require an electrical installation which incorporates flexibility and mechanical protection. This can be achieved by a conduit, tray or trunking installation.

In a block of purpose-built flats, all the electrical connections must be accessible from one flat without intruding upon the surrounding flats. A loop-in conduit system, in which the only connections are at the light switch and outlet positions, would meet this requirement.

For a domestic electrical installation an appropriate lighting scheme and multiple socket outlets for the connection of domestic appliances, all at a reasonable cost, are important factors which can usually be met by a PVC insulated and sheathed wiring system.

The final choice of a wiring system must rest with those designing the installation and those ordering the work, but whatever system is employed, good workmanship by competent persons is essential for compliance with the regulations. (IEE Regulations 134.1.1)

By definition a **competent person** is one who has the ability to perform a particular task properly.

Generally speaking an electrician will have the necessary skills to perform a wide range of electrical activities competently.

The HSE Regulation 16 states that persons 'must be competent to prevent danger .... so that the person themselves or others are not placed at risk due to a lack of skill when dealing with electrical equipment'.

The 17th Edition of the IEE Regulations also makes the following definitions relating to people.

An **ordinary person** is a person who is neither a skilled person nor an instructed person.

A **skilled person** is a person with technical knowledge or sufficient experience to be able to avoid the dangers which electricity may create.

An **instructed person** is a person adequately advised or supervised by skilled persons to be able to avoid the dangers which electricity may create.

Definition

By definition a *competent person* is one who has the ability to perform a particular task properly.

Key Fact

Definitions

A person can be described as
- ordinary
- competent
- instructed or
- skilled

depending upon that person's skill or ability.

Definition

An *ordinary person* is a person who is neither a skilled person nor an instructed person.

A *skilled person* is a person with technical knowledge or sufficient experience to be able to avoid the dangers which electricity may create.

An *instructed person* is a person adequately advised or supervised by skilled persons to be able to avoid the dangers which electricity may create.

Try This

Definitions

People may be described as an ordinary person, a skilled person, an instructed person or a competent person. Place people you know into each of these categories – for example yourself, your parents, your supervisor, etc.

The necessary skills can be acquired by an electrical trainee who has the correct attitude and dedication to his or her craft. NVQ Level 3 is 'skilled craft level' or the level required to be considered 'competent'.

Legal contracts

Before work commences, some form of legal contract should be agreed between the two parties, that is, those providing the work (e.g. the subcontracting electrical company) and those asking for the work to be carried out (e.g. the main building company).

A contract is a formal document which sets out the terms of agreement between the two parties. A standard form of building contract typically contains four sections:

1. The articles of agreement – this names the parties, the proposed building and the date of the contract period.

2. The contractual conditions – this states the rights and obligations of the parties concerned, for example whether there will be interim payments for work or a penalty if work is not completed on time.

3. The appendix – this contains details of costings, for example the rate to be paid for extras as day work, who will be responsible for defects, how much of the contract tender will be retained upon completion and for how long.

4. The supplementary agreement – this allows the electrical contractor to recoup any value-added tax paid on materials at interim periods.

In signing the contract, the electrical contractor has agreed to carry out the work to the appropriate standards in the time stated and for the agreed cost. The other party, say the main building contractor, is agreeing to pay the price stated for that work upon completion of the installation.

If a dispute arises the contract provides written evidence of what was agreed and will form the basis for a solution.

For smaller electrical jobs, a verbal contract may be agreed, but if a dispute arises there is no written evidence of what was agreed and it then becomes a matter of one person's word against another's.

Professional bodies supporting electrotechnical organizations

If you are reading this book I would guess that you are an electrical trainee working in one sector of the electrotechnical industry. You hope to eventually pass the City & Guilds 2330 Parts 2 and 3 qualifications, take your AM2 Practical Assessment and become a qualified electrician. Believe me, I do wish you well, because you are the future of the electrotechnical industry.

As a trainee, you are probably employed by an electrical company and attend your local college on either a 'Day Release' or 'Block Release' scheme. The combination of work and College will provide you with the skills you will need to become 'fully qualified'!

54

So, although you are doing all the work yourself, you are being sponsored or supported by the company that you work for, the JTL (JIB Training Limited) and the City & Guilds of London Institute to become professionally qualified as an electrician.

It is in this same way that the professional bodies support the electrotechnical industry. They provide a structure of help, support and guidance to the individual companies that make up the electrotechnical industry.

So let us look at some of the professional bodies which support the electrotechnical organizations like the company you work for.

The Institution of Engineering and Technology (IET)

- The IET was formed in spring 2006 by bringing together the Institution of Electrical Engineers (IEE) and the Institution of Incorporated Engineers (IIE)

- The IET is Europe's largest professional society for engineers

The Institution of Electrical Engineers (IEE)

- The IEE was established in 1871

- The IEE produces the IEE Wiring Regulations to BS 7671

- They also produce many other publications and provide training courses to help electricians, managers and supervisors to keep up to date with the changes in the relevant regulations

- The one Site Guide describes the 'requirements for electrical installations'

- Eight guidance notebooks are available

- The Electricians Guide to the Building Regulations clarifies the requirements for electrical operatives of the new Part P Regulations which came into effect on the 1st January 2005

- *Wiring Matters* is a quarterly magazine published by the IEE covering many of the topics which may trouble some of us in the electrotechnical industries

- All of these publications can be purchased by visiting the IET website at www.iet.org/shop or email: sales@iet.org.uk

The Electrical Contractors Association (ECA)

- The ECA was founded over 100 years ago and is a trade association representing electrotechnical companies

- Membership is made up of electrical contracting companies both large and small

- Customers employing an electrical contractor who has ECA membership are guaranteed that the work undertaken will meet all relevant regulations. If the work undertaken fails to meet the relevant standards, the ECA will arrange for the work to be rectified at no cost to the customer

- The work of the ECA member is regularly assessed by the Association's UKAS accredited inspection body

- Those electrotechnical companies which are members of the ECA are permitted to display the ECA logo on their company vehicles and stationery

- Further information can be found on the ECA website at www.eca. co.uk

The National Inspection Council for Electrical Installation Contracting (NICEIC)

- The NICEIC is an independent consumer safety organization, set up to protect users of electricity against the hazards of unsafe electrical installations

- It is the electrical industry's safety regulatory body

- The NICEIC publishes a list of approved contractors whose standard of work is regularly assessed by local area engineers

- Customers employing an electrical contractor who has NICEIC membership can be assured that the work carried out will meet all relevant standards. If the work undertaken fails to meet all relevant standards, the name of the electrical contractor will be removed from the 'NICEIC Approved List'

  Some work, such as local authority work, is only available to NICEIC approved contractors

- Further information can be found at www.niceic.org.uk

Trade Unions

- Trade Unions have a long history of representing workers in industry and commerce

- The relevant unions negotiate with employer organizations the pay and working conditions of their members

- The Trade Union which represents employees in the electrotechnical industry in the new millennium is called Amicus

- Through a network of local area offices the union offers advice and support for its members. They will also provide legal advice and representation if a member has a serious accident as a result of a health and safety issue or has a dispute with an employer

- Further information can be found at www.amicustheunion.org

Try This

Trade Organizations

Does the company you work for belong to a trade organization?

Why do they belong, and what are the advantages?

Do you belong to a Trade Union – if so, which one?

If not, why? Trade Union membership is often free while you are training.

Check your Understanding

When you have completed the questions, check out the answers at the back of this book.
Note: more than one multiple choice answers may be correct.

1. An electrotechnical company manufacturing 'white' goods would probably call itself:

   a. an electrical contracting company

   b. a panel building company

   c. a motor re-wind company

   d. an equipment and machine manufacturer.

2. An electrotechnical company installing lighting and power systems in buildings would probably call itself:

   a. an electrical contracting company

   b. a panel building company

   c. a motor re-wind company

   d. an equipment and machine manufacturer.

3. An electrotechnical company reconditioning transformers and electric motors would probably call itself:

   a. an electrical contracting company

   b. a panel building company

   c. a motor re-wind company

   d. an equipment and machine manufacturer.

4. An electrotechnical company manufacturing specialist switchgear, protection, monitoring and control equipment would probably call itself:

   a. an electrical contracting company

   b. a panel building company

   c. a motor re-wind company

   d. an equipment and machine manufacturer.

5. Emergency lighting and security systems

   a. make our roads and pavements safe for people

   b. monitor industrial processes

   c. ensure that a building is safe and secure

   d. provide a controlled environment for people in buildings.

6. Building management and control systems

   a. make our roads and pavements safe for people

   b. monitor industrial processes

   c. ensure that a building is safe and secure

   d. provide a controlled environment for people in buildings.

7. Electrical instrumentation systems
    a. make our roads and pavements safe for people
    b. monitor industrial processes
    c. ensure that a building is safe and secure
    d. provide a controlled environment for people in buildings.

8. Highway electrical systems
    a. make our roads and pavements safe for people
    b. monitor industrial processes
    c. ensure that a building is safe and secure
    d. provide a controlled environment for people in buildings.

9. A mature electrician having responsibility for the whole design and completion of small contracts will probably be called
    a. the contracts manager
    b. supervisor or foreman
    c. a skilled operative
    d. a mechanic fitter.

10. A person carrying out electrical work with a high degree of skill and competence under the guidance of a supervisor will probably be called
    a. the contracts manager
    b. supervisor or foreman
    c. a skilled operative
    d. a mechanic fitter.

11. A person responsible for a number of large electrical jobs on different sites will probably be called
    a. the contracts manager
    b. supervisor or foreman
    c. a skilled operative
    d. a mechanic fitter.

12. A skilled operative who has a core skill and qualification in mechanical engineering will probably be called
    a. the contracts manager
    b. supervisor or foreman
    c. a skilled operative
    d. a mechanic fitter.

13. A professional organization representing and supporting individual electrotechnical *employees* is
    a. The NICEIC
    b. The ECA
    c. The IEE
    d. Amicus, the Union.

14. A professional organization representing and supporting individual electrotechnical *companies* is
    a. The NICEIC
    b. The ECA
    c. The IEE
    d. Amicus, the Union.

15. Define what is meant by the electrotechnical industry.

16. Think about the person in your company who has the job title supervisor/foreman. What is his or her name, and what are his or her responsibilities at work.

17. Think about the person in your company who has the job title contracts manager or project manager. What is his or her name, and what are his or her responsibilities at work.

18. Think about the person in your company who might be called a skilled operative. What is his or her name, what does he or she do at work and what are his or her qualifications.

19. Use an annotated block diagram to show the structure of the staff in your electrotechnical company from manager to operative and trainee.

20. Does your electrotechnical company belong to the ECA or the NICEIC? Do the company vehicles display the ECA or NICEIC logo, if so:
    a. why does your company want to advertise the fact that it belongs to the ECA or NICEIC?
    b. what are the advantages of being a company member of the ECA or NICEIC?
    c. does the ECA or NICEIC Inspector call once a year, if so why, what does he or she do?

    You might want to ask your supervisor to help you with this question.

21. Define the meaning of the words 'ordinary person' in an electrotechnical context.

22. Define the meaning of the words 'competent person' in an electrotechnical context.

23. Define the meaning of the words 'skilled person' in an electrotechnical context.

24. Define the meaning of the words 'instructed person' in an electrotechnical context.

CH 3

Sources of technical information and communications

Unit 1 – Working effectively and safely in the electrotechnical environment – Outcome 3

Underpinning knowledge: when you have completed this chapter you should be able to:

- state the sources of technical information
- identify types of drawings and diagrams
- recognize BE EN 60617 symbols

Communications

When we talk about good communications we are talking about transferring information from one person to another both quickly and accurately. We do this by talking to other people, looking at drawings and plans and discussing these with colleagues from the same company and with other professionals who have an interest in the same project. The technical information used within our industry comes from many sources. The IEE Regulations (BS 7671) is the 'electrician's bible' and forms the basis of all our electrical design calculations and installation methods. British Standards, European Harmonised Standards and Codes of Practice provide detailed information for every sector of the electrotechnical industry, influencing all design and build considerations.

Sources of technical information

Equipment and accessories available to use in a specific situation can often be found in the very comprehensive manufacturer's catalogues and the catalogues of the major wholesalers that service the electrotechnical industries.

All of this technical information may be distributed and retrieved by using:

- conventional drawings and diagrams which we will look at in more detail below
- sketch drawing to illustrate an idea or the shape of say a bracket to hold a piece of electrical equipment
- the Internet can be used to download British Standards and Codes of Practice
- the Internet can also be used to download health and safety information from the health and safety executive at: www.gov.uk/hseor www.opsi.gov.uk
- CDs, DVDs, USB memory sticks and email can be used to communicate and store information electronically
- the facsimile (Fax) machine can be used to communicate with other busy professionals, information say about a project you are working on together.

If you are working at your company office with access to online computers, then technical information is only a fingertip or mouse click away. However, a construction site is a hostile environment for a laptop and so a hard copy of any data is preferable on site.

Let us now look at the types of drawings and diagrams which we use within our industry to communicate technical information between colleagues and other professionals. The type of diagram to be used in any particular situation is the one which most clearly communicates the desired information.

Site plans or layout drawings

These are scale drawings based upon the architect's site plan of the building and show the position of the electrical equipment which is to be installed. The electrical equipment is identified by a graphical symbol. The standard symbols used by the electrical contracting industry are those recommended by the British Standard EN 60617, *Graphical Symbols for Electrical Power, Telecommunications and Electronic Diagrams*. Some of the more common electrical installation symbols are given in Fig. 3.1.

The site plan or layout drawing will be drawn to a scale, smaller than the actual size of the building, so to find the actual measurement, you must measure the distance on the drawing and multiply by the scale.

For example, if the site plan is drawn to a scale of 1:100, then 10 mm on the site plan represents 1 m measured in the building.

The layout drawing or site plan of a small domestic extension is shown in Fig. 3.2. It can be seen that the mains intake position, probably a consumer unit, is situated in the storeroom which also contains one light controlled by a switch at the door. The bathroom contains one lighting point controlled by a one-way pull switch at the door. The kitchen has two doors and a switch is installed at each door to control the fluorescent luminaire. There are also three double sockets situated around the kitchen. The sitting room has a two-way switch at each door controlling the centre lighting point. Two wall lights with built-in switches are to be wired, one at each side of the window. Two double sockets and one switched socket are also to be installed in the sitting room. The bedroom has two lighting points controlled independently by two one-way switches at the door. The wiring diagrams and installation procedures for all these circuits can be found in later chapters.

Try This

Drawing

The next time you are on site ask your supervisor to show you the site plans. Ask him:

- how does the scale work
- put names to the equipment represented by British Standard symbols

As-fitted drawings

When the installation is completed a set of drawings should be produced which indicate the final positions of all the electrical equipment. As the building and electrical installation progresses, it is sometimes necessary to modify the positions of equipment indicated on the layout drawing because, for example, the position of a doorway has been changed. The layout drawings or site plans indicate the original intentions for the position of equipment, while the 'as-fitted' drawing indicates the actual positions of equipment upon completion of the contract.

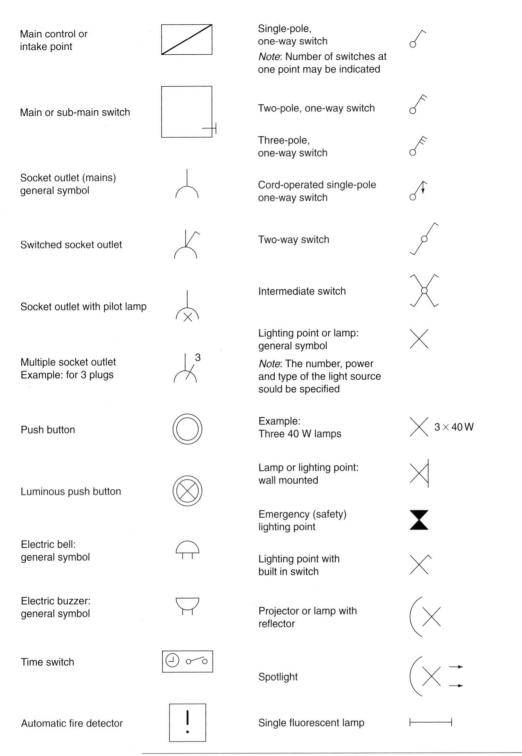

FIGURE 3.1
Some BS EN 60617 electrical installation symbols.

Try This

Drawings

Take a moment to clarify the difference between:

- layout drawings and
- as-fitted drawings.

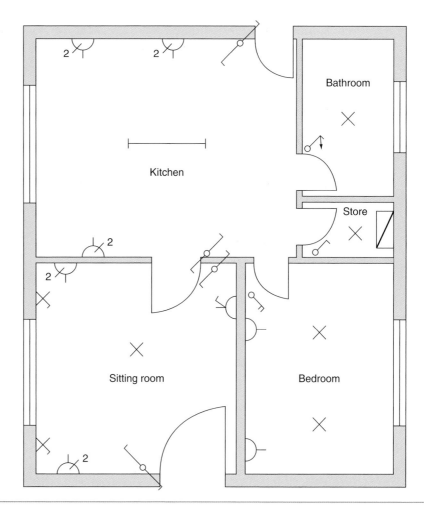

FIGURE 3.2
Layout drawing or site plan of a small electrical installation.

Detail drawings and assembly drawings

These are additional drawings produced by the architect to clarify some point of detail. For example, a drawing might be produced to give a fuller description of a suspended ceiling arrangement or the assembly arrangements of the metalwork for the suspended ceiling.

Location drawings

Location drawings identify the place where something is located. It might be the position of the manhole covers giving access to the drains. It might be the position of all water stop taps or the position of the emergency lighting fittings. This type of information may be placed on a blank copy of the architect's site plan or on a supplementary drawing.

Distribution cable route plans

On large installations there may be more than one position for the electrical supplies. Distribution cables may radiate from the site of the electrical mains intake position to other sub-mains positions. The site of the sub-mains and the route taken by the distribution cables may be shown on a blank copy of the architect's site plan or on the electricians 'as-fitted' drawings.

Block diagrams

A block diagram is a very simple diagram in which the various items or pieces of equipment are represented by a square or rectangular box. The purpose of the block diagram is to show how the components of the circuit relate to each other and, therefore, the individual circuit connections are not shown. Figure 3.3 shows the block diagram of a space heating control system.

Wiring diagrams

A wiring diagram or connection diagram shows the detailed connections between components or items of equipment. They do not indicate how a piece of equipment or circuit works. The purpose of a wiring diagram is to help someone with the actual wiring of the circuit. Figure 3.4 shows the wiring diagram for a space heating control system. Other wiring diagrams can be seen in Figs 4.4–4.7 of Chapter 14.

Circuit diagrams

A circuit diagram shows most clearly how a circuit works. All the essential parts and connections are represented by their graphical symbols. The purpose of a circuit diagram is to help our understanding of the circuit. It will be laid out as clearly as possible, without regard to the physical lay-out of the actual components and, therefore, it may not indicate the most convenient way to wire the circuit. Figure 3.5 shows the circuit diagram of our same space heating control system. Figures 5.1 and 5.2 in Chapter 5 are circuit diagrams.

Schematic diagrams

A schematic diagram is a diagram in outline of, for example, a motor starter circuit. It uses graphical symbols to indicate the interrelationship of the

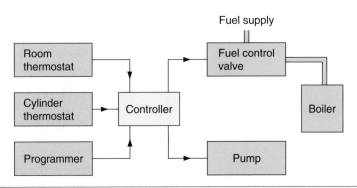

FIGURE 3.3

Block diagram – space heating control system (Honeywell Y. Plan).

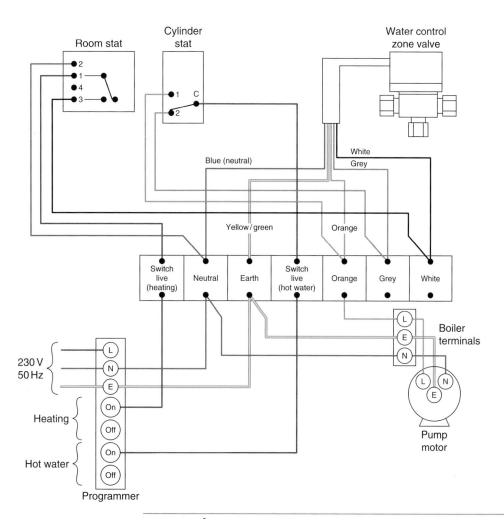

FIGURE 3.4

Wiring diagram – space heating control system (Honeywell Y. Plan).

electrical elements in a circuit. These help us to understand the working operation of the circuit but are not helpful in showing us how to wire the components. An electrical schematic diagram looks very like a circuit diagram. Figure 3.6 shows a schematic diagram.

Freehand working diagrams

Freehand working drawings or sketches are another important way in which we communicate our ideas. The drawings of the spring toggle bolt in Chapter 14 (Fig. 14.39) were done from freehand sketches. A freehand sketch may be done as an initial draft of an idea before a full working drawing is made. It is often much easier to produce a sketch of your ideas or intentions than to describe them or produce a list of instructions.

To convey the message or information clearly it is better to make your sketch large rather than too small. It should also contain all the dimensions necessary to indicate clearly the size of the finished object depicted by the sketch.

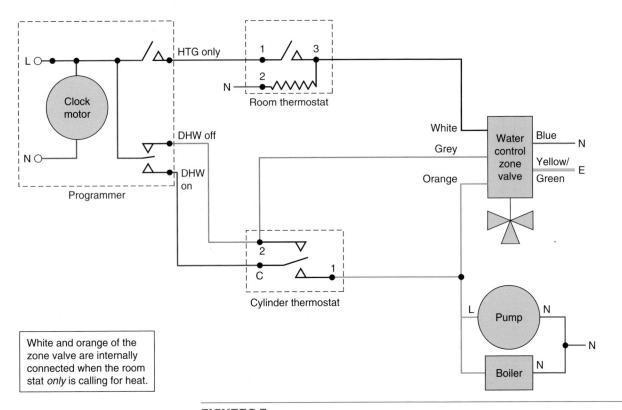

White and orange of the
zone valve are internally
connected when the room
stat *only* is calling for heat.

FIGURE 3.5

Circuit diagram – space heating control system (Honey well Y. plan).

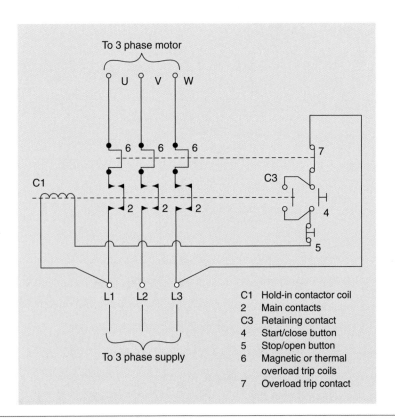

FIGURE 3.6

Schematic diagram – DOL motor starter.

All drawings and communications must be aimed at satisfying the client's wishes for the project. It is the client who will pay the final bill which, in turn, pays your wages. The detailed arrangements of what must be done to meet the client's wishes are contained in the client's specification documents and all your company's efforts must be directed at meeting the whole specification, but no more.

The positional reference system

A positional reference system can be used to mark exact positions in any space. It uses a simple grid reference system to mark out points in the space enclosed by the grid. It is easy to understand if we consider a specific example which I use when building prototype electronic circuits on matrix board. Matrix board is the insulated board full of holes into which we insert small pins and then attach the electronic components.

To set up the grid reference, count along the columns at the top of the board, starting from the left and then count down the rows. The position of point 4:3 would be 4 holes from the left and 3 holes down.

Prepare a matrix board, or any space for that matter, as follows:

- Turn the matrix board so that a manufactured straight edge is to the top and left-hand side

- Use a felt tip pen to mark the holes in groups of five along the top edge and down the left-hand edge as shown in Fig. 3.7.

The pins can then be inserted as required. Figure 3.7 shows a number of pin reference points. Counting from the left-hand side of the board there are 3:3, 3:16, 10:11, 18:3, 18:11, 25:3 and 25:16.

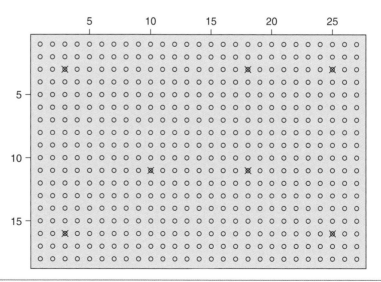

FIGURE 3.7

Positional reference system used to identify points on an electronic matrix board.

Telephone communications

Telephones today play one of the most important roles in enabling people to communicate with each other. You are never alone when you have a telephone. If there is a problem, you can ring your supervisor or foreman for help. The advantage of a telephone message over a written message is its speed; the disadvantage is that no record is kept of an agreement made over the telephone. Therefore, business agreements made on the telephone are often followed up by written confirmation.

When *taking* a telephone call, remember that you cannot be seen and, therefore, gestures and facial expressions will not help to make you understand. Always be polite and helpful when answering your company's telephone – you are your company's most important representative at that moment. Speak clearly and loud enough to be heard without shouting, sound cheerful and write down messages if asked. Always read back what you have written down to make sure that you are passing on what the caller intended.

Many companies now use standard telephone message pads such as that shown in Fig. 3.8 because they prompt people to collect all the relevant

FLASH-BANG ELECTRICAL   TELEPHONE MESSAGES

Date _____ Thurs 11 Aug. 08 _____ Time _09.30_

Message to _____ Dave Twem

Message from (Name) __ John Gall

(Address) _____ Bispham Site

_____ Blackpool.

(Telephone No.) _____ (01253) 123456

Message _____ Pick up Megger _from Jim on Saturday and take to Bispham_ _site on Monday._

Thanks

Message taken by _____ Dave Lan

FIGURE 3.8
Typical standard telephone message pad.

information. In this case, John Gall wants Dave Twem to pick up the Megger from Jim on Saturday and take it to the Bispham site on Monday. The person taking the call and relaying the message is Dave Low.

When *making* a telephone call, make sure you know what you want to say or ask. Make notes so that you have times, dates and any other relevant information ready before you make the call.

Check your Understanding

When you have completed the questions, check out the answers at the back of the book.

Note: more than one multiple choice answer may be correct.

1. A scale drawing which shows the original intention for the position of electrical equipment is called a:
   a. wiring diagram
   b. detail assembly drawing
   c. site plan or layout drawing
   d. as-fitted drawing.

2. The scale drawing which shows the actual position of the electrical equipment upon completion of the contract is called:
   a. wiring diagram
   b. detail assembly drawing
   c. site plan or layout drawing
   d. as-fitted drawing.

3. A scale drawing showing the position of equipment by graphical symbols is a description of a:
   a. block diagram
   b. site plan or layout diagram
   c. wiring diagram
   d. circuit diagram.

4. A diagram which shows the detailed connections between individual items of equipment is a description of a:
   a. block diagram
   b. site plan or layout diagram
   c. wiring diagram
   d. circuit diagram.

5. A diagram which shows most clearly how a circuit works, with all items represented by graphical symbols is a description of a:
   a. block diagram
   b. site plan or layout diagram
   c. wiring diagram
   d. circuit diagram.

6. A site plan has a scale of 1:100. From the scale drawing you can see that a socket outlet must be positioned on a wall 40 cm from the corner of the room in which you are standing. How far from the corner of the room would you actually measure to the centre of the socket's fixing position taking the scale into account:

    a. 0.4 m

    b. 4.0 m

    c. 10.0 m

    d. 40.0 m.

7. Sketch a site plan or layout drawing for the room which you normally use at college and indicate the position of all the electrical accessories in the room using BS EN 60617 symbols.

8. What methods could you use to find and store some information about :

    - Health and safety at work

    - British Standards

    - Electrical accessories and equipment.

9. What method would you use to let the office know that the materials you were expecting have not yet arrived.

10. What method would you use to send a long list of materials required for the job you are on to the wholesalers for later delivery to the site. Use bullet points.

11. What are the advantages and disadvantages of having sources of technical information on:

    a. some form of electronic storage system such as a CD, DVD or USB memory stick or

    b. hard copy such as a catalogue, drawings or On Site Guide.

    Would it make a difference if you were at the office or on a construction site?

12. State the advantages and disadvantages of:

    a. telephone messages

    b. written messages.

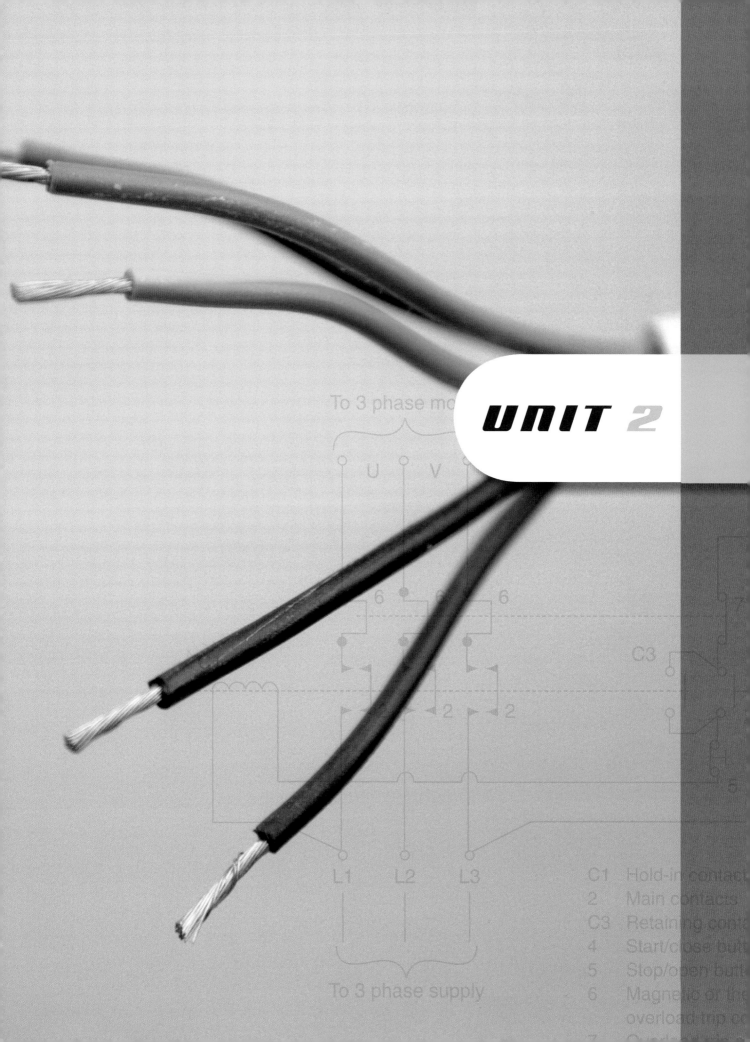

UNIT 2

To 3 phase mo...

U   V

6   6

C3

2   2

5

L1   L2   L3

C1   Hold-in contact
2   Main contacts
C3   Retaining con...
4   Start/close but...
5   Stop/open but...
6   Magnetic or t...
        overload trip ...

To 3 phase supply

CH 4

Basic electrotechnical units and theory

Unit 2 – Principles of electrotechnology – Outcome 1

Underpinning knowledge: when you have completed this chapter you should be able to:

- identify SI units and their symbols for various quantities such as current, voltage, resistance, force, mass, power and energy

- state the relationship between resistance, length and area

- state the relationship between force, mass, weight and turning force (Basic Mechanics)

- state the efficiency of a machine

- state the constructional features and principles of a simple alternator

- state the maximum rms and average value of an a.c. waveform

Units

Very early units of measurement were based on the things easily available – the length of a stride, the distance from the nose to the outstretched hand, the weight of a stone and the time-lapse of one day. Over the years, new units were introduced and old ones were modified. Different branches of science and engineering were working in isolation, using their own units, and the result was an overwhelming variety of units.

In all branches of science and engineering there is a need for a practical system of units which everyone can use. In 1960, the General Conference of Weights and Measures agreed to an international system called the Système International d'Unités (abbreviated to **SI units**).

SI units are based upon a small number of fundamental units from which all other units may be derived. Table 4.1 describes some of the basic units that we shall be using in our electrical studies.

Like all metric systems, SI units have the advantage that prefixes representing various multiples or sub-multiples may be used to increase or decrease the size of the unit by various powers of 10. Some of the more common prefixes and their symbols are shown in Table 4.2.

Definition

SI units are based upon a small number of fundamental units from which all other units may be derived.

80

Table 4.1 Basic SI Units

Quantity	Measure of	Basic unit	Symbol	Notes
Area	Length × length	Metre squared	m^2	
Current I	Electric current	Ampere	A	
Energy	Ability to do work	Joule	J	Joule is a very small unit 3.6×10^6 J = 1 kWh
Force	The effect on a body	Newton	N	
Frequency	Number of cycles	Hertz	Hz	mains frequency is 50 Hz
Length	Distance	Metre	m	
Mass	Amount of material	kilogram	kg	One metric tonne = 1000 kg
Magnetic flux Φ	Magnetic energy	Weber	Wb	
Magnetic flux density B	Number of lines of magnetic flux	Tesla	T	
Potential or pressure	Voltage	Volt	V	
Period T	Time taken to complete one cycle	Second	s	The 50 Hz mains supply has a period of 20 ms
Power	Rate of doing work	Watt	W	
Resistance	Opposition to current flow	Ohm	Ω	
Resistivity	Resistance of a sample piece of material	Ohm metre	ρ	Resistivity of copper is 17.5×10^{-9} Ωm
Temperature	Hotness or coldness	Kelvin	K	0°C = 273 K. A change of 1 K is the same as 1°C
Time	Time	Second	s	60 s = 1 min 60 min = 1 h
Weight	Force exerted by a mass	Kilogram	kg	1000 kg = 1 tonne

Note: A more detailed description can be found in this chapter.

Prefix	Symbol	Multiplication factor		
Table 4.2 Symbols and Multiples for Use with SI Units				
Mega	M	$\times 10^6$	or	$\times 1,000,000$
Kilo	K	$\times 10^3$	or	$\times 1000$
Hecto	H	$\times 10^2$	or	$\times 100$
Decca	da	$\times 10$	or	$\times 10$
Deci	D	$\times 10^{-1}$	or	$\div 10$
Centi	C	$\times 10^{-2}$	or	$\div 100$
Milli	M	$\times 10^{-3}$	or	$\div 1000$
Micro	μ	$\times 10^{-6}$	or	$\div 1,000,000$

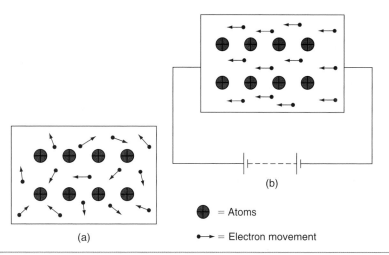

(b)

⊕ = Atoms

•— = Electron movement

(a)

FIGURE 4.1

Atoms and electrons on a material.

Basic circuit theory

All matter is made up of atoms which arrange themselves in a regular framework within the material. The atom is made up of a central, positively charged nucleus, surrounded by negatively charged electrons. The electrical properties of a material depend largely upon how tightly these electrons are bound to the central nucleus.

A **conductor** is a material in which the electrons are loosely bound to the central nucleus and are, therefore, free to drift around the material at random from one atom to another, as shown in Fig. 4.1(a). Materials which are good conductors include copper, brass, aluminium and silver.

An **insulator** is a material in which the outer electrons are tightly bound to the nucleus, so there are no free electrons to move around the material. Good insulating materials are PVC, rubber, glass and wood.

If a battery is attached to a conductor as shown in Fig. 4.1(b), the free electrons drift purposefully in one direction only. The free electrons close to the positive plate of the battery are attracted to it since unlike charges attract, and the free electrons near the negative plate will be repelled from it. For each electron entering the positive terminal of the battery, one will

Definition

A *conductor* is a material in which the electrons are loosely bound to the central nucleus and are, therefore, free to drift around the material at random from one atom to another, as shown in Fig. 4.1(a). Materials which are good conductors include copper, brass, aluminium and silver.

Definition

An *insulator* is a material in which the outer electrons are tightly bound to the nucleus, so there are no free electrons to move around the material. Good insulating materials are PVC, rubber, glass and wood.

be ejected from the negative terminal, so the number of electrons in the conductor remains constant.

The drift of electrons within a conductor is known as an **electric current**, measured in amperes and given the symbol I.

For a current to continue to flow, there must be a complete circuit for the electrons to move around. If the circuit is broken by opening a switch, for example, the electron flow and therefore the current will stop immediately.

To cause a current to flow continuously around a circuit, a driving force is required, just as a circulating pump is required to drive water around a central heating system. This driving force is the *electromotive force* (emf). Each time an electron passes through the source of emf, more energy is provided to send it on its way around the circuit.

An emf is always associated with energy conversion, such as chemical to electrical in batteries and mechanical to electrical in generators. The energy introduced into the circuit by the emf is transferred to the load terminals by the circuit conductors.

The **potential difference** (p.d.) is the change in energy levels measured across the load terminals. This is also called the volt drop or terminal voltage, since emf and p.d. are both measured in volts. **Resistance** every circuit offers some opposition to current flow, which we call the circuit *resistance*, measured in ohms (symbol Ω), to commemorate the famous German physicist Georg Simon Ohm, who was responsible for the analysis of electrical circuits.

Ohm's law

In 1826, Ohm published details of an experiment he had done to investigate the relationship between the current passing through and the potential difference between the ends of a wire. As a result of this experiment, he arrived at a law, now known as **Ohm's law**, which says that the current passing through a conductor under constant temperature conditions is proportional to the potential difference across the conductor. This may be expressed mathematically as

$$V = I \times R (V)$$

Transposing this formula, we also have

$$I = \frac{V}{R} (A) \quad \text{and} \quad R = \frac{V}{I} (\Omega)$$

Example 1

An electric heater, when connected to a 230 V supply, was found to take a current of 4 A. Calculate the element resistance.

$$R = \frac{V}{I}$$

Definition

The drift of electrons within a conductor is known as an *electric current*, measured in amperes and given the symbol I.

Definition

The *potential difference* (p.d.) is the change in energy levels measured across the load terminals. This is also called the volt drop or terminal voltage, since emf and p.d. are both measured in volts.

Definition

Every circuit offers some opposition to current flow, which we call the circuit *resistance*, measured in ohms (symbol Ω).

Definition

Ohm's law, which says that the current passing through a conductor under constant temperature conditions is proportional to the potential difference across the conductor.

$$\therefore R = \frac{230\text{ V}}{4\text{ A}} = 57.5\,\Omega$$

Example 2

The insulation resistance measured between phase conductors on a 400 V supply was found to be 2 MΩ. Calculate the leakage current.

$$I = \frac{V}{R}$$

$$\therefore I = \frac{400\text{ V}}{2 \times 10^6\,\Omega} = 200 \times 10^{-6}\text{ A} = 200\ \mu\text{A}$$

Example 3

When a 4 Ω resistor was connected across the terminals of an unknown d.c. supply, a current of 3 A flowed. Calculate the supply voltage.

$$V = I \times R$$
$$\therefore V = 3\text{A} \times 4\Omega = 12\text{V}$$

Resistivity

The resistance or opposition to current flow varies for different materials, each having a particular constant value. If we know the resistance of, say, 1 m of a material, then the resistance of 5 m will be five times the resistance of 1 m.

The **resistivity** (symbol ρ – the Greek letter 'rho') of a material is defined as the resistance of a sample of unit length and unit cross-section. Typical values are given in Table 4.3. Using the constants for a particular material we can calculate the resistance of any length and thickness of that material from the equation.

$$R = \frac{\rho l}{a}\,(\Omega)$$

where

ρ = the resistivity constant for the material (Ω m)

l = the length of the material (m)

a = the cross-sectional area of the material (m^2).

Definition

The *resistivity* (symbol ρ – the Greek letter 'rho') of a material is defined as the resistance of a sample of unit length and unit cross-section.

Table 4.3 Resistivity Values

Material	Resistivity (Ω m)
Silver	16.4×10^{-9}
Copper	17.5×10^{-9}
Aluminium	28.5×10^{-9}
Brass	75.0×10^{-9}
Iron	100.0×10^{-9}

Table 4.3 gives the resistivity of silver as $16.4 \times 10^{-9}\ \Omega$ m, which means that a sample of silver 1 m long and 1 m in cross-section will have a resistance of $16.4 \times 10^{-9}\ \Omega$.

Example 1

Calculate the resistance of 100 m of copper cable of 1.5 mm² cross-sectional area if the resistivity of copper is taken as $17.5 \times 10^{-9}\ \Omega$ m.

$$R = \frac{\rho l}{a}\ (\Omega)$$

$$\therefore R = \frac{17.5 \times 10^{-9}\ \Omega\ \text{m} \times 100\ \text{m}}{1.5 \times 10^{-6}\ \text{m}^2} = 1.16\ \Omega$$

Example 2

Calculate the resistance of 100 m of aluminium cable of 1.5 mm² cross-sectional area if the resistivity of aluminium is taken as $28.5 \times 10^{-9}\ \Omega$ m.

$$R = \frac{\rho l}{a}\ (\Omega)$$

$$\therefore R = \frac{28.5 \times 10^{-9}\ \Omega\ \text{m} \times 100\ \text{m}}{1.5 \times 10^{-6}\ \text{m}^2} = 1.9\ \Omega$$

The above examples show that the resistance of an aluminium cable is some 60% greater than a copper conductor of the same length and cross-section. Therefore, if an aluminium cable is to replace a copper cable, the conductor size must be increased to carry the rated current as given by the tables in Appendix 4 of the *IEE Regulations* and Appendix 6 of the *On Site Guide*.

The other factor which affects the resistance of a material is the temperature, and we will consider this later.

Try This

Resistance
- Take two 100 m lengths of singles cable (2 coils)
- Measure the resistance of 100 m of cable (1 coil)
  Value _____ Ω
- Join the two lengths together (200 m) and again measure the resistance
  Value _____ Ω
  Does this experiment prove resistance is proportional to length?
  If the resistance is doubled, it is proved! QED, Quo Erat Demonstration (Latin for it is proved).

Basic mechanics and machines

Mechanics is the scientific study of 'machines', where a machine is defined as a device which transmits motion or force from one place to another. An engine is one particular type of machine, an energy-transforming machine, converting fuel energy into a more directly useful form of work.

Most modern machines can be traced back to the five basic machines described by the Greek inventor Hero of Alexandria who lived about the time of Christ. The machines described by him were the wedge, the screw, the wheel and axle, the pulley and the lever. Originally they were used for simple purposes, to raise water and to move objects which man alone could not lift, but today their principles are of fundamental importance to our scientific understanding of mechanics. Let us now consider some fundamental mechanical principles and calculations.

Mass

Mass is a measure of the amount of material in a substance, such as metal, plastic, wood, brick or tissue, which is collectively known as a body. The mass of a body remains constant and can easily be found by comparing it on a set of balance scales with a set of standard masses. The SI unit of mass is the kilogram (kg).

Weight

Weight is a measure of the force which a body exerts on anything which supports it. Normally it exerts this force because it is being attracted towards the earth by the force of gravity.

For scientific purposes the weight of a body is *not* constant, because gravitational force varies from the equator to the poles; in space a body would be 'weightless' but here on earth under the influence of gravity a 1 kg mass would have a weight of approximately 9.81 N (see also the definition of 'force').

Speed

The feeling of speed is something with which we are all familiar. If we travel in a motor vehicle we know that an increase in speed would, excluding accidents, allow us to arrive at our destination more quickly. Therefore, **speed** is concerned with distance travelled and time taken. Suppose we were to travel a distance of 30 miles in 1 h; our speed would be an average of 30 miles/h:

$$\text{Speed} = \frac{\text{Distance (m)}}{\text{Time (s)}}$$

Velocity

In everyday conversation we often use the word velocity to mean the same as speed, and indeed the units are the same. However, for scientific purposes this is not acceptable since velocity is also concerned with direction. Velocity is speed in a given direction. For example, the speed of an aircraft might be 200 miles/h, but its velocity would be 200 miles/h in, say, a westerly direction. Speed is a scalar quantity, while velocity is a vector quantity.

$$\text{Velocity} = \frac{\text{Distance (m)}}{\text{Time (s)}}$$

Acceleration

When an aircraft takes off, it starts from rest and increases its velocity until it can fly. This change in velocity is called its acceleration. By definition, **acceleration** is the rate of change in velocity with time.

$$\text{Acceleration} = \frac{\text{Velocity}}{\text{Time}} = (\text{m/s}^2)$$

Example

If an aircraft accelerates from a velocity of 15 m/s to 35 m/s in 4 s, calculate its average acceleration.

Average velocity = 35 m/s – 15 m/s = 20 m/s

$$\text{Average acceleration} = \frac{\text{Velocity}}{\text{Time}} = \frac{20}{4} = 5 \text{ m/s}^2$$

Thus, the average acceleration is 5 m/s².

Force

Definition

Force The presence of a force can only be detected by its effect on a body. A force may cause a stationary object to move or bring a moving body to rest.

The presence of a force can only be detected by its effect on a body. A force may cause a stationary object to move or bring a moving body to rest. For example, a number of people pushing a broken-down motor car exert a force which propels it forward, but applying the motor car brakes applies a force on the brake drums which slows down or stops the vehicle. Gravitational force causes objects to fall to the ground. The apple fell from the tree on to Isaac Newton's head as a result of gravitational force. The standard rate of acceleration due to gravity is accepted as 9.81 m/s². Therefore, an apple weighing 1 kg will exert a force of 9.81 N since

$$\text{Force} = \text{Mass} \times \text{Acceleration (N)}$$

The SI unit of force is the newton, symbol N, to commemorate the great English scientist Sir Isaac Newton (1642–1727).

Example

A 50 kg bag of cement falls from a forklift truck while being lifted to a storage shelf. Determine the force with which the bag will strike the ground:

$$\text{Force} = \text{Mass} \times \text{Acceleration (N)}$$

$$\text{Force} = 50 \text{ kg} \times 9.81 \text{ m/s}^2 = 490.5 \text{ N}$$

A force can manifest itself in many different ways. Let us consider a few examples:

- 'Inertial force' is the force required to get things moving, to change direction or stop, like the motor car discussed above.

- 'Cohesive or adhesive force' is the force required to hold things together.

- 'Tensile force' is the force pulling things apart.

- 'Compressive force' is the force pushing things together.

- 'Friction force' is the force which resists or prevents the movement of two surfaces in contact.

Definition

- 'Inertial force' is the force required to get things moving, to change direction or stop, like the motor car discussed above.
- 'Cohesive or adhesive force' is the force required to hold things together.
- 'Tensile force' is the force pulling things apart.
- 'Compressive force' is the force pushing things together.
- 'Friction force' is the force which resists or prevents the movement of two surfaces in contact.
- 'Shearing force' is the force which moves one face of a material over another.
- 'Centripetal force' is the force acting towards the centre when a mass attached to a string is rotated in a circular path.
- 'Centrifugal force' is the force acting away from the centre, the opposite to centripetal force.
- 'Gravitational force' is the force acting towards the centre of the earth due to the effect of gravity.
- 'Magnetic force' is the force created by a magnetic field.
- 'Electrical force' is the force created by an electrical field.

Definition

Pressure or *stress* is a measure of the force per unit area.

- 'Shearing force' is the force which moves one face of a material over another.
- 'Centripetal force' is the force acting towards the centre when a mass attached to a string is rotated in a circular path.
- 'Centrifugal force' is the force acting away from the centre, the opposite to centripetal force.
- 'Gravitational force' is the force acting towards the centre of the earth due to the effect of gravity.
- 'Magnetic force' is the force created by a magnetic field.
- 'Electrical force' is the force created by an electrical field.

Pressure or stress

To move a broken-down motor car I might exert a force on the back of the car to propel it forward. My hands would apply a pressure on the body panel at the point of contact with the car. **Pressure** or stress is a measure of the force per unit area.

$$\text{Pressure or stress} = \frac{\text{Force}}{\text{Area}} \, (\text{N/m}^2)$$

Example 1

A young woman of mass 60 kg puts all her weight on to the heel of one shoe which has an area of 1 cm². Calculate the pressure exerted by the shoe on the floor (assuming the acceleration due to gravity to be 9.81 m/s²).

$$\text{Pressure} = \frac{\text{Force}}{\text{Area}} (\text{N/m}^2)$$

$$\text{Pressure} = \frac{60 \text{ kg} \times 9.81 \text{ m/s}^2}{1 \times 10^{-4} \text{m}^2} = 5886 \text{ kN/m}^2$$

Example 2

A small circus elephant of mass 1 tonne (1000 kg) puts all its weight on to one foot which has a surface area of 400 cm². Calculate the pressure exerted by the elephant's foot on the floor, assuming the acceleration due to gravity to be 9.81 m/s².

$$\text{Pressure} = \frac{\text{Force}}{\text{Area}} (\text{N/m}^2)$$

$$\text{Pressure} = \frac{1000 \text{ kg} \times 9.81 \text{ m/s}^2}{400 \times 10^{-4} \text{m}^2} = 245.3 \text{ kN/m}^2$$

These two examples show that the young woman exerts 24 times more pressure on the ground than the elephant. This is because her mass exerts a force over a much smaller area than the elephant's foot and is the reason why many wooden dance floors are damaged by high-heeled shoes.

Work done

Suppose a broken-down motor car was to be pushed along a road, work would be done on the car by applying the force necessary to move it along the road. Heavy breathing and perspiration would be evidence of the work done.

Definition

By definition, *work done* is dependent upon the force applied times the distance moved in the direction of the force.

By definition **Work done** is dependent upon the force applied times the distance moved in the direction of the force.

Work done = Force × Distance moved in the direction of the force (J)

The SI unit of work done is the newton metre or joule (symbol J). The joule is the preferred unit and it commemorates an English physicist, James Prescot Joule (1818–1889).

Example

A building hoist lifts ten 50 kg bags of cement through a vertical distance of 30 m to the top of a high-rise building. Calculate the work done by the hoist, assuming the acceleration due to gravity to be 9.81 m/s^2.

$$\text{Work done} = \text{Force} \times \text{Distance moved (J)}$$
$$\text{but force} = \text{Mass} \times \text{Acceleration (N)}$$
$$\therefore \text{Work done} = \text{Mass} \times \text{Acceleration} \times \text{Distance moved (J)}$$
$$\text{Work done} = 10 \times 50 \text{kg} \times 9.81 \text{m/s}^2 \times 30 \text{m}$$
$$\text{Work done} = 147.15 \text{kJ}$$

Power

Definition

By definition, *power* is the rate of doing work.

If one motor car can cover the distance between two points more quickly than another car, we say that the faster car is more powerful. It can do a given amount of work more quickly. By definition, **power** is the rate of doing work.

$$\text{Power} = \frac{\text{Work done}}{\text{Time taken}} \text{ (W)}$$

The SI unit of power, both electrical and mechanical, is the watt (symbol W). This commemorates the name of James Watt (1736–1819), the inventor of the steam engine.

Example 1

A building hoist lifts ten 50 kg bags of cement to the top of a 30 m high building. Calculate the rating (power) of the motor to perform this task in 60 s if the acceleration due to gravity is taken as 9.81 m/s^2.

$$\text{Power} = \frac{\text{Work done}}{\text{Time taken}} \text{ (W)}$$
$$\text{but Work done} = \text{Force} \times \text{Distance moved (J)}$$
$$\text{and Force} = \text{Mass} \times \text{Acceleration (N)}$$

By substitution,

$$\text{Power} = \frac{\text{Mass} \times \text{Acceleration} \times \text{Distance moved}}{\text{Time taken}} \text{ (W)}$$

$$\text{Power} = \frac{10 \times 50 \text{ kg} \times 9.81 \text{ m/s}^2 \times 30 \text{ m}}{60 \text{ s}}$$

$$\text{Power} = 2452.5 \text{ W}$$

The rating of the building hoist motor will be 2.45 kW.

Example 2

A hydroelectric power station pump motor working continuously during a 7 h period raises 856 tonnes of water through a vertical distance of 60 m. Determine the rating (power) of the motor, assuming the acceleration due to gravity is 9.81 m/s².

From Example 1,

$$\text{Power} = \frac{\text{Mass} \times \text{Acceleration} \times \text{Distance moved}}{\text{Time taken}} \text{(W)}$$

$$\text{Power} = \frac{856 \times 1000 \text{ kg} \times 9.81 \text{ m/s}^2 \times 60 \text{ m}}{7 \times 60 \times 60 \text{ s}}$$

$$\text{Power} = 20,000 \text{ W}$$

The rating of the pump motor is 20 kW.

Example 3

An electric hoist motor raises a load of 500 kg at a velocity of 2 m/s. Calculate the rating (power) of the motor if the acceleration due to gravity is 9.81 m/s².

$$\text{Power} = \frac{\text{Mass} \times \text{Acceleration} \times \text{Distance moved}}{\text{Time taken}} \text{(W)}$$

$$\text{but Velocity} = \frac{\text{Distance}}{\text{Time}} \text{(m/s)}$$

$$\therefore \text{Power} = \text{Mass} \times \text{Acceleration} \times \text{Velocity}$$
$$\text{Power} = 500 \text{kg} \times 9.81 \text{m/s}^2 \times 2 \text{m/s}$$
$$\text{Power} = 9810 \text{ W.}$$

The rating of the hoist motor is 9.81 kW.

Levers and turning force

A **lever** allows a heavy load to be lifted or moved by a small effort. Every time we open a door, turn on a tap or tighten a nut with a spanner, we exert a lever-action turning force. A **lever** is any rigid body which pivots or rotates about a fixed axis or fulcrum. The simplest form of lever is the crowbar, which is useful because it enables a person to lift a load at one end which is greater than the effort applied through his or her arm muscles at the other end. In this way the crowbar is said to provide a 'mechanical advantage'. A washbasin tap and a spanner both provide a mechanical advantage through the simple lever action. The mechanical advantage of a simple lever is dependent upon the length of lever on either side of the fulcrum. Applying the principle of turning forces to a lever, we obtain the formula:

Load force × Distance from fulcrum =
Effort force × Distance from fulcrum

This formula can perhaps better be understood by referring to Fig. 4.2. A small effort at a long distance from the fulcrum can balance a large load at a

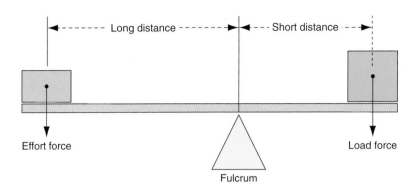

FIGURE 4.2
Turning forces of a simple lever.

short distance from the fulcrum. Thus a 'turning force' or 'turning moment' depends upon the distance from the fulcrum and the magnitude of the force.

Example

Calculate the effort required to raise a load of 500 kg when the effort is applied at a distance of five times the load distance from the fulcrum (assume the acceleration due to gravity to be 10 m/s^2).

$$Load force = Mass \times Acceleration (N)$$

$$Load force = 500 \, kg \times 10 \, m/s^2 = 5000 \, N$$

$$Load force \times Distance from fulcrum = Effort force \times Distance from fulcrum$$

$$5000 \, N \times 1 \, m = Effort force \times 5 \, m$$

$$\therefore Effort force = \frac{5000 \, N \times 1 \, m}{5 \, m} = 1000 \, N$$

Thus an effort force of 1000 N can overcome a load force of 5000 N using the mechanical advantage of this simple lever.

Simple machines

Our physical abilities in the field of lifting and moving heavy objects are limited. However, over the centuries we have used our superior intelligence to design tools, mechanisms and machines which have overcome this physical inadequacy. This concept is shown in Fig. 4.3.

By definition, a **machine** is an assembly of parts, some fixed, others movable, by which motion and force are transmitted. With the aid of a machine we are able to magnify the effort exerted at the input and lift or move large loads at the output.

Efficiency of any machine

In any machine the power available at the output is less than that which is put in because losses occur in the machine. The losses may result from friction in the bearings, wind resistance to moving parts, heat, noise or vibration.

Definition

By definition, a *machine* is an assembly of parts, some fixed, others movable, by which motion and force are transmitted. With the aid of a machine we are able to magnify the effort exerted at the input and lift or move large loads at the output.

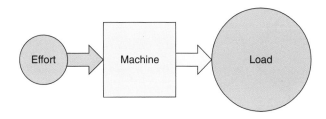

FIGURE 4.3
Simple machine concept.

Definition

The ratio of the output power to the input power is known as the *efficiency* of the machine. The symbol for efficiency is the Greek letter 'eta' (η). In general,

$$\eta = \frac{\text{Power output}}{\text{Power input}}$$

The ratio of the output power to the input power is known as the **efficiency** of the machine. The symbol for efficiency is the Greek letter 'eta' (η). In general,

$$\eta = \frac{\text{Power output}}{\text{Power input}}$$

Since efficiency is usually expressed as a percentage we modify the general formula as follows.

$$\eta = \frac{\text{Power output}}{\text{Power input}} \times 100$$

Example

A transformer feeds the 9.81 kW motor driving the mechanical hoist of the previous example. The input power to the transformer was found to be 10.9 kW. Find the efficiency of the transformer.

$$\eta = \frac{\text{Power output}}{\text{Power input}} \times 100$$

$$\eta = \frac{9.81\,\text{kW}}{10.9\,\text{kW}} \times 100 = 90\%$$

Thus the transformer is 90% efficient. Note that efficiency has no units, but is simply expressed as a percentage.

Electrical machines

Electrical machines are energy converters. If the machine input is mechanical energy and the output electrical energy then that machine is a generator, as shown in Fig. 4.4(a). Alternatively, if the machine input is electrical energy and the output mechanical energy then the machine is a motor, as shown in Fig. 4.4(b).

An electrical machine may be used as a motor or a generator, although in practice the machine will operate more efficiently when operated in the mode for which it was designed.

Simple a.c. generator or alternator

If a simple loop of wire is rotated between the poles of a permanent magnet, as shown in Fig. 4.5, the loop of wire will cut the lines of magnetic flux

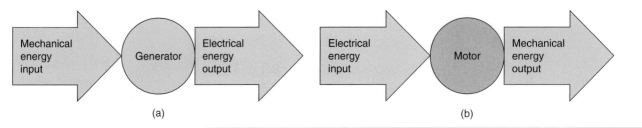

FIGURE 4.4
Electrical machines as energy converters.

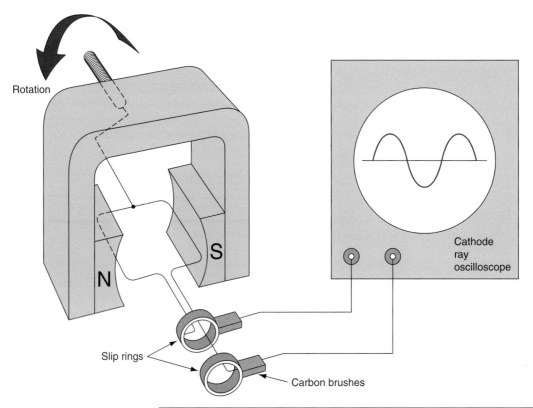

FIGURE 4.5
Simple a.c. generator or alternator.

Definition

Faraday's law which states that when a conductor cuts or is cut by a magnetic field, an emf is induced in that conductor.

between the north and south poles. This flux cutting will induce an emf in the wire by **Faraday's law** which states that *when a conductor cuts or is cut by a magnetic field, an emf is induced in that conductor*. If the generated emf is collected by carbon brushes at the slip rings and displayed on the screen of a cathode ray oscilloscope, the waveform will be seen to be approximately sinusoidal. Alternately changing, first positive and then negative, then positive again, giving an alternating output.

Simple d.c. generator or dynamo

If the slip rings of Fig. 4.5 are replaced by a single split ring, called a commutator, the generated emf will be seen to be in one direction, as shown in Fig. 4.6. The action of the commutator is to reverse the generated emf every half-cycle, rather like an automatic change-over switch. However,

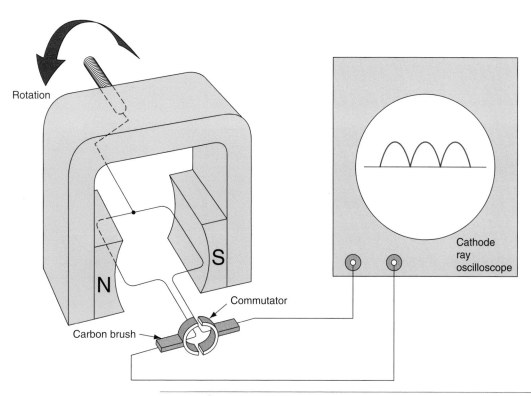

FIGURE 4.6
Simple d.c. generator or dynamo.

this simple arrangement produces a very bumpy d.c. output. In a practical machine, the commutator would contain many segments and many windings to produce a smoother d.c. output. Similar to the unidirectional battery supply shown in Fig. 4.7.

Alternating current theory

The supply which we obtain from a car battery is a unidirectional or d.c. supply, whereas the mains electricity supply is alternating or a.c. (see Fig. 4.7).

One of the reasons for using alternating supplies for the electricity mains supply is because we can very easily change the voltage levels by using a transformer which will only work on an a.c. supply.

The generated alternating supply at the power station is transformed up to 132,000 V, or more, for efficient transmission along the National Grid conductors.

Most electrical equipment makes use of alternating current supplies, and for this reason knowledge of alternating waveforms is necessary for all practising electricians.

When a coil of wire is rotated inside a magnetic field as shown in Fig. 4.5, a voltage is induced in the coil. The induced voltage follows a mathematical law known as the sinusoidal law and, therefore, we can say that a sine wave has been generated. Such a waveform has the characteristics displayed in Fig. 4.8.

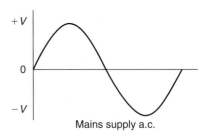

Battery supply d.c.

Mains supply a.c.

FIGURE 4.7

Unidirectional and alternating supply.

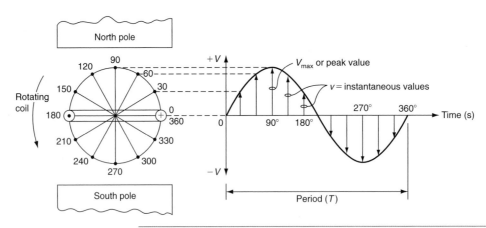

FIGURE 4.8

Characteristics of a sine wave.

In the United Kingdom we generate electricity at a frequency of 50 Hz and the time taken to complete each cycle is given by

$$T = \frac{1}{f}$$

$$\therefore T = \frac{1}{50 \, \text{Hz}} = 0.02 \, \text{s}$$

An alternating waveform is constantly changing from zero to a maximum, first in one direction, then in the opposite direction, and so the instantaneous values of the generated voltage are always changing. A useful description of the electrical effects of an a.c. waveform can be given by the maximum, average and rms values of the waveform.

The maximum or peak value is the greatest instantaneous value reached by the generated waveform. Cable and equipment insulation levels must be equal to or greater than this value.

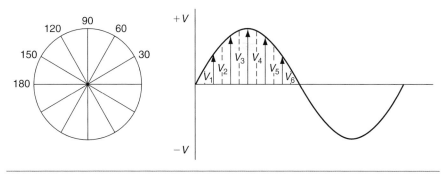

FIGURE 4.9

Sinusoidal waveform showing instantaneous values of voltage.

The average value is the average over one half-cycle of the instantaneous values as they change from zero to a maximum and can be found from the following formula applied to the sinusoidal waveform shown in Fig. 4.9.

$$V_{av} = \frac{V_1 + V_2 + V_3 + V_4 + V_5 + V_6}{6} = 0.637\,V_{max}$$

For any sinusoidal waveform the average value is equal to 0.637 of the maximum value.

The rms value is the square root of the mean of the individual squared values and is the value of an a.c. voltage which produces the same heating effect as a d.c. voltage. The value can be found from the following formula applied to the sinusoidal waveform shown in Fig. 4.9.

$$V_{rms} = \sqrt{\frac{V_1^2 + V_2^2 + V_3^2 + V_4^2 + V_5^2 + V_6^2}{6}}$$
$$= 0.7071\,V_{max}$$

For any sinusoidal waveform the rms value is equal to 0.7071 of the maximum value.

Example

The sinusoidal waveform applied to a particular circuit has a maximum value of 325.3 V. Calculate the average and rms value of the waveform.

$$\text{Average value } V_{av} = 0.637 \times V_{max}$$
$$\therefore V_{av} = 0.637 \times 325.3 = 207.2 \text{ V}$$
$$\text{rms value } V_{rms} = 0.7071 \times V_{max}$$
$$V_{rms} = 0.7071 \times 325.3 = 230 \text{ V}$$

When we say that the main supply to a domestic property is 230V, we really mean $230\,V_{rms}$. Such a waveform has an average value of about 207.2V and a maximum value of almost 325.3V but because the rms value gives the d.c. equivalent value we almost always give the rms value without identifying it as such.

Check your Understanding

When you have completed the questions, check out the answers at the back of the book.

Note: more than one multiple choice answer may be correct.

1. The SI unit of mass is the:
    a. kilogram or kg
    b. pound or lb
    c. metre or m
    d. millimetre or mm.

2. The SI unit of length is the:
    a. kilogram or kg
    b. pound or lb
    c. metre or m
    d. millimetre or mm.

3. The SI unit of time is the:
    a. minute or m
    b. second or s
    c. hour or h
    d. day or d.

4. The SI unit of electric current is the:
    a. ohm or Ω
    b. volt or V
    c. watt or W
    d. ampere or A.

5. The SI unit of resistance is the:
    a. ohm or Ω
    b. volt or V
    c. watt or W
    d. ampere or A.

6. The ampere is a measure of:
    a. potential difference
    b. power
    c. force
    d. electric current.

7. The watt is a measure of:
   a. potential difference
   b. power
   c. force
   d. electric current.

8. The volt is a measure of:
   a. potential difference
   b. power
   c. force
   d. electric current.

9. The Newton is a measure of:
   a. potential difference
   b. power
   c. force
   d. electric current.

10. Which of the following may be defined as 'a measure of the force which a body exerts on anything which supports it':
    a. acceleration
    b. force
    c. mass
    d. weight.

11. Which of the following may be defined as 'a measure of the amount of material in a substance':
    a. acceleration
    b. force
    c. mass
    d. weight.

12. Which of the following may be defined as 'may cause a stationary object to move or bring a moving body to rest':
    a. acceleration ·
    b. force
    c. mass
    d. weight

13. Which of the following may be defined as 'the force applied times the distance moved in the direction of the force':
    a. acceleration
    b. work done
    c. power
    d. velocity

14. Which of the following may be defined as 'the rate of doing work':
    a. acceleration
    b. work done
    c. power
    d. velocity

15. Which of the following may be defined as 'the speed in a given direction':
    a. acceleration
    b. work done
    c. power
    d. velocity

16. Briefly describe what we mean by 'a turning force' and give five practical examples of this effect.

17. Briefly define what we mean by a 'simple machine' and give five examples.

18. Briefly describe what we mean by 'the efficiency of a machine'.

19. Sketch the construction of a simple alternator and label all the parts.

20. State how an emf is induced in an alternator. Sketch and name the shape of the generated emf.

21. Calculate or state the average rms and maximum value of the domestic a.c. mains supply and show these values on a sketch of the mains supply.

Basic scientific concepts in electrotechnology

Unit 2 – Principles of electrotechnology – Outcome 2

Underpinning knowledge: when you have completed this chapter you should be able to:

- list the materials used for conductors and insulators
- describe the properties of conductor and insulating materials
- explain current flow in a series and parallel circuit
- describe how to connect voltmeters and ammeters
- describe the three effects of an electric current
- describe the flux patterns set up by permanent magnets, conductors and solenoids
- describe the construction of a basic transformer
- state the use of transformers on the National Grid

Properties of conductors and insulators

In Fig. 4.1 of the last chapter we looked at the atomic structure of materials. All materials are made up of atoms and electrons. What makes them different materials is the way in which the atoms and electrons are arranged and how strongly the electrons are attracted to the atoms.

A **conductor** is a material, usually a metal, in which the electrons are loosely bound to the central nucleus. These electrons can easily become 'free electrons' which allows heat and electricity to pass easily through the material.

An **insulator** is a material, usually a non-metal, in which the electrons are very firmly bound to the nucleus and, therefore, will not allow heat or electricity to pass through it.

Let us now define the terms and properties of some of the materials used in the electrotechnical industry.

Ferrous A word used to describe all metals in which the main constituent is iron. The word 'ferrous' comes from the Latin word *ferrum* meaning iron. Ferrous metals have magnetic properties. Cast iron, wrought iron and steel are all ferrous metals.

Non-ferrous Metals which *do not* contain iron are called non-ferrous. They are non-magnetic and resist rusting. Copper, aluminium, tin, lead, zinc and brass are examples of non-ferrous metals.

Alloy An alloy is a mixture of two or more metals. Brass is an alloy of copper and zinc, usually in the ratio 70–30% or 60–40%.

Corrosion The destruction of a metal by chemical action. Most corrosion takes place when a metal is in contact with moisture (see also mild steel and zinc).

Thermoplastic polymers These may be repeatedly warmed and cooled without appreciable changes occurring in the properties of the material. They are good insulators, but give off toxic fumes when burned. They have a flexible quality when operated up to a maximum temperature of 70°C but should not be flexed when the air temperature is near 0°C, otherwise they may crack. Polyvinylchloride (PVC) used for cable insulation is a thermoplastic polymer.

Thermosetting polymers Once heated and formed, products made from thermosetting polymers are fixed rigidly. Plug tops, socket outlets and switch plates are made from this material.

Rubber is a tough elastic substance made from the sap of tropical plants. It is a good insulator, but degrades and becomes brittle when exposed to sunlight.

Synthetic rubber is manufactured, as opposed to being produced naturally. Synthetic or artificial rubber is carefully manufactured to have all the

---

Definition

A *conductor* is a material, usually a metal, in which the electrons are loosely bound to the central nucleus. These electrons can easily become 'free electrons' which allows heat and electricity to pass easily through the material.

Definition

An *insulator* is a material, usually a non metal, in which the electrons are very firmly bound to the nucleus and, therefore, will not allow heat or electricity to pass through it.

Definition

Ferrous A word used to describe all metals in which the main constituent is iron.

Definition

Non-ferrous Metals which *do not* contain iron are called non-ferrous.

Definition

Alloy An alloy is a mixture of two or more metals.

Definition

Corrosion The destruction of a metal by chemical action.

Definition

Thermoplastic polymers These may be repeatedly warmed and cooled without appreciable changes occurring in the properties of the material.

Definition

Polyvinylchloride (PVC) used for cable insulation is a thermoplastic polymer.

100

good qualities of natural rubber – flexibility, good insulation and suitability for use over a wide range of temperatures.

Silicon rubber Introducing organic compounds into synthetic rubber produces a good insulating material which is flexible over a wide range of temperatures and which retains its insulating properties even when burned. These properties make it ideal for cables used in fire alarm installations such as FP200 cables.

Magnesium oxide The conductors of mineral insulated metal sheathed (MICC) cables are insulated with compressed magnesium oxide, a white chalk-like substance which is heat-resistant and a good insulator and lasts for many years. The magnesium oxide insulation, copper conductors and sheath, often additionally manufactured with various external sheaths to provide further protection from corrosion and weather, produce a cable designed for long-life and high-temperature installations. However, the magnesium oxide is very hygroscopic, which means that it attracts moisture and, therefore, the cable must be terminated with a special moisture-excluding seal, as shown in Fig. 6.3.

Copper

Copper is extracted from an ore which is mined in South Africa, North America, Australia and Chile. For electrical purposes it is refined to about 98.8% pure copper, the impurities being extracted from the ore by smelting and electrolysis. It is a very good conductor, is non-magnetic and offers considerable resistance to atmospheric corrosion. Copper toughens with work, but may be annealed, or softened, by heating to dull red before quenching.

Copper forms the largest portion of the alloy brass, and is used in the manufacture of electrical cables, domestic heating systems, refrigerator tubes and vehicle radiators. An attractive soft reddish brown metal, copper is easily worked and is also used to manufacture decorative articles and jewellery.

Aluminium

Aluminium is a grey-white metal obtained from the mineral bauxite which is found in the United States, Germany and the Russian Federation. It is a very good conductor, is non-magnetic, offers very good resistance to atmospheric corrosion and is notable for its extreme softness and lightness. It is used in the manufacture of power cables. The overhead cables of the National Grid are made of an aluminium conductor reinforced by a core of steel. Copper conductors would be too heavy to support themselves between the pylons. Lightness and resistance to corrosion make aluminium an ideal metal for the manufacture of cooking pots and food containers.

Aluminium alloys retain the corrosion resistance properties of pure aluminium with an increase in strength. The alloys are cast into cylinder heads and gearboxes for motorcars, and switch-boxes and luminaires for electrical installations. Special processes and fluxes have now been developed which allow aluminium to be welded and soldered.

Definition

Thermosetting polymers Once heated and formed, products made from thermosetting polymers are fixed rigidly. Plug tops, socket outlets and switch plates are made from this material.

Definition

Rubber is a tough elastic substance made from the sap of tropical plants.

Definition

Synthetic rubber is manufactured, as opposed to being produced naturally.

Definition

Silicon rubber Introducing organic compounds into synthetic rubber produces a good insulating material such as FP200 cables.

Definition

Magnesium oxide The conductors of mineral insulated metal sheathed (MICC) cables are insulated with compressed magnesium oxide

Try This

Materials

List 5 'good insulator' materials being used at your place of work.

Brass

Brass is a non-ferrous alloy of copper and zinc which is easily cast. Because it is harder than copper or aluminium it is easily machined. It is a good conductor and is highly resistant to corrosion. For these reasons it is often used in the electrical and plumbing trades. Taps, valves, pipes, electrical terminals, plug top pins and terminal glands for steel wire armour (SWA) and MI cables are some of the many applications.

Brass is an attractive yellow metal which is also used for decorative household articles and jewellery. The combined properties of being an attractive metal which is highly resistant to corrosion make it a popular metal for ships' furnishings.

Cast steel

Cast steel is also called tool steel or high-carbon steel. It is an alloy of iron and carbon which is melted in airtight crucibles and then poured into moulds to form ingots. These ingots are then rolled or pressed into various shapes from which the finished products are made. Cast steel can be hardened and tempered and is therefore ideal for manufacturing tools. Hammer heads, pliers, wire cutters, chisels, files and many machine parts are also made from cast steel.

Mild steel

Mild steel is also an alloy of iron and carbon but contains much less carbon than cast steel. It can be filed, drilled or sawn quite easily and may be bent when hot or cold, but repeated cold bending may cause it to fracture. In moist conditions corrosion takes place rapidly unless the metal is protected. Mild steel is the most widely used metal in the world, having considerable strength and rigidity without being brittle. Ships, bridges, girders, motorcar bodies, bicycles, nails, screws, conduit, trunking, tray and SWA are all made of mild steel.

Try This

Materials

List 5 'good conductor' materials being used at your place of work.

Zinc

Zinc is a non-ferrous metal which is used mainly to protect steel against corrosion and in making the alloy brass. Mild steel coated with zinc is sometimes called *galvanized steel*, and this coating considerably improves steel's resistance to corrosion. Conduit, trunking, tray, SWA, outside luminaires and electricity pylons are made of galvanized steel.

Resistors in series and parallel

In an electrical circuit resistors may be connected in series, in parallel, or in various combinations of series and parallel connections.

Series-connected resistors

In any series circuit a current I will flow through all parts of the circuit as a result of the potential difference supplied by a battery V_T. Therefore, we say that in a series circuit the current is common throughout that circuit.

When the current flows through each resistor in the circuit, R_1, R_2 and R_3 for example in Fig. 5.1, there will be a voltage drop across that resistor whose value will be determined by the values of I and R, since from Ohm's law $V = I \times R$. The sum of the individual voltage drops, V_1, V_2 and V_3 for example in Fig. 5.1, will be equal to the total voltage V_T.

We can summarize these statements as follows. For any series circuit, I is common throughout the circuit and

$$V_T = V_1 + V_2 + V_3 \tag{5.1}$$

Let us call the total circuit resistance R_T. From Ohm's law we know that $V = I \times R$ and therefore

$$
\begin{aligned}
&\text{Total voltage } V_T = I \times R_T \\
&\text{Voltage drop across } R_1 \text{ is } V_1 = I \times R_1 \\
&\text{Voltage drop across } R_2 \text{ is } V_2 = I \times R_2 \\
&\text{Voltage drop across } R_3 \text{ is } V_3 = I \times R_3
\end{aligned}
\tag{5.2}
$$

We are looking for an expression for the total resistance in any series circuit and, if we substitute equations (5.2) into equation (5.1) we have:

$$V_T = V_1 + V_2 + V_3$$

$$\therefore I \times R_T = I \times R_1 + I \times R_2 + I \times R_3$$

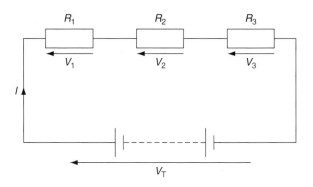

FIGURE 5.1

A series circuit.

Now, since I is common to all terms in the equation, we can divide both sides of the equation by I. This will cancel out I to leave us with an expression for the circuit resistance:

$$R_T = R_1 + R_2 + R_3$$

Note that the derivation of this formula is given for information only. Craft students need only state the expression $R_T = R_1 + R_2 + R_3$ for series connections.

Key Fact

Resistors

In a series circuit, total resistance $R_T = R_1 + R_2 + R_3$ ohms.

Parallel-connected resistors

In any parallel circuit, as shown in Fig. 5.2, the same voltage acts across all branches of the circuit. The total current will divide when it reaches a resistor junction, part of it flowing in each resistor. The sum of the individual currents, I_1, I_2 and I_3 for example in Fig. 5.2, will be equal to the total current I_T.

We can summarize these statements as follows. For any parallel circuit, V is common to all branches of the circuit and

$$I_T = I_1 + I_2 + I_3 \tag{5.3}$$

Let us call the total resistance R_T.

From Ohm's law we know, that $I = \frac{V}{R}$, and therefore

$$\begin{aligned} &\text{the total current } I_T = \frac{V}{R_T} \\ &\text{the current through } R_1 \text{ is } I_1 = \frac{V}{R_1} \\ &\text{the current through } R_2 \text{ is } I_2 = \frac{V}{R_2} \\ &\text{the current through } R_3 \text{ is } I_3 = \frac{V}{R_3} \end{aligned} \tag{5.4}$$

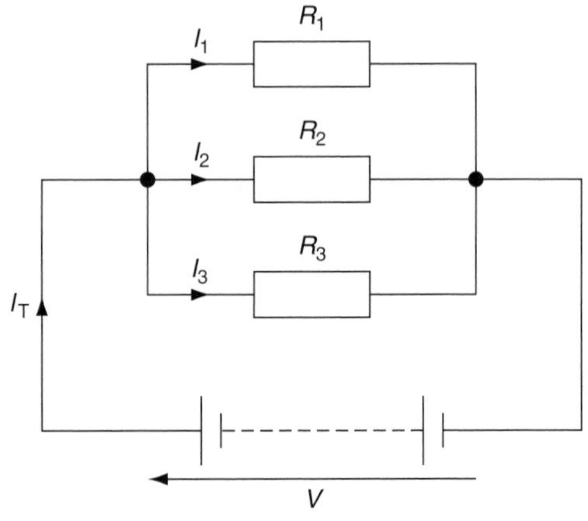

FIGURE 5.2
A parallel circuit.

We are looking for an expression for the equivalent resistance R_T in any *parallel* circuit and, if we substitute equations (5.4) into equation (5.3) we have:

$$I_T = I_1 + I_2 + I_3$$

$$\therefore \frac{V}{R_T} = \frac{V}{R_1} + \frac{V}{R_2} + \frac{V}{R_3}$$

Now, since V is common to all terms in the equation, we can divide both sides by V, leaving us with an expression for the circuit resistance:

$$\frac{1}{R_T} = \frac{1}{R_1} + \frac{1}{R_2} + \frac{1}{R_3}$$

Note that the derivation of this formula is given for information only. Craft students need only state the expression $1/R_T = 1/R_1 + 1/R_2 + 1/R_3$ for parallel connections.

Key Fact

Resistors

In a parallel circuit, total resistance can be found from $\dfrac{1}{R_T} = \dfrac{1}{R_1} + \dfrac{1}{R_2} + \dfrac{1}{R_3}$

Example 1

Three 6 Ω resistors are connected (a) in series (see Fig. 5.3), and (b) in parallel (see Fig. 5.4), across a 12 V battery. For each method of connection, find the total resistance and the values of all currents and voltages.

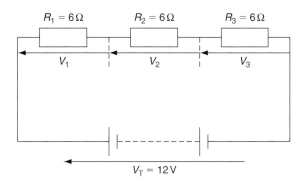

FIGURE 5.3
Resistors in series.

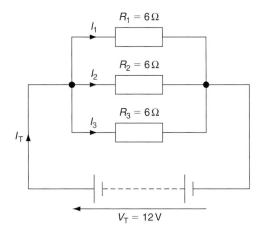

FIGURE 5.4
Resistors in parallel.

For any series connection

$$R_T = R_1 + R_2 + R_3$$
$$\therefore R_T = 6\,\Omega + 6\,\Omega + 6\,\Omega = 18\,\Omega$$

Total current $I_T = \dfrac{V_T}{R_T}$

$$\therefore I_T = \frac{12\,V}{18\,\Omega} = 0.67\,A$$

The voltage drop across R_1 is

$$V_1 = I_T \times R_1$$
$$\therefore V_1 = 0.67\,A \times 6\,\Omega = 4\,V$$

The voltage drop across R_2 is

$$V_2 = I_T \times R_2$$
$$\therefore V_2 = 0.67\,A \times 6\,\Omega = 4\,V$$

The voltage drop across R_3 is

$$V_3 = I_T \times R_3$$
$$\therefore V_3 = 0.67\,A \times 6\,\Omega = 4\,V$$

For any parallel connection,

$$\frac{1}{R_T} = \frac{1}{R_1} + \frac{1}{R_2} + \frac{1}{R_3}$$
$$\therefore \frac{1}{R_T} = \frac{1}{6\,\Omega} + \frac{1}{6\,\Omega} + \frac{1}{6\,\Omega}$$
$$\frac{1}{R_T} = \frac{1+1+1}{6\,\Omega} = \frac{3}{6\,\Omega}$$
$$R_T = \frac{6\,\Omega}{3} = 2\,\Omega$$

Total current $I_T = \dfrac{V_T}{R_T}$

$$\therefore I_T = \frac{12\,V}{2\,\Omega} = 6\,A$$

The current flowing through R_1 is

$$I_1 = \frac{V_T}{R_1}$$
$$\therefore I_1 = \frac{12\,V}{6\,\Omega} = 2\,A$$

The current flowing through R_2 is

$$I_2 = \frac{V_T}{R_2}$$
$$\therefore I_2 = \frac{12\,V}{6\,\Omega} = 2\,A$$

The current flowing through R_3 is

$$I_3 = \frac{V_T}{R_3}$$
$$\therefore I_3 = \frac{12\,V}{6\,\Omega} = 2\,A$$

Series and parallel combinations

The most complex arrangement of series and parallel resistors can be simplified into a single equivalent resistor by combining the separate rules for series and parallel resistors.

Example 2

Resolve the circuit shown in Fig. 5.5 into a single resistor and calculate the potential difference across each resistor.

By inspection, the circuit contains a parallel group consisting of R_3, R_4 and R_5 and a series group consisting of R_1 and R_2 in series with the equivalent resistor for the parallel branch.

Consider the parallel group. We will label this group R_P. Then

$$\frac{1}{R_P} = \frac{1}{R_3} + \frac{1}{R_4} + \frac{1}{R_5}$$

$$\frac{1}{R_P} = \frac{1}{2\,\Omega} + \frac{1}{3\,\Omega} + \frac{1}{6\,\Omega}$$

$$\frac{1}{R_P} = \frac{3+2+1}{6\,\Omega} = \frac{6}{6\,\Omega}$$

$$R_P = \frac{6\,\Omega}{6} = 1\,\Omega$$

Figure 5.5 may now be represented by the more simple equivalent shown in Fig. 5.6.

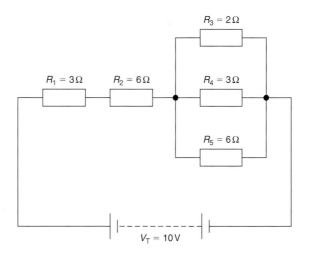

FIGURE 5.5

A series/parallel circuit.

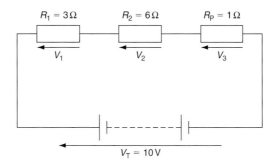

FIGURE 5.6

Equivalent series circuit.

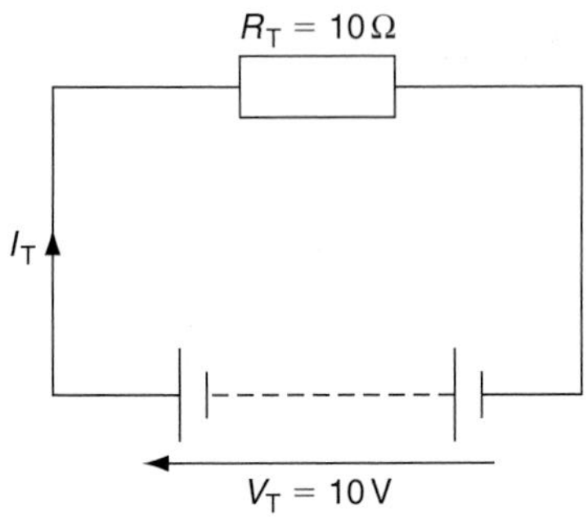

$R_T = 10\,\Omega$

I_T

$V_T = 10\,V$

FIGURE 5.7

Single equivalent resistor for Fig. 5.5.

Since all resistors are now in series,

$$R_T = R_1 + R_2 + R_P$$
$$\therefore R_T = 3\,\Omega + 6\,\Omega + 1\,\Omega = 10$$

Thus, the circuit may be represented by a single equivalent resistor of value 10 Ω as shown in Fig. 5.7. The total current flowing in the circuit may be found by using Ohm's law:

$$I_T = \frac{V_T}{R_T} = \frac{10\,V}{10\,\Omega} = 1\,A$$

The potential differences across the individual resistors are

$$V_1 = I_T \times R_1 = 1\,A \times 3\,\Omega = 3\,V$$
$$V_2 = I_T \times R_2 = 1\,A \times 6\,\Omega = 6\,V$$
$$V_P = I_T \times R_P = 1\,A \times 1\,\Omega = 1\,V$$

Since the same voltage acts across all branches of a parallel circuit the same p.d. of 1 V will exist across each resistor in the parallel branch R_3, R_4 and R_5.

Example 3

Determine the total resistance and the current flowing through each resistor for the circuit shown in Fig. 5.8.

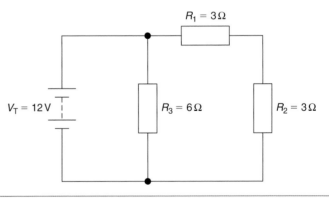

$R_1 = 3\,\Omega$

$V_T = 12\,V$

$R_3 = 6\,\Omega$

$R_2 = 3\,\Omega$

FIGURE 5.8

A series/parallel circuit for Example 3.

By inspection, it can be seen that R_1 and R_2 are connected in series while R_3 is connected in parallel across R_1 and R_2. The circuit may be more easily understood if we redraw it as in Fig. 5.9.

For the series branch, the equivalent resistor can be found from

$$R_S = R_1 + R_2$$
$$\therefore R_S = 3\,\Omega + 3\,\Omega = 6\,\Omega$$

Figure 5.9 may now be represented by a more simple equivalent circuit, as in Fig. 5.10.

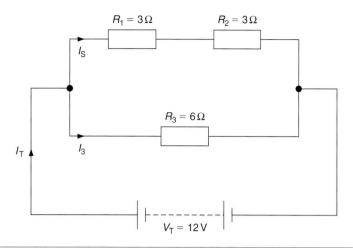

FIGURE 5.9

Equivalent circuit for Example 2.

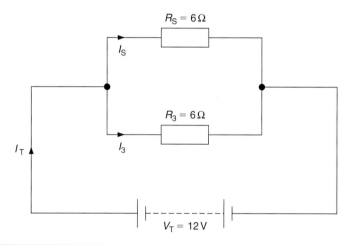

FIGURE 5.10

Simplified equivalent circuit for Example 2.

Since the resistors are now in parallel, the equivalent resistance may be found from

$$\frac{1}{R_\mathrm{T}} = \frac{1}{R_\mathrm{S}} + \frac{1}{R_3}$$

$$\therefore \frac{1}{R_\mathrm{T}} = \frac{1}{6\ \Omega} + \frac{1}{6\ \Omega}$$

$$\frac{1}{R_\mathrm{T}} = \frac{1+1}{6\ \Omega} = \frac{2}{6\ \Omega}$$

$$R_\mathrm{T} = \frac{6\ \Omega}{2} = 3\ \Omega$$

The total current is

$$I_\mathrm{T} = \frac{V_\mathrm{T}}{R_\mathrm{T}} = \frac{12\ \mathrm{V}}{3\ \Omega} = 4\ \mathrm{A}$$

Let us call the current flowing through resistor R_3 I_3.

$$\therefore I_3 = \frac{V_\mathrm{T}}{R_3} = \frac{12\ \mathrm{V}}{6\ \Omega} = 2\ \mathrm{A}$$

Let us call the current flowing through both resistors R_1 and R_2, as shown in Fig. 5.9, I_S.

$$\therefore I_S = \frac{V_T}{R_S} = \frac{12\,V}{6\,\Omega} = 2\,A$$

Measuring volts and amps

The type of instrument to be purchased for general use in the electrotechnical industries is a difficult choice because there are so many different types on the market and every manufacturer's representative is convinced that his company's product is the best. However, most instruments can be broadly grouped under two general headings: those having *analogue* and those with *digital* displays.

Analogue meters or instruments

Analogue meters have a pointer moving across a calibrated scale. They are the only choice when a general trend or variation in value is to be observed. Hi-fi equipment often uses analogue displays to indicate how power levels vary with time, which is more informative than a specific value. Red or danger zones can be indicated on industrial instruments. The fuel gauge on a motor car often indicates full, half full or danger on an analogue display which is much more informative than an indication of the exact number of litres of petrol remaining in the tank.

These meters are only accurate when used in the calibrated position – usually horizontally.

Most meters using an analogue scale incorporate a mirror to eliminate parallax error. The user must look straight at the pointer on the scale when taking readings and the correct position is indicated when the pointer image in the mirror is hidden behind the actual pointer. That is the point at which a reading should be taken from the appropriate scale of the instrument.

Digital meters or instruments

Digital meters provide the same functions as analogue meters but they display the indicated value using a seven-segment LED to give a numerical value of the measurement. Modern digital meters use semiconductor technology to give the instrument a very high-input impedance, typically about $10\,M\Omega$ and, therefore, they are ideal for testing most electrical or electronic circuits.

The choice between an analogue and a digital display is a difficult one and must be dictated by specific circumstances. However, if you are an electrician or service engineer intending to purchase a new instrument, I think on balance that a good-quality digital multimeter such as that shown in Fig. 5.11 would be best. Having no moving parts, digital meters tend to be more rugged and, having a very high-input impedance, they are ideally suited to testing all circuits that an electrician might work on in his daily work.

Definition

Analogue meters have a pointer moving across a calibrated scale.

Definition

Digital meters provide the same functions as analogue meters but they display the indicated value using a seven-segment LED to give a numerical value of the measurement.

FIGURE 5.11

Digital multimeter suitable for testing electrical and electronic circuits.

The multimeter

Multimeters are designed to measure voltage, current or resistance. Before taking measurements the appropriate volt, ampere or ohm scale should be selected. To avoid damaging the instrument it is good practice first to switch to the highest value on a particular scale range. For example, if the 10 A scale is first selected and a reading of 2.5 A is displayed, we then know that a more appropriate scale would be the 3 A or 5 A range. This will give a more accurate reading which might be, say, 2.49 A. When the multimeter is used as an ammeter to measure current it must be connected in series with the test circuit, as shown in Fig. 5.12(a). When used as a voltmeter the multimeter must be connected in parallel with the component, as shown in Fig. 5.12(b).

When using a commercial multirange meter as an ohmmeter for testing electronic components, care must be exercised in identifying the positive terminal. The red terminal of the meter, identifying the positive input for testing voltage and current, usually becomes the negative terminal when the meter is used as an ohmmeter because of the way the internal battery is connected to the meter movement. Check the meter manufacturers handbook before using a multimeter to test electronic components.

The three effects of an electric current

When an electric current flows in a circuit it can have one or more of the following three effects: *heating, magnetic* or *chemical.*

HEATING EFFECT

The movement of electrons within a conductor, which is the flow of an electric current, causes an increase in the temperature of the conductor. The amount of heat generated by this current flow depends upon the type and dimensions of the conductor and the quantity of current flowing. By changing these variables, a conductor may be operated hot and used as the heating element of a fire, or be operated cool and used as an electrical installation conductor.

The heating effect of an electric current is also the principle upon which a fuse gives protection to a circuit. The fuse element is made of a metal with a low melting point and forms a part of the electrical circuit. If an excessive current flows, the fuse element overheats and melts, breaking the circuit.

MAGNETIC EFFECT

Whenever a current flows in a conductor a magnetic field is set up around the conductor like an extension of the insulation. The magnetic field increases with the current and collapses if the current is switched off. A conductor carrying current and wound into a solenoid produces a magnetic field very similar to a permanent magnet, but has the advantage of being switched on and off by any switch which controls the circuit current.

The magnetic effect of an electric current is the principle upon which electric bells, relays, instruments, motors and generators work.

Definition

When an electric current flows in a circuit it can have one or more of the following three effects: *heating, magnetic* or *chemical.*

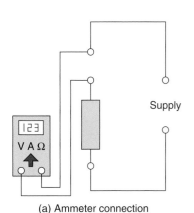

(a) Ammeter connection

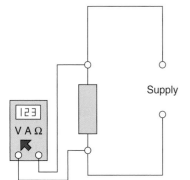

(b) Voltmeter connection

FIGURE 5.12

Using a multimeter (a) as an ammeter and (b) as a voltmeter.

CHEMICAL EFFECT

When an electric current flows through a conducting liquid, the liquid is separated into its chemical parts. The conductors which make contact with the liquid are called the anode and cathode. The liquid itself is called the electrolyte, and the process is called *electrolysis*.

Electrolysis is an industrial process used in the refining of metals and electroplating. It was one of the earliest industrial applications of electric current. Most of the aluminium produced today is extracted from its ore by electrochemical methods. Electroplating serves a double purpose by protecting a base metal from atmospheric erosion and also giving it a more expensive and attractive appearance. Silver and nickel plating has long been used to enhance the appearance of cutlery, candlesticks and sporting trophies.

An anode and cathode of dissimilar metal placed in an electrolyte can react chemically and produce an emf. When a load is connected across the anode and cathode, a current is drawn from this arrangement, which is called a cell. A battery is made up of a number of cells. It has many useful applications in providing portable electrical power, but electrochemical action can also be undesirable since it is the basis of electrochemical corrosion which rots our motor cars, industrial containers and bridges.

Magnetism

The Greeks knew as early as 600 BC that a certain form of iron ore, now known as magnetite or lodestone, had the property of attracting small pieces of iron. Later, during the Middle Ages, navigational compasses were made using the magnetic properties of lodestone. Small pieces of lodestone attached to wooden splints floating in a bowl of water always came to rest pointing in a north–south direction. The word lodestone is derived from an old English word meaning 'the way', and the word magnetism is derived from Magnesia, the place where magnetic ore was first discovered.

Iron, nickel and cobalt are the only elements which are attracted strongly by a magnet. These materials are said to be *ferromagnetic*. Copper, brass, wood, PVC and glass are not attracted by a magnet and are, therefore, described as *non-magnetic*.

Some basic rules of magnetism

1. Lines of magnetic flux have no physical existence, but they were introduced by Michael Faraday (1791–1867) as a way of explaining the magnetic energy existing in space or in a material. They help us to visualize and explain the magnetic effects. The symbol used for magnetic flux is the Greek letter Φ (phi) and the unit of magnetic flux is the weber (symbol Wb), pronounced 'veber', to commemorate the work of the German physicist Wilhelm Weber (1804–1891).

2. Lines of magnetic flux always form closed loops.

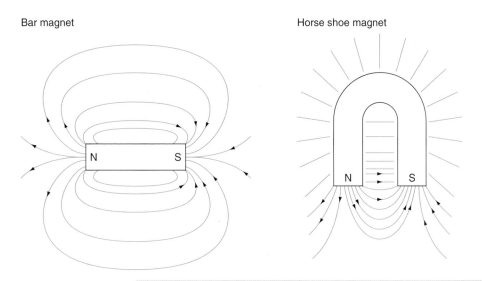

3. Lines of magnetic flux behave like stretched elastic bands, always trying to shorten themselves.

4. Lines of magnetic flux never cross over each other.

5. Lines of magnetic flux travel along a magnetic material and always emerge out of the 'north pole' end of the magnet.

6. Lines of magnetic flux pass through space and non-magnetic materials undisturbed.

7. The region of space through which the influence of a magnet can be detected is called the **magnetic field** of that magnet.

8. The number of lines of magnetic flux within a magnetic field is a measure of the flux density. Strong magnetic fields have a high-flux density. The symbol used for flux density is B, and the unit of flux density is the tesla (symbol T), to commemorate the work of the Croatian-born American physicist Nikola Tesla (1857–1943).

9. The places on a magnetic material where the lines of flux are concentrated are called the **magnetic poles**.

10. Like poles repel; unlike poles attract. These two statements are sometimes called the 'first laws of magnetism' and are shown in Fig. 5.14.

Magnetic fields

If a permanent magnet is placed on a surface and covered by a piece of paper, iron filings can be shaken on to the paper from a dispenser. Gently tapping the paper then causes the filings to take up the shape of the magnetic field surrounding the permanent magnet. The magnetic fields around a permanent magnet are shown in Figs 5.13 and 5.14.

Definition

The region of space through which the influence of a magnet can be detected is called the *magnetic field* of that magnet.

Definition

The places on a magnetic material where the lines of flux are concentrated are called the *magnetic poles*.

Definition

Like poles repel; unlike poles attract. These two statements are sometimes called the 'first laws of magnetism' and are shown in Fig. 5.14.

Bar magnet

Horse shoe magnet

FIGURE 5.13

Magnetic fields around a permanent magnet.

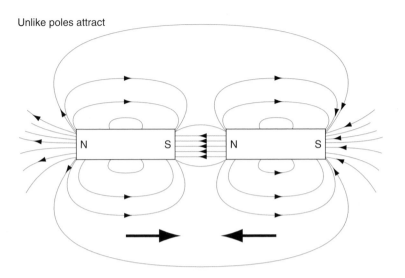

Unlike poles attract

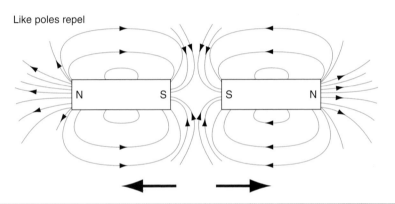

Like poles repel

FIGURE 5.14
The first laws of magnetism.

Electromagnetism

Electricity and magnetism have been inseparably connected since the experiments by Oersted and Faraday in the early nineteenth century. An electric current flowing in a conductor produces a magnetic field 'around' the conductor which is proportional to the current. Thus a small current produces a weak magnetic field, while a large current will produce a strong magnetic field. The magnetic field 'spirals' around the conductor, as shown in Fig. 5.15 and its direction can be determined by the 'dot' or 'cross' notation and the 'screw rule'. To do this, we think of the current as being represented by a dart or arrow inside the conductor. The dot represents current coming towards us when we would see the point of the arrow or dart inside the conductor. The cross represents current going away from us when we would see the flights of the dart or arrow. Imagine a corkscrew or screw being turned so that it will move in the direction of the current. Therefore, if the current was coming out of the paper, as shown in Fig. 5.15(a), the magnetic field would be spiralling anticlockwise around the conductor. If the current was going into the paper, as shown by Fig. 5.15(b), the magnetic field would spiral clockwise around the conductor.

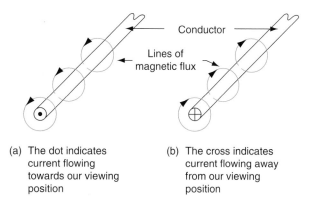

(a) The dot indicates current flowing towards our viewing position

(b) The cross indicates current flowing away from our viewing position

FIGURE 5.15
Magnetic fields around a current carrying conductor.

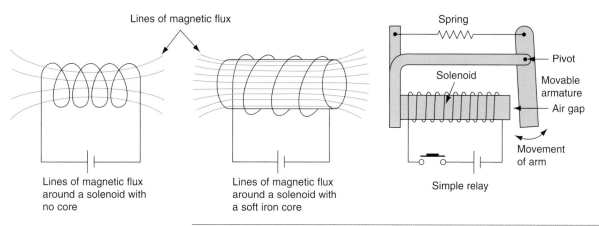

Lines of magnetic flux around a solenoid with no core

Lines of magnetic flux around a solenoid with a soft iron core

Simple relay

FIGURE 5.16
The solenoid and one practical application: the relay.

A current flowing in a *coil* of wire or solenoid establishes a magnetic field which is very similar to that of a bar magnet. Winding the coil around a soft iron core increases the flux density because the lines of magnetic flux concentrate on the magnetic material. The advantage of the electromagnet when compared with the permanent magnet is that the magnetism of the electromagnet can be switched on and off by a functional switch controlling the coil current. This effect is put to practical use in the electrical relay as used in a motor starter or alarm circuit. Figure 5.16 shows the structure and one application of the solenoid.

Electrical Transformers

A transformer is an electrical machine without moving parts, which is used to change the value of an alternating voltage.

A transformer will only work on an alternating supply, it will not normally work from a D.C. supply such as a battery.

- A transformer such as that shown in Fig. 5.17 consists of two coils called the primary and secondary coils or windings, wound on to a common core. The iron core of the transformer is not solid but made up of very thin sheets called laminations, to improve efficiency.

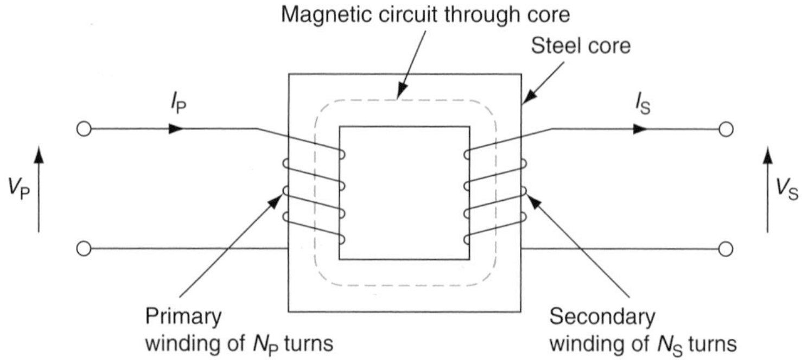

Magnetic circuit through core

Steel core

I_P

I_S

V_P

V_S

Primary
winding of N_P turns

Secondary
winding of N_S turns

FIGURE 5.17

A simple transformer.

- An alternating voltage applied to the primary winding establishes an alternating magnetic flux in the core.

- The magnetic flux in the core causes a voltage to be induced in the secondary winding of the transformers.

- The voltage in both the primary and secondary windings is proportional to the number of turns.

- This means that if you increase the number of secondary turns you will increase the output voltage. This has an application in power distribution.

- Alternatively, reducing the number of secondary turns will reduce the output voltage. This is useful for low voltage supplies such as domestic bell transformers. Because it has no moving parts, a transformer can have a very high efficiency. Large power transformers, used on electrical distribution systems, can have an efficiency of better than 90%.

FIGURE 5.18

Typical oil filled power transformer.

Large power transformers need cooling to take the heat generated away from the core. This is often achieved by totally immersing the core and windings in insulating oil. A sketch of an oil immersed transformer can be seen in Fig. 5.18.

Very small transformers are used in electronic applications. Small transformers are used as isolating transformers in shaver sockets and can also be used to supply separated extra low voltage (SELV) sources. Equipment supplied from a SELV source may be installed in a bathroom or shower-room, provided that it is suitably enclosed and protected from the ingress of moisture. This includes equipment such as water heaters, pumps for showers and whirlpool baths.

Try This

Have you seen any transformers in action?

Were they big or small – what were they being used for?

Have you been close up to a transmission tower, perhaps when you were walking in the countryside?

Electrical power on the National Grid

Electricity is generated in large modern Power Stations at 25 kV (25,000 volts). It is then transformed up to 132 kV or 270 kV for transmission to other parts of the country on the National Grid network. This is a network of overhead conductors suspended on transmission towers which link together the Power Stations and the millions of users of electricity.

Raising the voltage to these very high values reduces the losses on the transmission network. 66 kV or 33 kV are used for secondary transmission lines and then these high voltages are reduced to 11 kV at local sub-stations for distribution to end users such as factories, shops and houses at 400 V and 230 V.

The ease and efficiency of changing the voltage levels is only possible because we generate an a.c. supply. Transformers are then used to change the voltage levels to those which are appropriate. Very high voltages for transmission, lower voltages for safe end use. This would not be possible if a d.c. supply was generated.

Figure 5.19 shows a simplified diagram of electricity distribution.

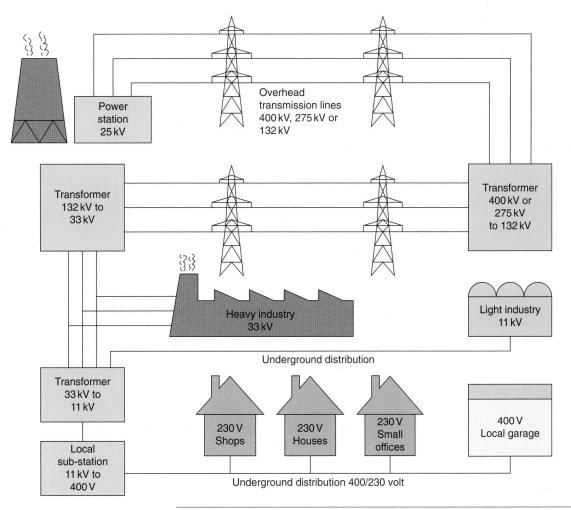

FIGURE 5.19

Simplified diagram of the distribution of electricity from power station to consumer.

Check Your Understanding

When you have completed the questions, check out the answers at the back of the book.
Note: more than one multiple choice answer may be correct.

1. Good conductor materials are:
    a. copper
    b. PVC
    c. brass
    d. wood.

2. Good insulator materials are:
    a. copper
    b. PVC
    c. brass
    d. wood.

3. A good conductor material:
    a. has lots of free electrons
    b. has no free electrons
    c. may be made of copper
    d. may be made of plastic.

4. A good insulator material:
    a. has lots of free electrons
    b. has no free electrons
    c. may be made of copper
    d. may be made of plastic.

5. In a series circuit
    a. the current is 'common' to all resistors
    b. the voltage is 'common' to all resistors
    c. $R_T = R_1 + R_2$
    d. $\dfrac{1}{R_T} = \dfrac{1}{R_1} + \dfrac{1}{R_2}$.

6. In a parallel circuit
    a. the current is 'common' to all resistors
    b. the voltage is 'common' to all resistors
    c. $R_T = R_1 + R_2$
    d. $\dfrac{1}{R_T} = \dfrac{1}{R_1} + \dfrac{1}{R_2}$.

7. The current taken by a 10 Ω resistor when connected to a 230 V supply will be:

    a. 2.3A

    b. 10A

    c. 23A

    d. 230A.

8. The resistance of a kettle element which takes 12A from a 230A main supply is:

    a. 2.88 Ω

    b. 5.00 Ω

    c. 12.24 Ω

    d. 19.16 Ω.

9. A 12 ohm filament lamp was found to be taking a current of 2 A at full brilliance. The voltage across the lamp under these conditions is:

    a. 6V

    b. 12V

    c. 24V

    d. 48V.

10. Current flowing through a solenoid sets up a magnetic flux. If an iron core is added to the solenoid while the current is maintained at a constant value, the magnetic flux will:

    a. remain constant

    b. totally collapse

    c. decrease in strength

    d. increase in strength.

11. Resistors of 6 Ω and 3 Ω are connected in series. The combined resistance value will be:

    a. 2.0 Ω

    b. 3.6 Ω

    c. 6.3 Ω

    d. 9.0 Ω.

12. Resistors of 6 Ω and 3 Ω are connected in parallel. The combined resistance value will be:

    a. 2.0 Ω

    b. 3.6 Ω

    c. 6.3 Ω

    d. 9.0 Ω.

13. Resistors of 20 Ω, 40 Ω and 60 Ω are connected in series. The total resistance value will be:

    a. 10.9 Ω

    b. 20.0 Ω

    c. 60.6 Ω

    d. 120 Ω.

14. Resistors of 20 Ω, 40 Ω and 60 Ω are connected in parallel. The total resistance value will be:

    a. 10.9 Ω

    b. 20.0 Ω

    c. 60.0 Ω

    d. 120 Ω.

15. Two identical resistors are connected in series across a 24 V battery. The voltage drop across each resistor will be:

    a. 2 V

    b. 6 V

    c. 12 V

    d. 24 V.

16. Two identical resistors are connected in parallel across a 24 V battery. The voltage drop across each resistor will be:

    a. 2 V

    b. 6 V

    c. 12 V

    d. 24 V.

17. Electricity is generated in a modern power station at:

    a. 230 V

    b. 400 V

    c. 25 kV

    d. 132 kV.

18. Electricity is distributed on the National Grid at:

   a. 230 V

   b. 400 V

   c. 25 kV

   d. 132 kV.

19. Describe with the aid of a simple diagram how the atoms and electrons behave in a material said to be a good conductor of electricity.

20. Describe, with the aid of a simple diagram, how the atoms and electrons behave in a material said to be a good insulator.

21. List five materials which are used as good conductors in the electrotechnical industry.

22. List five materials which are used as good insulators in the electrotechnical industry.

23. Sketch a simple circuit of two resistors connected in series across a battery and explain how the current flows in this circuit.

24. Sketch a simple circuit of two resistors connected in parallel across a battery and explain how the current flows in this circuit.

25. Sketch a simple circuit to show how a voltmeter and ammeter would be connected into the circuit to measure total voltage and total current.

26. Describe the advantage of using an a.c. supply for the National Grid rather than a d.c. supply.

27. Sketch the construction of a simple transformer and label the primary and secondary windings. Why is the metal core of the transformer laminated? How do we cool a big power transformer?

28. List five practical applications for a transformer – for example, a shaver socket.

29. Describe the three effects of an electric current.

30. Sketch the magnetic flux patterns:

   a. around a simple bar magnet

   b. a horseshoe magnet

   c. explain the action and state one application for a solenoid.

Basic electrical circuits and cables

Unit 2 – Principles of electrotechnology – Outcome 3

Underpinning knowledge: when you have completed this chapter you should be able to:

- state the component parts of electrical cables

- state the component parts of an electrical circuit

- determine appropriate wiring systems

- identify the requirements of 'protective equipotential bonding'

- list exposed conductive parts and extraneous conductive parts

- list basic principles of 'basic protection, 'fault protection' and 'overload protection'

Electrical cables

In Chapters 4 and 5 we looked at the science behind conductors and insulators. In this chapter, we will look at a practical application for that science, electrical cables.

Most cables can be considered to be constructed in three parts: the **conductor** which must be of a suitable cross-section to carry the load current; the **insulation**, which has a colour or number code for identification; and the **outer sheath** which may contain some means of providing protection from mechanical damage.

The conductors of a cable are made of either copper or aluminium and may be stranded or solid. Solid conductors are only used in fixed wiring installations and may be shaped in larger cables. Stranded conductors are more flexible and conductor sizes from 4.0 to 25 mm^2 contain seven strands. A 10 mm^2 conductor, for example, has seven 1.35 mm diameter strands which collectively make up the 10 mm^2 cross-sectional area of the cable. Conductors above 25 mm^2 have more than seven strands, depending upon the size of the cable. Flexible cords have multiple strands of very fine wire, as fine as one strand of human hair. This gives the cable its very flexible quality.

Definition

Cables can be considered to be constructed in three parts: the *conductor* which must be of a suitable cross-section to carry the load current; the *insulation*, which has a colour or number code for identification; and the *outer sheath* which may contain some means of providing protection from mechanical damage.

New wiring colours

Twenty-eight years ago the United Kingdom agreed to adopt the European colour code for flexible cords, that is, brown for live or phase conductor, blue for the neutral conductor and green combined with yellow for earth conductors. However, no similar harmonization was proposed for nonflexible cables used for fixed wiring. These were to remain as red for live or phase conductor, black for the neutral conductor and green combined with yellow for earth conductors.

On 31 March 2004, the IEE published Amendment No. 2 to BS 7671: 2001 which specified new cable core colours for all fixed wiring in UK electrical installations. These new core colours will 'harmonize' the United Kingdom with the practice in mainland Europe.

Fixed cable core colours up to 2006

- *Single-phase* supplies red line conductors, black neutral conductors, and green combined with yellow for earth conductors.

- *Three-Phase* supplies red, yellow and blue line conductors, black neutral conductors and green combined with yellow for earth conductors.

These core colours must not be used after 31 March 2006.

New (harmonized) fixed cable core colours

- *Single-phase* supplies brown line conductors, blue neutral conductors and green combined with yellow for earth conductors (just like flexible cords).

- *Three-phase* supplies brown, black and grey line conductors, blue neutral conductors and green combined with yellow for earth conductors.

Cable core colours from 31st of March 2004 onwards.

Extensions or alterations to existing *single-phase* installations do not require marking at the interface between the old and new fixed wiring colours. However, a warning notice must be fixed at the consumer unit or distribution fuse board which states:

> *Caution – this installation has wiring colours to two versions of BS 7671. Great care should be taken before undertaking extensions, alterations or repair that all conductors are correctly identified.*

Alterations to *three-phase* installations must be marked at the interface L1, L2, L3 for the lines and N for the neutral. Both new and old cables must be marked. These markings are preferred to coloured tape and a caution notice is again required at the distribution board. Appendix 7 of BS 7671: 2008 deals with harmonized cable core colours.

PVC insulated and sheathed cables

Domestic and commercial installations use this cable, which may be clipped direct to a surface, sunk in plaster or installed in conduit or trunking. It is the simplest and least expensive cable. Figure 6.1 shows a sketch of a twin and earth cable.

The conductors are covered with a colour-coded PVC insulation and then contained singly or with others in a PVC outer sheath.

PVC/SWA cable

PVC insulated steel wire armour cables are used for wiring underground between buildings, for main supplies to dwellings, rising sub-mains and industrial installations. They are used where some mechanical protection of the cable conductors is required.

The conductors are covered with colour-coded PVC insulation and then contained either singly or with others in a PVC sheath (see Fig. 6.2). Around this sheath is placed an armour protection of steel wires twisted along the length of the cable, and a final PVC sheath covering the steel wires protects them from corrosion. The armour sheath also provides the circuit protective

Safety First

PVC cables

- PVC cables should not be installed when the surrounding temperature is below 0°C.
- The PVC insulation becomes brittle at low temperatures and may be damaged during installation.
- IEE Regulation 522.1.2.

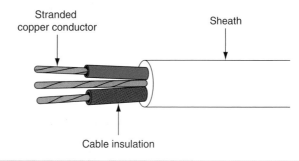

FIGURE 6.1

A twin and earth PVC insulated and sheathed cable.

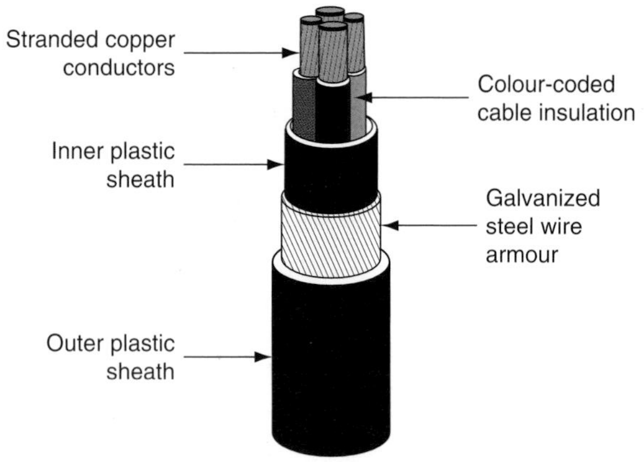

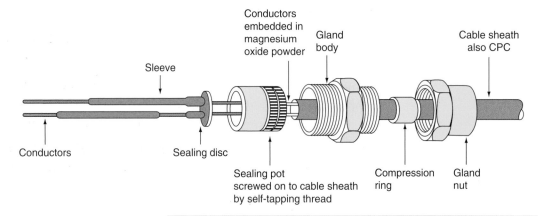

FIGURE 6.2

A four-core PVC/SWA cable.

FIGURE 6.3

MI cable with terminating seal and gland.

conductor (CPC) and the cable is simply terminated using a compression gland.

MI cable

A mineral insulated (MI) cable has a seamless copper sheath which makes it waterproof and fire- and corrosion-resistant. These characteristics often make it the only cable choice for hazardous or high-temperature installations such as oil refineries and chemical works, boiler houses and furnaces, petrol pump and fire alarm installations.

The cable has a small overall diameter when compared to alternative cables and may be supplied as bare copper or with a PVC oversheath. It is colour-coded orange for general electrical wiring, white for emergency lighting or red for fire alarm wiring. The copper outer sheath provides the CPC, and the cable is terminated with a pot and sealed with compound and a compression gland (see Fig. 6.3).

The copper conductors are embedded in a white powder, magnesium oxide, which is non-ageing and non-combustible, but which is hygroscopic, which means that it readily absorbs moisture from the surrounding air, unless adequately terminated. The termination of an MI cable is a complicated process requiring the electrician to demonstrate a high level of practical skill and expertise for the termination to be successful.

FP 200 cable

FP 200 cable is similar in appearance to an MI cable in that it is a circular tube, or the shape of a pencil, and is available with a red or white sheath. However, it is much simpler to use and terminate than an MI cable.

The cable is available with either solid or stranded conductors that are insulated with 'insudite' a fire resistant insulation material. The conductors are then screened, by wrapping an aluminium tape around the insulated conductors, that is, between the insulated conductors and the outer sheath. This aluminium tape screen is applied metal side down and in contact with the bare CPC.

The sheath is circular and made of a robust thermoplastic low smoke, zero halogen material.

FP 200 is available in 2, 3, 4, 7, 12 and 19 cores with a conductor size range from 1.0 to 4.0 mm. The core colours are: two core, red and black, three core, red, yellow and blue and four core, black, red, yellow and blue.

The cable is as easy to use as a PVC insulated and sheathed cable. No special terminations are required, the cable may be terminated through a grommet into a knock out box or terminated through a simple compression gland.

The cable is a fire resistant cable, primarily intended for use in fire alarms and emergency lighting installations or it may be embedded in plaster.

High-voltage power cables

The cables used for high-voltage power distribution require termination and installation expertise beyond the normal experience of a contracting electrician. The regulations covering high-voltage distribution are beyond the scope of the IEE regulations for electrical installations. Operating at voltages in excess of 33 kV and delivering thousands of kilowatts, these cables are either suspended out of reach on pylons or buried in the ground in carefully constructed trenches.

High-voltage overhead cables

Suspended from cable towers or pylons, overhead cables must be light, flexible and strong.

The cable is constructed of stranded aluminium conductors formed around a core of steel stranded conductors (see Fig. 6.4). The aluminium conductors carry the current and the steel core provides the tensile strength required to suspend the cable between pylons. The cable is not insulated since it is placed out of reach and insulation would only add to the weight of the cable.

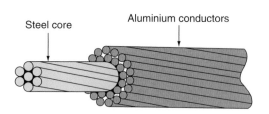

Steel core   Aluminium conductors

FIGURE 6.4

132 kV overhead cable construction.

Component parts of an electrical circuit

For a piece of electrical equipment to work efficiently and effectively it must be correctly connected to an electrical circuit. So what is an electrical circuit?

An electrical circuit has the following five components as shown in Fig. 6.5:

- a source of electrical energy. This might be a battery giving a d.c. (direct current) supply or the mains supply which is a.c. (alternating current)

- a source of circuit protection. This might be a fuse or circuit breaker which will protect the circuit from 'overcurrent'

- the circuit conductors or cables. These carry voltage and current to power the load

- a means to control the circuit. This might be a simple on/off switch but it might also be a dimmer or a thermostat

- and a load. This is something which needs electricity to make it work. It might be an electric lamp, an electrical appliance, an electric motor or an i-pod.

Choosing an appropriate wiring system

An electrical installation is made up of many different electrical circuits, lighting circuits, power circuits, single-phase domestic circuits and three-phase industrial or commercial circuits.

Whatever the type of circuit, the circuit conductors are contained within cables or enclosures.

Part 5 of the IEE Regulations tells us that electrical equipment and materials must be chosen so that they are suitable for the installed conditions, taking into account temperature, the presence of water, corrosion, mechanical damage, vibration or exposure to solar radiation. Therefore, PVC insulated and sheathed cables are suitable for domestic installations but for a cable requiring mechanical protection and suitable for burying underground, a PVC/SWA cable would be preferable. These two types of cable are shown in Figs 6.1 and 6.2 of this chapter.

MI cables are waterproof, heatproof and corrosion resistant with some mechanical protection. These qualities often make it the only cable choice

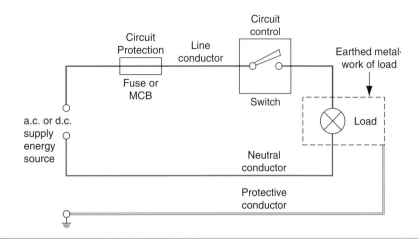

FIGURE 6.5
Component parts of an electric circuit.

for hazardous or high-temperature installations such as oil refineries, chemical works, boiler houses and petrol pump installations. An MI cable with terminating gland and seal is shown in Fig. 6.3.

We will be looking in detail at all of the different wiring systems and circuits later in this book in Chapter 14.

Electricity supplies

Electricity is generated in modern power stations at 25 kV and fed through transformers to the consumer over a complex network of cables known as the national grid system. This is a network of cables, mostly at a very high voltage, suspended from transmission towers, linking together the 175 power stations and millions of consumers. There are approximately 5,000 miles of high-voltage transmission lines in England and Wales, running mostly through the countryside.

Man-made structures erected in rural areas often give rise to concern, but every effort is made to route the overhead lines away from areas where they might spoil a fine view. There is full consultation with local authorities and interested parties as to the route which lines will take. Farmers are paid a small fee for having transmission line towers on their land. Over the years many different tower designs and colours have been tried, but for the conditions in the United Kingdom, galvanized steel lattice towers are considered the least conspicuous and most efficient.

For those who consider transmission towers unsightly, the obvious suggestion might be to run all cables underground. In areas of exceptional beauty this is done, but underground cables are about 16 times more expensive than the equivalent overhead lines. The cost of running the largest lines underground is about £3 million per mile compared with about £200,000 overhead. On long transmission lines the losses can be high, but by raising the operating voltage and therefore reducing the current for a given power, the I^2R losses are reduced, the cable diameter is reduced and the overall efficiency of transmission is increased. In order to standardize equipment, standard voltages are used. These are:

- 400 and 275 kV for the super grid
- 132 kV for the original grid
- 66 and 33 kV for secondary transmission
- 11 kV for high-voltage distribution
- 400 V for commercial consumer supplies
- 230 V for domestic consumer supplies.

A diagrammatic representation showing the distribution of power from the power station to the consumer was given in Chapter 5 in Fig. 5.19.

All local distribution in the United Kingdom is by underground cables from sub-stations placed close to the load centre and supplied at 11 kV. Transformers in these local sub-stations reduce the voltage to 400 V, three-phase and neutral distributor cables connect this supply to consumers. Connecting to one-phase and neutral of a three-phase 400 V supply gives a 230 V single-phase supply suitable for domestic consumers.

When single-phase loads are supplied from a three-phase supply, the load should be 'balanced' across the phases. That is, the load should be equally distributed across the three phases so that each phase carries approximately the same current. This prevents any one phase being overloaded.

Safe electrical installations

The provision of a safe electrical system is fundamental to the whole concept of using electricity in and around buildings safely. The electrical installation as a whole must be protected against overload and short circuit damage and the people using the installation must be protected against electric shock. An installation that meets the requirements of the IEE Wiring Regulations – Requirements for Electrical Installations, will be so protected. The method most universally used in the United Kingdom to provide for the safe use of electrical energy is protective equipotential bonding coupled with automatic disconnection of the supply by fuses or miniature circuit breakers (MCBs). So, let us look at these essential safety elements.

The consumer's mains equipment is normally fixed close to the point at which the supply cable enters the building. To meet the requirements of the IEE Regulations it must provide:

- protection against electric shock (Chapter 41)
- protection against overcurrent (Chapter 43)
- isolation and switching (Chapter 53).

Protection against electric shock, both 'basic protection' and 'fault protection' is provided by insulating and placing live parts out of reach in suitable

Definition

To provide *overcurrent protection* it is necessary to provide a device which will disconnect the supply automatically before the overload current can cause a rise in temperature which would damage the installation. A fuse or MCB would meet this requirement.

Definition

An *isolator* is a mechanical device which is operated manually and is provided so that the whole of the installation, one circuit or one piece of equipment may be cut off from the live supply.

Definition

The switching of electrically operated equipment in normal service is referred to as *functional switching*.

Definition

Earthing is the connection of the exposed conductive parts of an electrical installation to the main protective earthing terminal of the installation.

Definition

Bonding is the linking together of the exposed or extraneous metal parts of an electrical installation for the purpose of safety.

---

enclosures, earthing and bonding metal work and providing fuses or circuit breakers so that the supply is automatically disconnected under fault conditions.

To provide **overcurrent protection** it is necessary to provide a device which will disconnect the supply automatically before the overload current can cause a rise in temperature which would damage the installation. A fuse or MCB would meet this requirement.

An **isolator** is a mechanical device which is operated manually and is provided so that the whole of the installation, one circuit or one piece of equipment may be cut off from the live supply. In addition, a means of switching off for maintenance or emergency switching must be provided. A switch may provide the means of isolation, but an isolator differs from a switch in that it is intended to be opened when the circuit concerned is not carrying current. Its purpose is to ensure the safety of those working on the circuit by making dead those parts which are live in normal service. One device may provide both isolation and switching provided that the characteristics of the device meet the Regulations for both functions. The switching of electrically operated equipment in normal service is referred to as **functional switching**.

Circuits are controlled by switchgear which is assembled so that the circuit may be operated safely under normal conditions, isolated automatically under fault conditions, or isolated manually for safe maintenance. These requirements are met by good workmanship carried out by competent persons and the installation of approved British Standard materials such as switches, isolators, fuses or circuit breakers. (IEE Regulation 131.1.1). The equipment belonging to the supply authority is sealed to prevent unauthorized entry, because if connection were made to the supply before the meter, the energy used by the consumer would not be recorded on the meter. Figure 6.6 shows the connections and equipment at a domestic service position.

Protective electrical bonding to earth

The purpose of the bonding regulations is to keep all the exposed metalwork of an installation at the same earth potential as the metalwork of the electrical installation, so that no currents can flow and cause an electric shock. For a current to flow there must be a difference of potential between two points, but if the points are joined together there can be no potential difference. This bonding or linking together of the exposed metal parts of an installation is known as 'protective equipotential bonding' and gives protection against electric shock.

Let us now define some of the important new words as they apply to electrical installations.

Earthing is the connection of the exposed conductive parts of an electrical installation to the main protective earthing terminal of the installation.

Bonding is the linking together of the exposed or extraneous metal parts of an electrical installation for the purpose of safety.

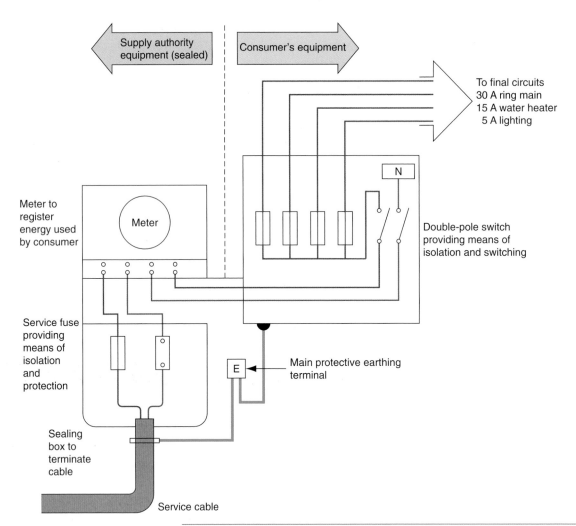

FIGURE 6.6

Simplified diagram of connections and equipment at a domestic service position.

Definition

Exposed conductive parts are the metalwork of the electrical installation. The conduit, trunking, metal boxes and equipment that make up the electrical installation.

Definition

Extraneous conductive parts are the other metal parts which do not form a part of the electrical installation. The structural steelwork of the building, gas, water and central heating pipes and radiators.

Exposed conductive parts are the metalwork of the electrical installation. The conduit, trunking, metal boxes and equipment that make up the electrical installation.

Extraneous conductive parts are the other metal parts which do not form a part of the electrical installation. The structural steelwork of the building, gas, water and central heating pipes and radiators.

Basic protection is protection against electric shock under fault free conditions and is provided by insulating live parts in accordance with section 416 of the IEE Regulations.

Fault protection is protection against electric shock under single fault conditions and is provided by protective equipotential bonding and automatic disconnection of the supply (by a fuse or MCB) in accordance with IEE Regulations 411.3 to 6.

Protection from electric shock is provided by basic protection and fault protection.

Protective equipotential bonding is equipotential bonding for the purpose of safety.

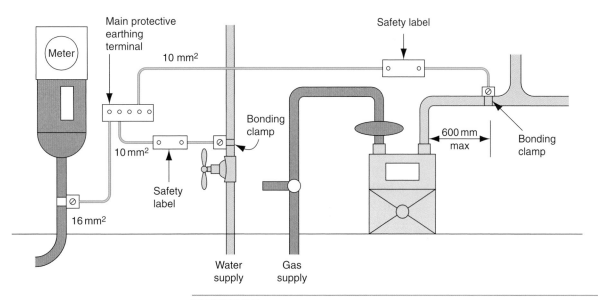

FIGURE 6.7
Protective equipotential bonding of gas and water supplies.

133

Try This

Memory Aid
Writing out important definitions helps you to remember them.

Protective equipotential bonding

Where earthed electrical equipment may come into contact with the metalwork of other services, they too must be effectively connected to the main protective earthing terminal of the installation (IEE Regulation 411.3.1.2).

Other services are described as:

- main water pipes
- main gas pipes
- other service pipes and ducting
- central heating and air conditioning systems
- exposed metal parts of the building structure
- lightning protective conductors

Protective equipotential bonding should be made to gas and water services at their point of entry into the building, as shown in Fig. 6.7, using insulated bonding conductors of not less than half the cross-section of the incoming main earthing conductor. The minimum permitted size is $6\,mm^2$ but the cross-section need not exceed $25\,mm^2$ (IEE Regulation 544.1.2). The bonding clamp must be fitted on the consumer's side of the gas meter between the outlet union, before any branch pipework but within $600\,mm$ of the meter 544.1.3.

A permanent label must also be fixed in a visible position at or near the point of connection of the bonding conductor with the words 'Safety Electrical Connection – Do Not Remove' (IEE Regulation 514.13.1). Supplementary bonding is described in Chapter 14 of this book.

Electrical shock and overload protection

Electric shock is normally caused either by touching a conductive part that is normally live, or by touching an exposed conductive part made live by a fault. The touch voltage curve in Fig. 6.8 shows that a person in contact with 230 V must be released from this danger in 40 ms if harmful effects are to be avoided. Similarly, a person in contact with 400V must be released in 15 ms to avoid being harmed.

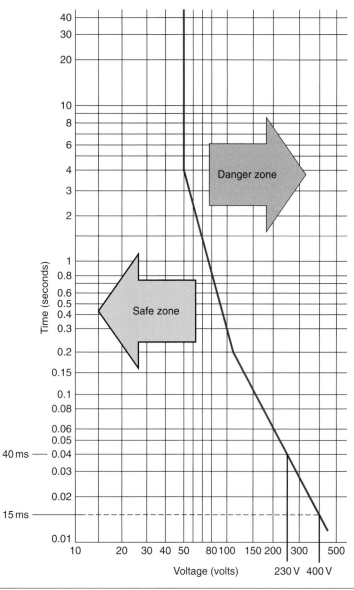

FIGURE 6.8

Touch voltage curve.

In general, protection against touching live parts is achieved by insulating live parts and called 'basic protection'. Protection against touching something made live as a result of a fault, and called 'fault protection' (IEE Regulation 131.2.2) is achieved by protective equipotential bonding and automatic disconnection of the supply in the event of a fault occurring. Separated extra low-voltage supplies (SELV) provide protection against both 'basic' and 'fault' protection.

Part 4 of the IEE Regulations deals with the application of protective measures for safety and Chapter 53 with the regulations for switching devices or switchgear required for protection, isolation and switching of a consumer's installation.

The consumer's main switchgear must be readily accessible to the consumer and be able to:

- isolate the complete installation from the supply
- protect against overcurrent
- cut off the current in the event of a serious fault occurring.

Protection against overcurrent

Excessive current may flow in a circuit as a result of an overload or a short circuit. An overload or overcurrent is defined as a current which exceeds the rated value in an otherwise healthy circuit. A short circuit is an overcurrent resulting from a fault of negligible impedance between live conductors having a difference in potential under normal operating conditions. Overload currents usually occur in a circuit because it is abused by the consumer or because it has been badly designed or modified by the installer. Short circuits usually occur as a result of an accident which could not have been predicted before the event.

An overload may result in currents of two or three times the rated current flowing in the circuit, while short circuit currents may be hundreds of times greater than the rated current. In both cases, the basic requirement for protection is that the circuit should be interrupted before the fault causes a temperature rise which might damage the insulation, terminations, joints or the surroundings of the conductors. If the device used for overload protection is also capable of breaking a prospective short circuit current safely, then one device may be used to give protection from both faults (Regulation 432.1). Devices which offer protection from overcurrent are:

- semi-enclosed fuses manufactured to BS 3036
- cartridge fuses manufactured to BS 1361 and BS 1362
- high breaking capacity fuses (HBC fuses) manufactured to BS 88
- MCBs manufactured to BS EN 60898.

We will look at overcurrent protection, fuses and MCBs in more detail later in this book in Chapter 12.

Definition

An *overload* or *overcurrent* is defined as a current which exceeds the rated value in an otherwise healthy circuit. A short circuit is an overcurrent resulting from a fault of negligible impedance.

Key Fact

Fuses

How does a fuse work?
- under fault conditions excessive current flows
- the fuse element gets hot
- the fuse element melts
- disconnecting the circuit it protected.

Safety First

Bonding

- bonding clamps must be of an approved type
- must be fitted to clean pipework
- must be tight and secure
- must have a visible label
- IEE Regulation 514.13.1.

When you have completed the questions, check out the answers at the back of the book.
Note: more than one multiple choice answer may be correct.

1. An electrical cable is made up of three parts which are:
   a. conduction, convection and radiation
   b. conductor, insulation and outer sheath
   c. heating, magnetic and chemical
   d. conductors and insulators.

2. An appropriate wiring method for a domestic installation would be a:
   a. metal conduit installation
   b. trunking and tray installation
   c. PVC cables
   d. PVC/SWA cables.

3. An appropriate wiring method for an underground feed to a remote building would be a:
   a. metal conduit installations
   b. trunking and tray installation
   c. PVC cables
   d. PVC/SWA cables.

4. An appropriate wiring method for a high-temperature installation in a boiler house is:
   a. metal conduit installation
   b. trunking and tray installation
   c. FP200 cables
   d. MI cables.

5. The cables suspended from the transmission towers of the National Grid Network are made from:
   a. copper and brass
   b. copper with PVC insulation
   c. aluminium and steel
   d. aluminium and porcelain.

6. An appropriate wiring system for a three-phase industrial installation would be:
   a. PVC cables
   b. PVC conduit
   c. one which meets the requirements of Part 2 of the IEE Regulations
   d. one which meets the requirements of Part 5 of the IEE Regulations.

7. A current which exceeds the rated value in an otherwise healthy circuit is one definition of:

   a. earthing

   b. bonding

   c. overload

   d. short circuit.

8. An overcurrent resulting from a fault of negligible impedance is one definition of:

   a. earthing

   b. bonding

   c. overload

   d. short circuit.

9. The connection of the exposed conductive parts of an installation to the main protective earthing terminal of the installation is one definition of:

   a. earthing

   b. protective equipotential bonding

   c. overload

   d. short circuit.

10. The linking together of the exposed or extraneous conductive parts of an installation for the purpose of safety is one definition of:

    a. earthing

    b. protective equipotential bonding

    c. exposed conductive parts

    d. extraneous conductive parts.

11. The conduit and trunking parts of the electrical installation are:

    a. earth conductors

    b. bonding conductors

    c. exposed conductive parts

    d. extraneous conductive parts.

12. The gas, water and central heating pipes of the building, not forming a part of the electrical installation are called:

    a. earthing conductors

    b. bonding conductors

    c. exposed conductive parts

    d. extraneous conductive parts.

13. The protection provided by insulating the live parts of the electrical installation is called:

    a. overload protection

    b. short circuit protection

    c. basic protection

    d. fault protection.

14. The protection provided by protective equipotential bonding and automatic disconnection of the supply is called:

    a. overload protection

    b. short circuit protection

    c. basic protection

    d. fault protection.

15. Produce a quick coloured sketch of a PVC insulated and sheathed cable and name the parts.

16. Produce a quick coloured sketch of a PVC/SWA cable and name the parts.

17. Produce a quick sketch of an electric circuit and name the five component parts.

18. Give an example of a device or accessory for each component part. For example, the supply might be from the a.c. mains or a battery.

19. In your own words state the meaning of circuit overload and short circuit protection. What will provide this type of protection?

20. State the purpose of earthing and earth protection. What do we do to achieve it and why do we do it?

21. In your own words state the meaning of exposed and extraneous conductive parts and give examples of each.

22. In your own words state the meaning of earthing and bonding. What type of cables and equipment would an electrician use to achieve earthing and bonding on an electrical installation.

23. In your own words state what we mean by 'basic protection' and how is it achieved.

24. In your own words state what we mean by 'fault protection' and how it is achieved.

Tools and equipment used for electrotechnical applications

Unit 2 – Principles of electrotechnology – Outcome 4

Underpinning knowledge: when you have completed this chapter, you should be able to:

- state the application of hand and power tools

- state that electrically operated tools must undergo inspection checks

- state the need for safe handling and storage of tools and equipment

- identify what we mean by 'good housekeeping'

Hand tools

A craftsman earns his living by hiring out his skills or selling products made using his skills and expertise. He shapes his environment, mostly for the better, improving the living standards of himself and others.

Tools extend the limited physical responses of the human body and therefore good quality, sharp tools are important to a craftsman. An electrician is no less a craftsman than a wood carver. Both must work with a high degree of skill and expertise and both must have sympathy and respect for the materials, which they use. Modern electrical installations using new materials are lasting longer than 50 years. Therefore they must be properly installed. Good design, good workmanship by competent persons and the use of proper materials are essential if the installation is to comply with the relevant regulations, (IEE Regulation 134.1.1) and reliably and safely meet the requirements of the customer for over half a century.

An electrician must develop a number of basic craft skills particular to his own trade, but he also requires some of the skills used in many other trades. An electrician's tool-kit will reflect both the specific and general nature of the work.

The basic tools required by an electrician are those used in the stripping and connecting of conductors.

These are pliers, side cutters, knife and an assortment of screwdrivers, as shown in Fig. 7.1.

The tools required in addition to these basic implements will depend upon the type of installation work being undertaken. When wiring new houses

140

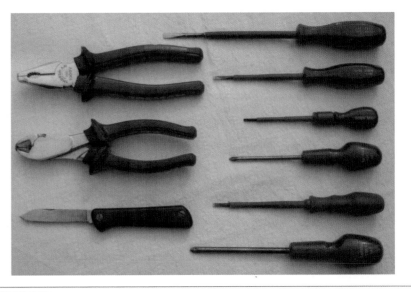

FIGURE 7.1
The tools used for making electrical connections.

or rewiring old ones, the additional tools required are those usually associated with a bricklayer and joiner. Examples are shown in Fig. 7.2.

When working on industrial installations, installing conduit and trunking, the additional tools required by an electrician would more normally be those associated with a fitter or sheet-metal fabricator, and examples are shown in Fig. 7.3.

Where special tools are required, for example, those required to terminate mineral insulated (MI) cables or the bending and cutting tools for conduit

FIGURE 7.2
Some additional tools required by an electrician engaged in house wiring.

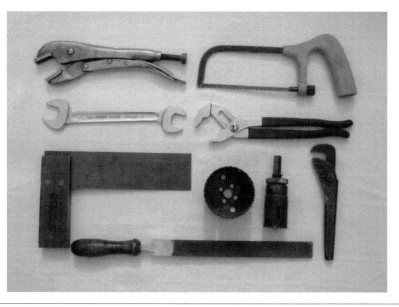

FIGURE 7.3
Some additional tools required by an electrician engaged in industrial installations.

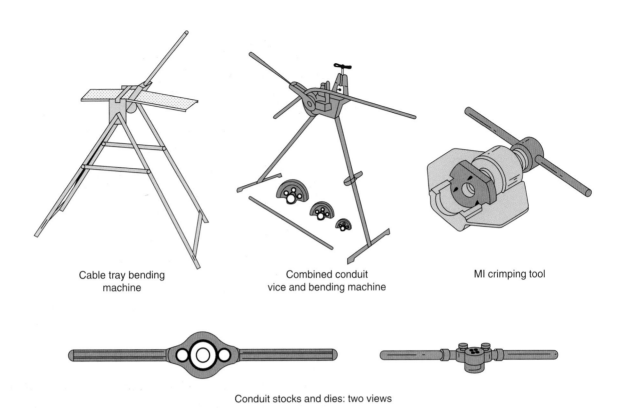

Cable tray bending machine

Combined conduit vice and bending machine

MI crimping tool

Conduit stocks and dies: two views

FIGURE 7.4

Some special tools required by an electrician engaged in industrial installations.

142

and cable trays as shown in Fig. 7.4, they will often be provided by an employer but most hand tools are provided by the electrician himself.

In general, good-quality tools last longer and stay sharper than those of inferior quality, but tools are very expensive to buy. A good set of tools can be assembled over the training period if the basic tools are bought first and the extended tool-kit acquired one tool at a time.

Another name for an installation electrician is a 'journeyman' electrician and, as the name implies, an electrician must be mobile and prepared to carry his tools from one job to another. Therefore, a good toolbox is an essential early investment, so that the right tools for the job can be easily transported.

Tools should be cared for and maintained in good condition if they are to be used efficiently and remain serviceable. Screwdrivers should have a flat squared off end and wood chisels should be very sharp. Access to a grind-stone will help an electrician to maintain his tools in first-class condition. Additionally, wood chisels will require sharpening on an oilstone to give them a very sharp edge.

Electrical power tools

Portable electrical tools can reduce much of the hard work for any trades-man and increase his productivity. Electrical tools should be maintained

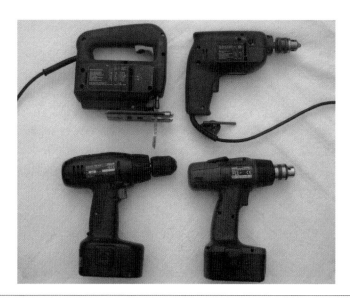

FIGURE 7.5
Electrical hand tools.

in a good condition and be appropriate for the purpose for which they are used. Many construction sites now insist on low voltage or battery tools being used which can further increase safety without any loss of productivity. Some useful electrical tools are shown in Fig. 7.5.

Electric drills are probably used most frequently of all electrical tools. They may be used to drill metal or wood. Wire brushes are made which fit into the drill chuck for cleaning the metal. Variable-speed electric drills, which incorporate a vibrator, will also drill brick and concrete as easily as wood when fitted with a masonry drill bit.

Hammer drills give between two and three thousand impacts per minute and are used for drilling concrete walls and floors.

Cordless electric drills are also available which incorporate a rechargeable battery, usually in the handle. They offer the convenience of electric drilling when an electrical supply is not available or if an extension cable is impractical.

Angle grinders are useful for cutting chases in brick or concrete. The discs are interchangeable. Silicon carbide discs are suitable for cutting slate, marble, tiles, brick and concrete, and aluminium oxide discs for cutting iron and steel such as conduit and trucking.

Jigsaws can be fitted with wood or metal cutting blades. With a wood cutting blade fitted they are useful for cutting across floorboards and skirting boards or any other application where a padsaw would be used. With a metal cutting blade fitted, they may be used to cut trunking.

When a lot of trunking work is to be undertaken, an electric nibbler is a worthwhile investment. This nibbles out the sheet metal, is easily controllable and is one alternative to the jigsaw.

Safety First

Power Tools

Many construction sites now only allow
- Low voltage tools
- PAT tested tools
- battery powered tools.

All tools must be used safely and sensibly. Cutting tools should be sharpened and screwdrivers ground to a sharp square end on a grindstone.

It is particularly important to check that the plug top and cables of hand held electrically powered tools and extension leads are in good condition. Damaged plug tops and cables must be repaired before you use them. All electrical power tools of 110V and 230V must be inspected and tested with a portable appliance tester (PAT) in accordance with the company's procedures, but probably at least once each year. PAT testing tests the quality of the insulation resistance and the earth continuity. Inspection checks the condition of the plug top, fuse and lead.

Tools and equipment that are left lying about in the workplace can become damaged or stolen and may also be the cause of people slipping, tripping or falling. Tidy up regularly and put power tools back in their boxes. You personally may have no control over the condition of the workplace in general, but keeping your own work area clean and tidy is the mark of a skilled and conscientious craftsman.

Safety First

Site Safety

The Health and Safety at Work Act makes *you* responsible for

- following the site safety rules
- demonstrating safe working practice
- Wearing or using appropriate PPEc.

Try This

Power Tools

- Look at the power tools at work
- do they have a PAT Test label
- Is it 'in date.'

Safe working practice

Every year thousands of people have accidents at their place of work despite the legal requirements laid down by the Health and Safety Executive. Many people recover quickly but an accident at work can result in permanent harm or even death.

At the very least, injuries hurt individuals. They may prevent you from doing the things you enjoy in your spare-time and they cost a lot of money, to you in loss of earnings and to your employer in loss of production and possibly damage to equipment. Your place of work may look harmless but it can be dangerous.

If there are five or more people employed by your company then the company must have its own safety policy as described in Chapter 1. This must spell out the organization and arrangements which have been put in place to ensure that you and your workmates are working in a safe place.

Your employer must also have carried out an assessment on the risks to your health and safety in the place where you are working. You should be told about the safety policy and risk assessment, for example, you may have been given a relevant leaflet when you started work. We will discuss Risk Assessment in some detail in Chapter 8 of this book.

You have a responsibility under the Health and Safety at Work Act to:

- learn how to work safely and to follow company procedures of work;

- obey all safety rules, notices and signs;

- not interfere with or misuse anything provided for safety;

- report anything that seems damaged, faulty or dangerous;

- behave sensibly, not play practical jokes and not distract other people at work;

- walk sensibly and not run around the workplace;

- use the prescribed walkways;

- drive only those vehicles for which you have been properly trained and passed the necessary test;

- not wear jewellery which could become caught in moving parts if you are using machinery at work;

- always wear appropriate clothing and PPE if necessary.

Common causes of accidents at work

Slips, trips and falls are still the major causes of accidents at work. To help prevent them:

- keep work areas clean and tidy;

- keep walkways clear;

- do not leave objects tools and equipment lying around blocking up walkways;

- clean up spills or wet patches on the floor straight away.

Manual handling, that is moving objects by hand, may result in strains, sprains and trap injury pains.

To help prevent them:

- use a mechanical aid to move heavy objects, such as a sack truck or flat bed truck

- only lift and carry what you can manage easily

- wear gloves to avoid rough or sharp edges

146

- use a good manual lifting technique which is discussed in the next chapter.

When using equipment, machines and some tools, such as angle grinders, are difficult to use.

To help prevent injuries:

- wear goggles

- wear appropriate PPE

- make sure tools and equipment are in good condition and carry an 'in date' PAT test label.

Badly stored equipment can become unstable and fall on to someone. To prevent this:

- stack equipment sensibly and securely

- stack heavy objects low down

- stack objects so they can be reached without stretching or reaching over.

Fire safety was discussed earlier in this book in Chapter 1.

Electricity and its safe use is what the electrotechnical industry and the regulations are all about.

To prevent electrical accidents always use the 'safe isolation procedure' before any work begins as described in Chapter 1.

Carrying out the bullet pointed activities listed above, which help to prevent the causes of accidents, can all be considered to be 'good housekeeping' because they individually contribute to a safer work environment.

Personal hygiene

In the work environment, dirt, thermal roof insulation and contact with chemicals and cleaning fluids may make you feel ill or cause unpleasant skin complaints. Therefore, you should always:

- wear appropriate personal protective equipment;

- wash your hands after using the toilet, after work and before you eat a meal, using soap and water or appropriate cleaners;

- dry your hands with the towel or dryer provided; do not use rags or your clothes;

- use barrier creams or latex gloves when they are provided to protect your skin;

- obtain medical advice about any skin complaint such as rashes, blisters or ulcers and tell your supervisor of the problems being experienced.

- The workplace must be a safe place to work or where particular hazards exist, they must be clearly identified and appropriate PPE provided.

We will look at the hazards associated with working above ground and good manual handling techniques in Chapter 8 of this book.

Check your Understanding

When you have completed the questions, check out the answers at the back of the book.
Note: more than one multiple choice answer may be correct.

1. Which hand tools would you use for terminating conductors in a junction box?
    a. a pair of side cutters or knife
    b. a screwdriver
    c. a wood chisel and saw
    d. a tenon saw.

2. Which hand tools would you use for removing cable insulation?
    a. a pair of side cutters or knife
    b. a screwdriver
    c. a wood chisel and saw.
    d. a tenon saw.

3. Which hand tools would you use to cut across a floorboard before lifting?
    a. a pair or side cutters or knife
    b. a screwdriver
    c. a wood chisel and saw
    d. a tenon saw.

4. Which hand tools would you use to cut and remove a notch in a floor joist?
    a. a pair of side cutters or knife
    b. a screwdriver
    c. a wood chisel and saw
    d. a tenon saw.

5. PAT testing is carried out on:
    a. hand tools
    b. domestic appliances only
    c. work electrical tools
    d. electrical equipment e.g. 110V transformers.

6. When PAT testing a 110V electric drill we are testing the:
    a. efficiency of the drill
    b. earth continuity of the drill
    c. hammer action of the drill
    d. insulation resistance of the drill.

7. For each type of tool shown in Figs 7.1 to 7.5, name the tool or piece of equipment and state one application (what you would use it for) for each.

8. Describe 'good practice' when handling and storing hand and power tools.

9. State five points that a PAT inspection and test checks on each appliance.

10. State what you would do with electrical tools on site when you have finished using them in order to make sure:

    a. they remain in good condition

    b. they are available the next time you want to use them.

11. Briefly describe that we mean by 'good housekeeping' on site.

12. State some of the actions that you could take at work that would make the work environment safer and that could be considered 'good housekeeping'.

13. Slips, trips and falls are the most common cause of accidents at work. What can you do at work to reduce the possibility of an accident being caused by a slip, trip or fall?

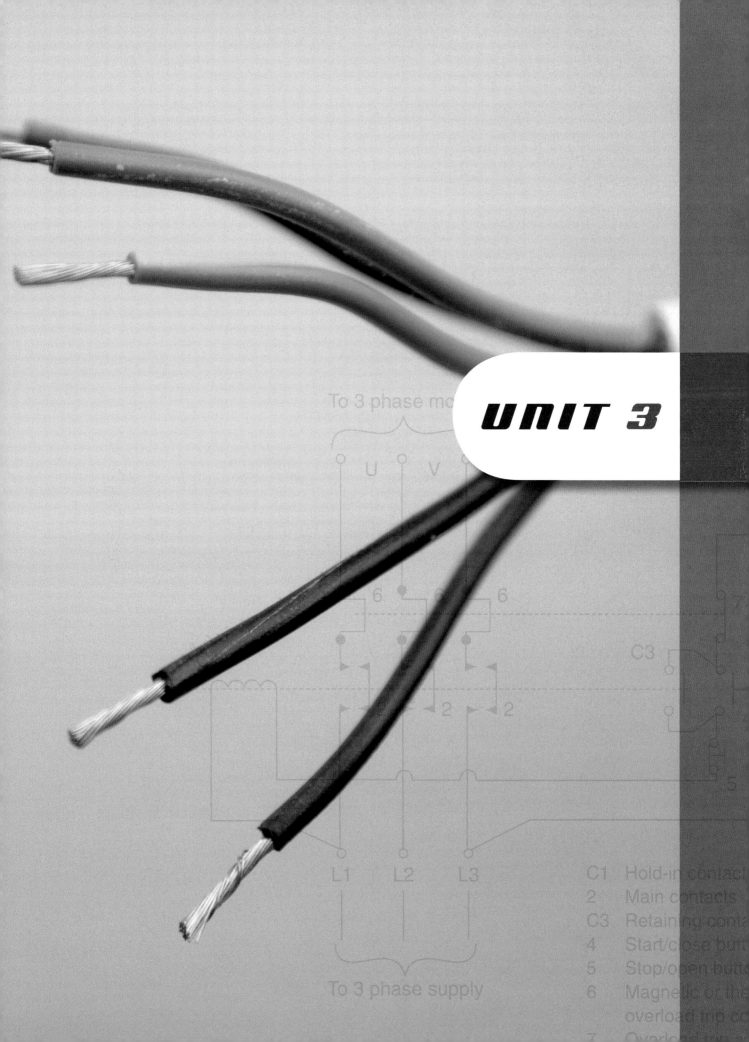

UNIT 3

To 3 phase motor

U   V

6   6   6

C3

2   2

L1   L2   L3

C1   Hold-in contact
2   Main contacts
C3   Retaining contact
4   Start/close button
5   Stop/open button
6   Magnetic or thermal
     overload trip coil

To 3 phase supply

Safe systems of working

Unit 3 – Application of health and safety and electrical principles – Outcome 1

Underpinning knowledge: when you have completed this chapter you should be able to:

- state the health and safety risks, precautions and procedures in the workplace
- define hazard and risks
- list the five stages of a risk assessment
- list the common categories of risk
- list the precautions taken to control risks
- state the rules for manual handling
- state the safety requirements when working at height
- list a safe electrical isolation procedure
- state why there is the need to leave the workplace clean and tidy
- state the procedure for disposing of waste materials

Health and safety risks, precautions and procedures

In Chapter 1 of this book, we looked at some of the health and safety rules and regulations. In particular, we now know that the Health and Safety at Work Act is the most important piece of recent legislation, because it places responsibilities for safety at work on both employers and employees. This responsibility is enforceable by law. We know what the regulations say about the control of substances, which might be hazardous to our health at work, because we briefly looked at the COSHH Regulations 2002 in Chapter 1. We also know that if there is a risk to health and safety at work our employer must provide personal protective equipment (PPE) free of charge, for us to use so that we are safe at work. The law is in place, we all apply the principles of health and safety at work and we always wear the appropriate PPE, so what are the risks? Well, getting injured at work is not a pleasant subject to think about but each year about 300 people in Great Britain lose their lives at work. In addition, there are about 158,000 non-fatal injuries reported to the Health and Safety Executive (HSE) each year and an estimated 2.2 million people suffer ill health caused by, or made worse by, work. It is a mistake to believe that these things only happen in dangerous occupations such as deep-sea diving, mining and quarrying, fishing industry, tunnelling and fire fighting or that it only happens in exceptional circumstances such as would never happen in your workplace. This is not the case. Some basic thinking and acting beforehand could have prevented most of these accident statistics from happening.

The most common categories of risk and causes of accidents at work are:

- slips, trips and falls
- manual handling, that is moving objects by hand
- using equipment, machinery or tools
- storage of goods and materials which then become unstable
- fire
- electricity
- mechanical handling.

Precautions taken to control risks:

- eliminate the cause
- substitute a procedure or product with less risk
- enclose the dangerous situation
- put guards around the hazard
- use safe systems of work
- supervise, train and give information to staff
- if the hazard cannot be removed or minimized then provide PPE.

Let us now look at the application of some of the procedures that make the workplace a safer place to work, but first I want to explain what I mean, when I use the words hazard and risk.

Hazard and risk

A **hazard** is something with the 'potential' to cause harm, for example, chemicals, electricity or working above ground.

A **risk** is the 'likelihood' of harm actually being done.

Competent persons are often referred to in the Health and Safety at Work Regulations, but who is 'competent'? For the purposes of the Act, a competent person is anyone who has the necessary technical skills, training and expertise to safely carry out the particular activity. Therefore, a competent person dealing with a hazardous situation reduces the risk.

Think about your workplace and at each stage of what you do, think about what might go wrong. Some simple activities may be hazardous. Here are some typical activities where accidents might happen.

Typical activity	Potential hazard
Receiving materials	Lifting and carrying
Stacking and storing	Falling materials
Movement of people	Slips, trips and falls
Building maintenance	Working at heights or in confined spaces
Movement of vehicles	Collisions

How high are the risks? Think about what might be the worst result; is it a broken finger or someone suffering permanent lung damage or being killed? How likely is it to happen? How often is that type of work carried out and how close do people get to the hazard? How likely is it that something will go wrong?

How many people might be injured if things go wrong? Might this also include people who do not work for your company?

Employers of more than five people must document the risks at work and the process is known as **hazard risk assessment**.

Hazard risk assessment: The procedure

The Management of Health and Safety at Work Regulations 1999 tells us that employers must systematically examine the workplace, the work activity and the management of safety in the establishment through a process of risk assessments. A record of all significant risk assessment findings must be kept in a safe place and be made available to an HSE inspector if required. Information based on the risk assessment findings must be communicated to relevant staff and if changes in work behaviour patterns are recommended in the interests of safety, then they must be put in place.

So risk assessment must form a part of any employer's robust policy of health and safety. However, an employer only needs to 'formally' assess the significant risks. He is not expected to assess the trivial and minor types of household risks. Staff are expected to read and to act upon these formal risk assessments, and they are unlikely to do so enthusiastically if the file is full of trivia. An assessment of risk is nothing more than a careful examination of what, in your work, could cause harm to people. It is a record that shows whether sufficient precautions have been taken to prevent harm.

The HSE recommends five steps to any risk assessment.

STEP 1

Look at what might reasonably be expected to cause harm. Ignore the trivial and concentrate only on significant hazards that could result in serious harm or injury. Manufacturer's data sheets or instructions can also help you spot hazards and put risks in their true perspective.

STEP 2

Decide who might be harmed and how. Think about people who might not be in the workplace all the time – cleaners, visitors, contractors or maintenance personnel. Include members of the public or people who share the workplace. Is there a chance that they could be injured by activities taking place in the workplace?

STEP 3

Evaluate what is the risk arising from an identified hazard. Is it adequately controlled or should more be done? Even after precautions have been put in place, some risk may remain. What you have to decide, for each significant hazard, is whether this remaining risk is low, medium or high. First of all, ask yourself if you have done all the things that the law says you have got to do. For example, there are legal requirements on the prevention of access to dangerous machinery. Then ask yourself whether generally accepted industry standards are in place, but do not stop there – think for yourself, because the law also says that you must do what is reasonably practicable to keep the workplace safe. Your real aim is to make all risks small by adding precautions, if necessary.

If you find that something needs to be done, ask yourself:

- Can I get rid of this hazard altogether?

- If not, how can I control the risk so that harm is unlikely?

Only use PPE when there is nothing else that you can reasonably do.

If the work that you do varies a lot, or if there is movement between one site and another, select those hazards which you can reasonably foresee, the ones that apply to most jobs and assess the risks for them. After that, if you spot any unusual hazards when you get on site, take what action seems necessary.

STEP 4

Record your findings and say what you are going to do about risks that are not adequately controlled. If there are fewer than five employees you do not need to write anything down but if there are five or more employees,

Key Fact

Definition

- Hazard is something that might cause harm
- Risk is the chance that harm will be done.

the significant findings of the risk assessment must be recorded. This means writing down the more significant hazards and assessing if they are adequately controlled and recording your most important conclusions. Most employers have a standard risk assessment form which they use such as that shown in Fig. 8.1 but any format is suitable. The important thing is to make a record.

HAZARD RISK ASSESSMENT	FLASH-BANG ELECTRICAL CO.
For Company name or site: _____ Address: _____ _____	Assessment undertaken by: _____ Signed: _____ Date: _____
STEP 5 Assessment review date: _____	
STEP 1 List the hazards here	STEP 2 Decide who might be harmed
STEP 3 Evaluate (what is) the risk – is it adequately controlled? State risk level as low, medium or high	STEP 4 Further action – what else is required to control any risk identified as medium or high?

FIGURE 8.1
Hazard risk assessment standard form.

There is no need to show how the assessment was made, provided you can show that:

1. a proper check was made

2. you asked those who might be affected

3. you dealt with all obvious and significant hazards

4. the precautions are reasonable and the remaining risk is low

5. you informed your employees about your findings.

Risk assessments need to be *suitable* and *sufficient*, not perfect. The two main points are:

1. Are the precautions reasonable?

2. Is there a record to show that a proper check was made?

File away the written assessment in a dedicated file for future reference or use. It can help if an HSE inspector questions the company's precautions or if the company becomes involved in any legal action. It shows that the company has done what the law requires.

STEP 5

Review the assessments from time to time and revise them if necessary.

Completing a risk assessment

When completing a risk assessment such as that shown in Fig. 8.1, do not be over complicated. In most firms in the commercial, service and light industrial sector, the hazards are few and simple. Checking them is common sense but necessary.

STEP 1

List only hazards which you could reasonably expect to result in significant harm under the conditions prevailing in your workplace. Use the following examples as a guide:

- Slipping or tripping hazards (e.g. from poorly maintained or partly installed floors and stairs)

- Fire (e.g. from flammable materials you might be using, such as solvents)

- Chemicals (e.g. from battery acid)

- Moving parts of machinery (e.g. blades)

- Rotating parts of hand tools (e.g. drills)

- Accidental discharge of cartridge operated tools

- High pressure air from airlines (e.g. air powered tools)

- Pressure systems (e.g. steam boilers)

- Vehicles (e.g. fork lift trucks)

- Electricity (e.g. faulty tools and equipment)
- Dust (e.g. from grinding operations or thermal insulation)
- Fumes (e.g. from welding)
- Manual handling (e.g. lifting, moving or supporting loads)
- Noise levels too high (e.g. machinery)
- Poor lighting levels (e.g. working in temporary or enclosed spaces)
- Low temperatures (e.g. working outdoors or in refrigeration plant)
- High temperatures (e.g. working in boiler rooms or furnaces).

STEP 2

Decide who might be harmed, do not list individuals by name. Just think about groups of people doing similar work or who might be affected by your work:

- Office staff
- Electricians
- Maintenance personnel
- Other contractors on site
- Operators of equipment
- Cleaners
- Members of the public.

Pay particular attention to those who may be more vulnerable, such as:

- staff with disabilities
- visitors
- young or inexperienced staff
- people working in isolation or enclosed spaces.

STEP 3

Calculate what is the risk – is it adequately controlled? Have you already taken precautions to protect against the hazards which you have listed in Step 1? For example:

- have you provided adequate information to staff?
- have you provided training or instruction?

Do the precautions already taken

- meet the legal standards required?
- comply with recognized industrial practice?
- represent good practice?
- reduce the risk as far as is reasonably practicable?

Safety First

Safety Procedures

- Hazard risk assessment is *an essential part* of any health and safety management system
- The aim of the planning process is to minimize risk
- HSE Publication, HSG (65).

159

If you can answer 'yes' to the above points then the risks are adequately controlled, but you need to state the precautions you have put in place. You can refer to company procedures, company rules, company practices, etc., in giving this information. For example, if we consider there might be a risk of electric shock from using electrical power tools, then the risk of a shock will be *less* if the company policy is to portable appliance test (PAT) all power tools each year and to fit a label to the tool showing that it has been tested for electrical safety. If the stated company procedure is to use battery drills whenever possible, or 110V drills when this is not possible, and to *never* use 230V drills, then this again will reduce the risk. If a policy such as this is written down in the company Safety Policy Statement, then you can simply refer to the appropriate section of the Safety Policy Statement and the level of risk will be low. (Note: PAT testing is described in Advanced Electrical Installation Work.)

STEP 4

Further action – what more could be done to reduce those risks, which were found to be inadequately controlled?

You will need to give priority to those risks that affect large numbers of people or which could result in serious harm. Senior managers should apply the principles below when taking action, if possible in the following order:

1. Remove the risk completely

2. Try a less risky option

3. Prevent access to the hazard (e.g. by guarding)

4. Organize work differently in order to reduce exposure to the hazard

5. Issue PPE

6. Provide welfare facilities (e.g. washing facilities for removal of contamination and first aid).

Any hazard identified by a risk assessment as *high risk* must be brought to the attention of the person responsible for health and safety within the company. Ideally, in Step 4 of the risk assessment you should be writing 'No further action is required. The risks are under control and identified as low risk'.

The assessor may use as many standard hazard risk assessment forms, such as that shown in Fig. 8.1, as the assessment requires. Upon completion they should be stapled together or placed in a plastic wallet and stored in the dedicated file.

You might like to carry out a risk assessment on a situation you are familiar with at work, using the standard form of Fig. 8.1, or your employer's standard forms. Alternatively, you might like to complete the visual display unit (VDU) workstation risk assessment checklist given in the next section.

We all use computers, and you might find it interesting to carry out a risk assessment of the computer workstation you use most, either at home, work or college, just for fun and to get an idea of how to carry out a risk assessment.

VDU operation hazards

Those who work at supermarket checkouts, assemble equipment or components, or work for long periods at a VDU and keyboard can be at risk because of the repetitive nature of the work. The hazard associated with these activities is a medical condition called 'upper limb disorders'. The term covers a number of related medical conditions.

Health and Safety (Display Screen Equipment) Regulations 1992

To encourage employers to protect the health of their workers and reduce the risks associated with VDU work, the HSE have introduced the Health and Safety (Display Screen Equipment) Regulations 1992. The regulations came into force on 1 January 1993, and employers who use standard office VDUs must show that they have taken steps to comply with the regulations.

So who is affected by the regulations? The regulations identify employees who use VDU equipment as 'users' if they:

- use a VDU more or less continuously on most days

- use a VDU more or less continuously for periods of an hour or more each day

- need to transfer information quickly to or from the screen

- need to apply high levels of attention or concentration to information displayed on a screen

- are very dependent upon VDUs or have little choice about using them.

All VDU users must be trained to use the equipment safely and protect themselves from upper limb disorders, temporary eyestrain, headaches, fatigue and stress.

To comply with the regulations an employer must:

- train users of VDU equipment and those who will carry out a risk assessment

- carry out a workstation risk assessment

- plan changes of activities or breaks for users

- provide eye and eyesight testing for users

- make sure new workstations comply with the regulations in the future

- give users information on the above.

USER TRAINING

Good user training will normally cover the following topics:

- the operating hazards and risks as described above
- the importance of good posture and changing position as shown in Fig. 8.2
- how to adjust furniture to avoid risks
- how to organize the workstation to avoid awkward or repeated stretching movements
- how to avoid reflections and glare on the monitor screen
- how to organize working routines so that there is a change of activity or a break
- how to adjust and clean the monitor screen
- how a user might contribute to a workstation risk assessment
- who to contact if problems arise.

When carrying out user training, the trainer might want to consider using a video, a computer-based training programme, discussions or seminars or the HSE employee leaflet *Working with VDUs* which can be obtained from the address given in Appendix B.

WORKSTATION RISK ASSESSMENT

A simple way to carry out a workstation risk assessment is to use a checklist such as that shown later in this section. Users can work through the checklist themselves. They know what the problems at their workstation are and whether they are comfortable or not. A trainer/assessor should then check the completed checklist and resolve the problems which the user cannot solve. For example, users may not know how the adjustment mechanism actually operates on their chair – a shorter user may benefit from a foot-rest as shown in Fig. 8.2, or a document holder may be more convenient for word processing users as shown in Fig. 8.3.

BREAKS

Breaking up long spells of display screen work helps to prevent fatigue and upper limb problems. Where possible encourage VDU users to carry

FIGURE 8.2

Examples of good posture when using VDU equipment.

out other tasks such as taking telephone calls, filing and photocopying. Otherwise, plan for users to take breaks away from the VDU screen if possible. The length of break required is not fixed by the law; the time will vary depending upon the work being done. Breaks should be taken before users become tired and short frequent breaks are better than longer infrequent ones.

EYE AND EYESIGHT TESTING

VDU users and those who are to become users of VDU equipment can request an eye and eyesight test that is free of charge to them. If the test shows that they need glasses specifically to carry out their VDU work, then their employer must pay for a basic pair of frames and lenses. Users are also entitled to further tests at regular intervals but if the user's normal glasses are suitable for VDU work, then the employer is not required to pay for them.

WORKSTATIONS

Make sure that new workstations comply with the regulations when:

- major changes to the workstation display screen equipment, furniture or software are made

- new users start work or change workstations

- workstations are re-sited

- the nature of the work changes considerably.

Users, trainers and assessors should focus on those aspects which have changed. For example:

- if the location of the workstation has changed, is the lighting adequate, is lighting or sunlight now reflecting off the display unit?

- different users have different needs – replacing a tall user with a short user may mean that a footrest is required

- users working from a number of source documents will need more desk space than users who are word processing.

A risk assessment should always be carried out on a new workstation or when a new operator takes over a workstation. Some questions cannot be answered until a user has had an opportunity to try the workstation. For example, does the user find the layout comfortable to operate, are there reflections on the screen at different times of the day as the sun moves around the building?

To be comfortable the operator should adjust the chair and equipment so that:

- Arms are horizontal and eyes are roughly at the height of the top of the VDU casing.

- Hands can rest on the work surface in front of the keyboard with fingers outstretched over the keys.

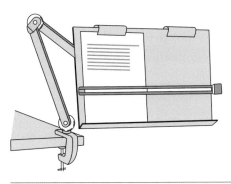

FIGURE 8.3
A document holder typically used by a word processing VDU operator.

163

- Feet are placed flat on the floor – too much pressure on the backs of legs and knees may mean that a footrest is needed.

- The small of the back is supported by the chair. The back should be held straight with the shoulders relaxed.

The arms on the chair or obstructions under the desk must not prevent the user from getting close enough to the keyboard comfortably.

INFORMATION

Good employers, who comply with the Display Screen Equipment Regulations, should let their employees know what care has been taken to reduce the risk to their health and safety at work. Users should be given information on:

- the health and safety relating to their particular workstations

- the risk assessments carried out and the steps taken to reduce risks

- the recommended break times and changes in activity to reduce risks

- the company procedures for obtaining eye and eyesight tests.

This information might be communicated to workers by:

- telling staff, for example, as part of an induction programme

- circulating a booklet or leaflet to relevant staff

- putting the information on a notice board

- using a computer-based information system, provided staff are trained in their use.

VDU workstation risk assessment checklist

Using a checklist such as that shown below or the more extensive checklist shown in the HSE book 'VDUs, An Easy Guide to the Regulations' is one way to assess workstation risks. You do not have to, but many employers find it a convenient method.

Risk factors are grouped under five headings and to each question the user should initially give a simple yes/no response. A 'yes' response means that no further action is necessary but a 'no' response will indicate that further follow-up action is required to reduce or eliminate risks to a user.

1. *Is the display screen image clear?*

    1.1 Are the characters readable?   Y/N

    1.2 Is the image free of flicker and movement?   Y/N

    1.3 Are brightness and contrast adjustable?   Y/N

    1.4 Does the screen swivel and tilt?   Y/N

    1.5 Is the screen free from glare and reflections?   Y/N

2. *Is the keyboard comfortable?*

    2.1 Is the keyboard tiltable?   Y/N

    2.2 Can you find a comfortable keyboard position?   Y/N

    2.3 Is there enough space to rest your hands in front of the keyboard?   Y/N

    2.4 Are the characters on the keys easily readable?   Y/N

3. *Does the furniture fit the work and user?*

    3.1 Is the work surface large enough?   Y/N

    3.2 Is the surface free of reflections?   Y/N

    3.3 Is the chair stable?   Y/N

    3.4 Do the adjustment mechanisms work?   Y/N

    3.5 Are you comfortable?   Y/N

4. *Is the surrounding environment risk free?*

    4.1 Is there enough room to change position and vary movement?   Y/N

    4.2 Are levels of light, heat and noise comfortable?   Y/N

    4.3 Does the air feel comfortable in terms of temperature and humidity?   Y/N

5. Is the software user friendly?

    5.1 Can you comfortably use the software?   Y/N

    5.2 Is the software suitable for the work task?   Y/N

    5.3 Have you had enough training?   Y/N

A copy of all risk assessments carried out should be placed in a dedicated file which can then be held by the trainer/assessor or other responsible person.

A copy of the full checklist can be found in the publication 'VDUs, an Easy Guide to the Regulations'. Other relevant publications include 'Display Screen Equipment Work and Guidance on Regulations L26' and 'Industry Advisory (General) leaflet IND(G) 36(L) 1993 Working with VDUs'. These and other health and safety publications are available from the HSE; the address is given in Appendix B.

Safe manual handling

Definition

Manual handling is lifting, transporting or supporting loads by hand or by bodily force.

Manual handling is lifting, transporting or supporting loads by hand or by bodily force. The load might be any heavy object, a printer, a VDU, a box of tools or a stepladder. Whatever the heavy object is, it must be moved thoughtfully and carefully, using appropriate lifting techniques if personal pain and injury are to be avoided. *Many people hurt their back, arms and*

165

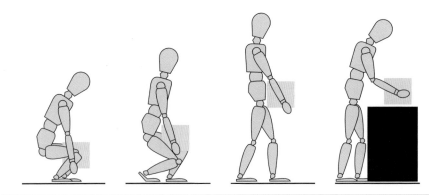

FIGURE 8.4
Correct manual lifting and carrying procedure.

feet, and over one-third of all 3-day reported injuries submitted to the HSE each year are the result of manual handling.

When lifting heavy loads, correct lifting procedures must be adopted to avoid back injuries. Figure 8.4 demonstrates the technique. Do not lift objects from the floor with the back bent and the legs straight as this causes excessive stress on the spine. Always lift with the back straight and the legs bent so that the powerful leg muscles do the lifting work. Bend at the hips and knees to get down to the level of the object being lifted, positioning the body as close to the object as possible. Grasp the object firmly and, keeping the back straight and the head erect, use the leg muscles to raise in a smooth movement. Carry the load close to the body. When putting the object down, keep the back straight and bend at the hips and knees, reversing the lifting procedure. A bad lifting technique will result in sprains, strains and pains. *There have been too many injuries over the years resulting from bad manual handling techniques. The problem has become so serious that the HSE has introduced new legislation* under the Health and Safety at Work Act 1974, the Manual Handling Operations Regulations 1992. Publications such as *Getting to Grips with Manual Handling* can be obtained from HSE Books; the address and Infoline are given in Appendix B.

Where a job involves considerable manual handling, employers must now train employees in the correct lifting procedures and provide the appropriate equipment necessary to promote the safe manual handling of loads.

Consider some 'good practice' when lifting loads.

- Do not lift the load manually if it is more appropriate to use a mechanical aid. Only lift or carry what you can easily manage.
- Always use a trolley, wheelbarrow or truck such as those shown in Fig. 8.5 when these are available.
- Plan ahead to avoid unnecessary or repeated movement of loads.
- Take account of the centre of gravity of the load when lifting – the weight acts through the centre of gravity.
- Never leave a suspended load unsupervised.
- Always lift and lower loads gently.
- Clear obstacles out of the lifting area.

Safety First

Lifting

- bend your legs
- keep your back straight
- use the leg muscles to raise the weight in a smooth movement

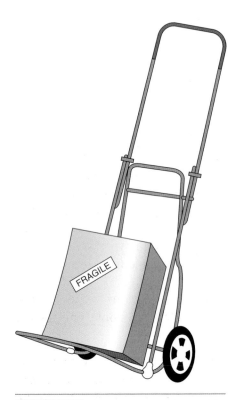

FIGURE 8.5
Always use a mechanical aid to transport a load when available.

- Use the manual lifting techniques described above and avoid sudden or jerky movements.

- Use gloves when manual handling to avoid injury from rough or sharp edges.

- Take special care when moving loads wrapped in grease or bubble-wrap.

- Never move a load over other people or walk under a suspended load.

Safe working above ground

Working above ground level creates added dangers and slows down the work rate of the electrician. New Work at Height Regulations came into force on the 6 April 2005. Every precaution should be taken to ensure that the working platform is appropriate for the purpose and in good condition.

LADDERS

The term ladder is generally taken to include step ladders and trestles. The use of ladders for working above ground level is only acceptable for access and work of short duration (Work at Height Regulations: 2005).

It is advisable to inspect the ladder before climbing it. It should be straight and firm. All rungs and tie rods should be in position and there should be no cracks in the stiles. The ladder should not be painted since the paint may be hiding defects.

Extension ladders should be erected in the closed position and extended one section at a time. Each section should overlap by at least the number of rungs indicated below:

- Ladder up to 4.8 m length – 2 rungs overlap

- Ladder up to 6.0 m length – 3 rungs overlap

- Ladder over 6.0 m length – 4 rungs overlap.

The angle of the ladder to the building should be in the proportion 4 up to 1 out or 75° as shown in Fig. 8.6. The ladder should be lashed at the top and bottom when possible to prevent unwanted movement and placed on firm and level ground. Footing is only considered effective for ladders smaller than 6 m and manufactured securing devices should always be considered. When ladders provide access to a roof or working platform the ladder must extend at least 1.05 m or 5 rungs above the landing place.

Short ladders may be carried by one person resting the ladder on the shoulder, but longer ladders should be carried by two people, one at each end, to avoid accidents when turning corners.

Long ladders or extension ladders should be erected by two people as shown in Fig. 8.7. One person stands on or 'foots' the ladder, while the other person lifts and walks under the ladder towards the walls. When the ladder is upright it can be positioned in the correct place, at the correct angle and secured before being climbed.

167

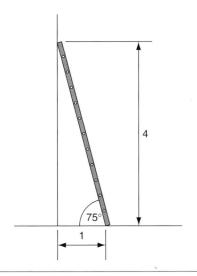

FIGURE 8.6
A correctly erected ladder.

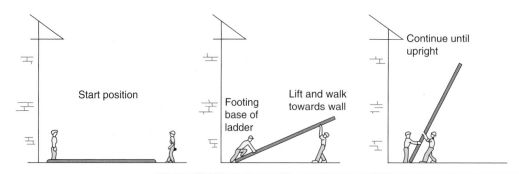

FIGURE 8.7

Correct procedure for erecting long or extension ladder.

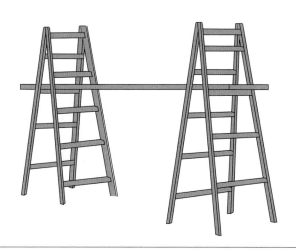

FIGURE 8.8

A trestle scaffold.

TRESTLE SCAFFOLD

Figure 8.8 shows a trestle scaffold. Two pairs of trestles spanned by scaffolding boards provide a simple working platform. The platform must be at least two boards or 450 mm wide. At least one-third of the trestle must be above the working platform. If the platform is more than 2 m above the ground, toeboards and guardrails must be fitted, and a separate ladder provided for access. The boards which form the working platform should be of equal length and not overhang the trestles by more than four times their own thickness. The maximum span of boards between trestles is:

- 1.3 m for boards 40 mm thick
- 2.5 m for boards 50 mm thick.

Trestles which are higher than 3.6 m must be tied to the building to give them stability. Where anyone can fall more than 4.5 m from the working platform, trestles may not be used.

MOBILE SCAFFOLD TOWERS

Mobile scaffold towers may be constructed of basic scaffold components or made from light alloy tube. The tower is built up by slotting the sections together until the required height is reached. A scaffold tower is shown in Fig. 8.9.

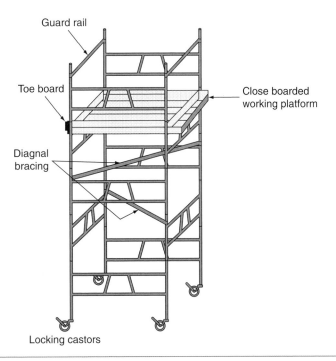

FIGURE 8.9
A mobile scaffold tower.

If the working platform is above 2 m from the ground it must be closed boarded and fitted with guardrails and toeboards. When the platform is being used, all four wheels must be locked. The platform must not be moved unless it is clear of tools, equipment and workers and should be pushed at the base of the tower and not at the top.

The stability of the tower depends upon the ratio of the base width to tower height. A ratio of base to height of 1:3 gives good stability. Outriggers can be used to increase stability by effectively increasing the base width. If outriggers are used then they must be fitted diagonally across all four corners of the tower and not on one side only. The tower must not be built more than 12 m high unless it has been specially designed for that purpose. Any tower higher than 9 m should be secured to the structure of the building to increase stability.

Access to the working platform of a scaffold tower should be by a ladder securely fastened vertically to the inside of the tower. Ladders must never be leaned against a tower since this might push the tower over.

Secure electrical isolation

Electric shock occurs when a person becomes part of the electrical circuit. The level or intensity of the shock will depend upon many factors, such as age, fitness and the circumstances in which the shock is received. The lethal level is approximately 50 mA, above which muscles contract, the heart flutters and breathing stops. A shock above the 50 mA level is therefore fatal unless the person is quickly separated from the supply. Below 50 mA only an unpleasant tingling sensation may be experienced or you may be thrown across a room or shocked enough to fall from a roof or ladder, but the resulting fall may lead to serious injury.

169

To prevent people receiving an electric shock accidentally, all circuits contain protective devices. All exposed metal is earthed; fuses and miniature circuit breakers (MCBs) are designed to trip under fault conditions, and residual current devices (RCDs) are designed to trip below the fatal level as described in Chapter 6.

Construction workers and particularly electricians do receive electric shocks, usually as a result of carelessness or unforeseen circumstances. As an electrician working on electrical equipment you must always make sure that the equipment is switched off or electrically isolated before commencing work. Every circuit must be provided with a means of isolation (IEE Regulation 132.15). When working on portable equipment or desk top units it is often simply a matter of unplugging the equipment from the adjacent supply. Larger pieces of equipment and electrical machines may require isolating at the local isolator switch before work commences. To deter anyone from re-connecting the supply while work is being carried out on equipment, a sign 'Danger – Electrician at Work' should be displayed on the isolator and the isolation 'secured' with a small padlock or the fuses removed so that no one can re-connect whilst work is being carried out on that piece of equipment. The Electricity at Work Regulations 1989 are very specific at Regulation 12(1) that we must ensure the disconnection and separation of electrical equipment from every source of supply and that this disconnection and separation is secure. Where a test instrument or voltage indicator is used to prove the supply dead, Regulation 4(3) of the Electricity at Work Regulations 1989 recommends that the following procedure is adopted.

1. First connect the test device such as that shown in Fig. 8.10 to the supply which is to be isolated. The test device should indicate main's voltage.

2. Next, isolate the supply and observe that the test device now reads zero volts.

3. Then connect the same test device to a known live supply or proving unit such as that shown in Fig. 8.11 to 'prove' that the tester is still working correctly.

4. Finally secure the isolation and place warning signs; only then should work commence.

The test device being used by the electrician must incorporate safe test leads which comply with the Health and Safety Executive Guidance Note 38 on electrical test equipment. These leads should incorporate barriers to prevent the user touching live terminals when testing and incorporating a protective fuse and be well insulated and robust, such as those shown in Fig. 8.12.

To isolate a piece of equipment or individual circuit successfully, competently, safely and in accordance with all the relevant regulations, we must follow a procedure such as that given by the flow diagram of Fig. 8.13. Start at the top and work down the flow diagram.

Definition

Electrical isolation We must ensure the disconnection and separation of electrical equipment from every source of supply and that this disconnection and separation is secure.

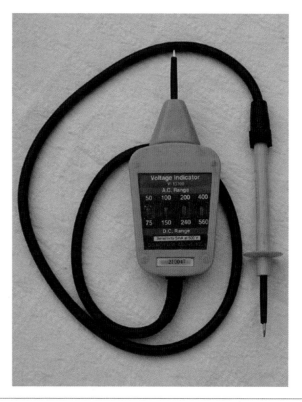

FIGURE 8.10
Typical voltage indicator.

Battery

PROVING UNIT PU2

MARTINDALE
TESTING
DEVICES
VI-13700/1
VI-16200

50-500V
AC/DC

CAUTION ⚠

READ INSTRUCTIONS
BEFORE OPERATING

6LF22 9V MN1604

Alpha
ELECTRONICS
LIMITED
MANCHESTER M29 0QA
TEL: (0942) 873434

SWITCH ON

Insert probe ends of testing device into
a.c./d.c. output sockets and while holding
testing device apply a light pressure in
direction of sockets

SWITCH OFF

Withdrawal of testing device instantly de-
energises PU2

PROOF TESTING

With probes of testing device inserted into
sockets APPLY SUFFICIENT PRESSURE
for good electrical contact

CHECK △ INDICATES AT ALL TIMES
during testing. Replace battery if, in normal
use, it does not illuminate.

Check ALL NEON LAMPS located within
the testing device illuminate for duration of
PROOF TEST

FIGURE 8.11
Voltage proving unit.

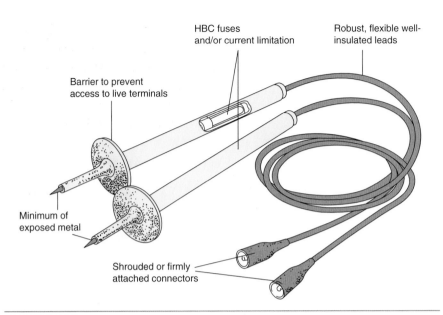

HBC fuses
and/or current limitation

Robust, flexible well-
insulated leads

Barrier to prevent
access to live terminals

Minimum of
exposed metal

Shrouded or firmly
attached connectors

FIGURE 8.12
Recommended type of test probe and leads.

When the heavy outlined amber boxes are reached, pause and ask yourself whether everything is satisfactory up to this point. If the answer is 'yes', move on. If the answer is 'no', go back as indicated by the diagram.

LIVE TESTING

The Electricity at Work Regulations 1989 at Regulation 4(3) tells us that it is preferable that supplies be made dead before work commences. However, it does acknowledge that some work, such as fault finding and testing, may require the electrical equipment to remain energized. Therefore, if the fault finding and testing can only be successfully carried out live then the person carrying out the fault diagnosis must:

- be trained so that they understand the equipment and the potential hazards of working live and can, therefore, be deemed 'competent' to carry out that activity

- only use approved test equipment

- set up appropriate warning notices and barriers so that the work activity does not create a situation dangerous to others.

While live testing may be required by workers in the electrotechnical industries in order to find the fault, live repair work must not be carried out. The individual circuit or piece of equipment must first be isolated before work commences in order to comply with the Electricity at Work Regulations 1989.

Disposing of waste

We have said many times in this book so far, that having a good attitude to health and safety, working conscientiously and neatly, keeping passageways clear and regularly tidying up the workplace is the sign of a good and competent craftsman. But what do you do with the rubbish that the working environment produces? Well, all the packaging material for electrical fittings and accessories usually goes into either your employer's skip or the skip on site designated for that purpose. All the off-cuts of conduit, trunking and tray also go into the skip. In fact, most of the general site debris will probably go into the skip and the waste disposal company will take the skip contents to a designated local council land fill area for safe disposal.

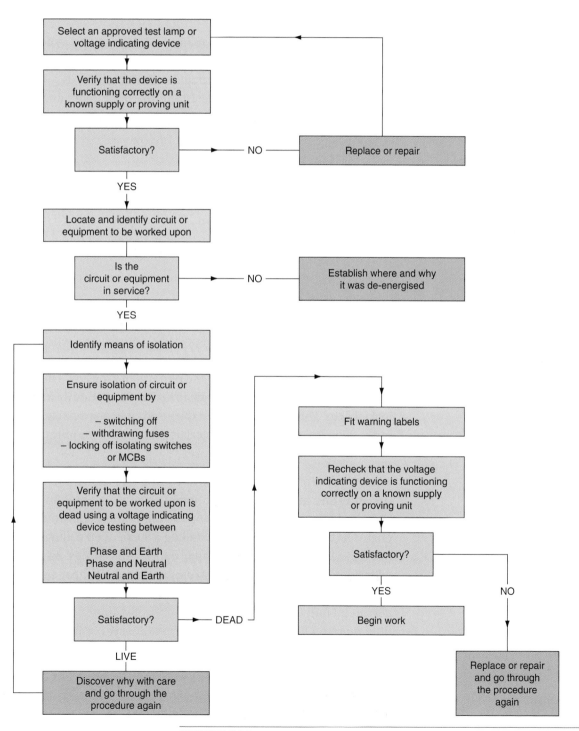

FIGURE 8.13
Flowchart for a secure isolation procedure.

The part coils of cable and any other reuseable leftover lengths of conduit, trunking or tray will be taken back to your employer's stores area. Here it will be stored for future use and the returned quantities deducted from the costs allocated to that job.

What goes into the skip for normal disposal into a land fill site is usually a matter of common sense. However, some substances require special consideration and disposal. We will now look at asbestos and large quantities of used fluorescent tubes.

Asbestos is a mineral found in many rock formations. When separated it becomes a fluffy, fibrous material with many uses. It was used extensively in the construction industry during the 1960s and 1970s for roofing material, ceiling and floor tiles, fire resistant board for doors and partitions, for thermal insulation and commercial and industrial pipe lagging.

In the buildings where it was installed some 40 years ago, when left alone, it does not represent a health hazard, but those buildings are increasingly becoming in need of renovation and modernization. It is in the dismantling and breaking up of these asbestos materials that the health hazard increases. Asbestos is a serious health hazard if the dust is inhaled. The tiny asbestos particles find their way into delicate lung tissue and remain embedded for life, causing constant irritation and eventually, serious lung disease.

Working with asbestos materials is not a job for anyone in the electro-technical industry. If asbestos is present in situations or buildings where you are expected to work, it should be removed by a specialist contractor before your work commences. Specialist contractors, who will wear fully protective suits and use breathing apparatus, are the only people who can safely and responsibly carry out the removal of asbestos. They will wrap the asbestos in thick plastic bags and store them temporarily in a covered and locked skip. This material is then disposed of in a special land fill site with other toxic industrial waste materials and the site monitored by the local authority for the foreseeable future.

There is a lot of work for electrical contractors in my part of the country, updating and improving the lighting in government buildings and schools. This work often involves removing the old fluorescent fittings, hanging on chains or fixed to beams and installing a suspended ceiling and an appropriate number of recessed modular fluorescent fittings. So what do we do with the old fittings? Well, the fittings are made of sheet steel, a couple of plastic lampholders, a little cable, a starter and ballast. All of these materials can go into the ordinary skip. However, the fluorescent tubes contain a little mercury and fluorescent powder with toxic elements, which cannot be disposed of in the normal land fill sites.

New Hazardous Waste Regulations were introduced in July 2005 and under these Regulations lamps and tubes are classified as hazardous. While each lamp contains only a small amount of mercury, vast numbers of lamps and tubes are disposed of in the United Kingdom every year resulting in a significant environmental threat.

The environmentally responsible way to dispose of fluorescent lamps and tubes is to recycle them.

The process usually goes like this:

- your employer arranges for the local electrical wholesaler to deliver a plastic used lamp waste container of an appropriate size for the job

- expired lamps and tubes are placed whole into the container, which often has a grating inside to prevent the tubes breaking when being transported

- when the container is full of used lamps and tubes, you telephone the electrical wholesaler and ask them to pick up the filled container and deliver it to one of the specialist recycling centres

- your electrical company will receive a 'Duty of Care Note' and full recycling documents which ought to be filed safely as proof that the hazardous waste was recycled safely

- the charge is approximately 50p for each 1800 mm tube and this cost is passed on to the customer through the final account.

175

Check your Understanding

When you have completed the questions, check out the answers at the back of the book.
Note: more than one multiple choice answer may be correct.

1. Two of the most common categories of risk and causes of accidents at work are:
    a. slips, trips and falls
    b. put guards around the hazard
    c. manual handling
    d. use safe systems of work.

2. Two of the most common precautions taken to control risks are:
    a. slips, trips and falls
    b. put guards around the hazard
    c. manual handling
    d. use safe systems of work throuit questions.

3. Something which has the potential to cause harm is one definition of:
    a. health and safety
    b. risk
    c. competent person
    d. hazard.

4. The chances of harm actually being done is one definition of:
    a. electricity
    b. risk
    c. health and safety
    d. hazard.

5. A competent person dealing with a hazardous situation:
    a. must wear appropriate PPE
    b. display a health and safety poster
    c. reduces the risk
    d. increases the risk.

6. Employers of companies employing more than five people must:
    a. become a member of the NICEIC
    b. provide PPE if appropriate
    c. carry out a hazard risk assessment
    d. display a health and safety poster.

7. There are five parts to a hazard risk assessment procedure. Identify one from the list below:

   a. wear appropriate PPE

   b. notify the HSE that you intend to carry out a risk assessment

   c. list the hazards and who might be harmed

   d. substitute a procedure with less risk.

8. Lifting, transporting or supporting heavy objects by hand or bodily force is one definition of:

   a. working at height

   b. a mobile scaffold tower

   c. a sack truck

   d. manual handling.

9. When working above ground for long periods of time the most appropriate piece of equipment to use would be:

   a. a ladder

   b. a trestle scaffold

   c. a mobile scaffold tower

   d. a pair of sky hooks.

10. The most appropriate piece of equipment to use for gaining access to a permanent scaffold would be:

   a. a ladder

   b. a trestle scaffold

   c. a mobile scaffold tower

   d. a pair of sky hooks.

11. The Electricity at Work Regulations tell us that 'we must ensure the disconnection and separation of electrical equipment from every source of supply and the separation must be secure'. A procedure to comply with this Regulation is called:

   a. work at height

   b. a hazard risk assessment

   c. a safe isolation procedure

   d. a workstation risk assessment.

12. The Electricity at Work Regulations absolutely forbid the following work activity:

    a. working at height

    b. testing live electrical systems

    c. live repair work on electrical circuits

    d. working without the appropriate PPE.

13. 'Good housekeeping' at work is about:

    a. cleaning up and putting waste in the skip

    b. working safely

    c. making the tea and collecting everyone's lunch

    d. putting tools and equipment away after use.

14. List five common categories of risk.

15. List five common precautions which might be taken to control risk.

16. Use bullet points to list the main stages involved in lifting a heavy box from the floor, carrying it across a room and placing it on a worktop, using a safe manual handling technique.

17. Describe a safe manual handling technique for moving a heavy electric motor out of the stores, across a yard and into the back of a van for delivery to site.

18. Use bullet points to list a step-by-step safe electrical isolation procedure for isolating a circuit in a three-phase distribution fuse board.

19. Use bullet points to list each stage in the erection and securing of a long extension ladder. Identify all actions which would make the ladder safe to use.

20. Describe how you would use a mobile scaffold tower to re-lamp all the light fittings in a supermarket. Use bullet points to give a step-by-step account of re-lamping the first two fittings.

21. What is a proving unit used for?

22. The HSE Guidance Note GS 38 tells us about suitable test probe leads. Use a sketch to identify the main recommendations.

23. State how you would deal with the following materials when you are cleaning up at the end of the job:

- pieces of conduit and tray
- cardboard packaging material
- empty cable rolls
- half full cable rolls
- bending machines for conduit and tray
- your own box of tools
- your employer's power tools
- 100 old fluorescent light fittings
- 200 used fluorescent tubes.

CH 9

Using technical information

Unit 3 – Application of health and safety and electrical principles – Outcome 2

Underpinning knowledge: when you have completed this chapter you should be able to:

- state methods of distributing technical information
- list the types of technical information
- state who uses technical information
- state types of 'on-site' documentation
- state how to receive materials on site
- state the importance of presenting the 'right image'

In Chapter 3 of this book we looked at *sources* of technical information and *specific types* of technical information, such as site plans and layout drawings. We said that:

Technical information can be received or distributed by

- written reports
- conventional scale drawings and information sheets
- the Internet and email
- CD, DVD or Memory Stick
- Facsimile (Fax machine).

In this chapter we are going to look at how we use technical information so this might be a good time to revise the work done in Chapter 3.

Types of technical information

Technical information is communicated to electrotechnical personnel in lots of different ways. It comes in the form of:

- *Specifications* – these are details of the client's requirements, usually drawn up by an architect. For example, the specification may give information about the type of wiring system to be employed or detail the type of luminaires or other equipment to be used.

- *Manufacturer's data* – if certain equipment is specified, let's say a particular type of luminaire or other piece of equipment, then the manufacturer's data sheet will give specific instructions for their assembly and fixing requirements. It is always good practice to read the data sheet before fitting the equipment. A copy of the data sheet should also be placed in the job file for the client to receive when the job is completed.

- *Reports and schedules* – a report is the written detail of something that has happened or the answer to a particular question asked by another professional person or the client. It might be the details of why an employee is to be disciplined or a report of some problem on site.

If the report is internal to the organization, a handwritten report is acceptable, but if the final report will go outside the organization, then it must be more formal and typed.

A schedule gives information about a programme or timetable of work, details of when certain events will take place. For example, when the electricians will start to do the 'first fix' and how many days it will take. A bar chart is an easy to understand schedule of work that shows how different activities interact on a project. Figure 9.1 shows a bar chart or schedule of work where activity A takes 2 days to complete and activity B starts at the same time as activity A but carries on for 8 days etc.

- *User instructions* – give information about the operation of a piece of equipment. Manufacturers of equipment provide 'user instructions'

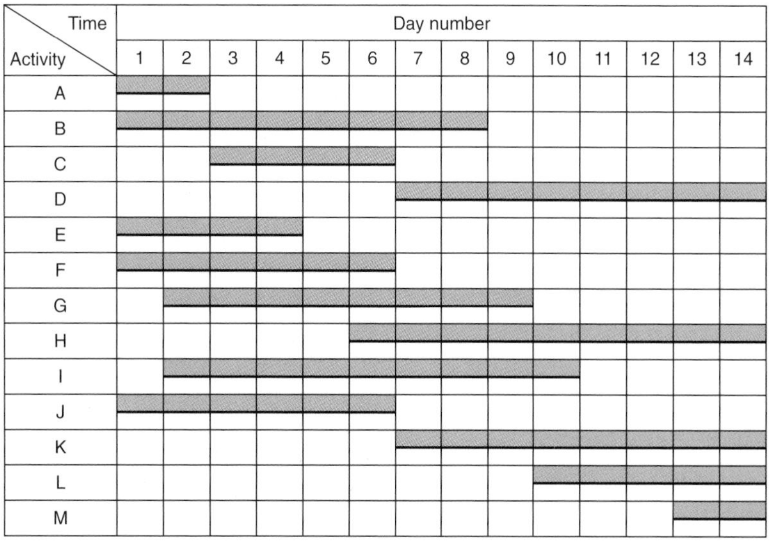
Time
Activity
Day number
1
2
3
4
5
6
7
8
9
10
11
12
13
14
A
B
C
D
E
F
G
H
I
J
K
L
M

FIGURE 9.1
A simple bar chart or schedule of work.

and a copy should be placed in the job file for the client to receive when the project is handed over.

- *Job sheets and Time sheets* – give 'on site' information. Job sheets give information about what is to be done and are usually issued by a manager to an electrician. Time sheets are a record of where an individual worker has been spending his time, which job and for how long. This information is used to make up individual wages and to allocate company costs to a particular job. We will look at these again later under the sub-heading 'on-site documentation'.

Those who need or use technical information

Technical information is required by many of the professionals involved in any electrotechnical activity, so who are the key people?

- The Operative – in our case this will be the skilled electricians actually on site, doing the job for the electrotechnical company.
- The Supervisor – he may have overall responsibility for a number of electricians on site and will need the 'big picture'.
- The Contractor – the main contractor takes on the responsibility of the whole project for the client. The main contractor may take on a sub-contractor to carry out some part of the whole project. On a large construction site the electrical contractor is usually the sub-contractor.
- Site Agent – he will be responsible for the smooth running of the whole project and for bringing the contract to a conclusion on schedule and within budget. The site agent may be nominated by the architect.
- Customer or Client – they also are the people ordering the work to be done. They will pay the final bill that pays everyone's wages.

On-site documentation

A lot of communications between and within larger organizations take place by completing standard forms or sending internal memos. Written messages have the advantage of being 'auditable'. An auditor can follow the paperwork trail to see, for example, who was responsible for ordering certain materials.

When completing standard forms, follow the instructions given and ensure that your writing is legible. Do not leave blank spaces on the form, always specifying 'not applicable' or 'N/A' whenever necessary. Sign or give your name and the date as asked for on the form. Finally, read through the form again to make sure you have answered all the relevant sections correctly.

Internal memos are forms of written communication used within an organization; they are not normally used for communicating with customers or suppliers. Figure 9.2 shows the layout of a typical standard memo form used by Dave Twem to notify John Gall that he has ordered the hammer drill.

Letters provide a permanent record of communications between organizations and individuals. They may be handwritten for internal use but formal business letters give a better impression of the organization if they are type-written. A letter should be written using simple concise language, and the tone of the letter should always be polite even if it is one of complaint. Always include the date of the correspondence. The greeting on a formal letter should be 'Dear Sir/Madam' and concluded with 'Yours faithfully'. A less formal greeting would be 'Dear Mr Smith' and concluded 'Yours sincerely'. Your name and status should be typed below your signature.

Delivery notes

When materials are delivered to site, the person receiving the goods is required to sign the driver's '**delivery note**'. This record is used to confirm that goods have been delivered by the supplier, who will then send out an invoice requesting payment, usually at the end of the month.

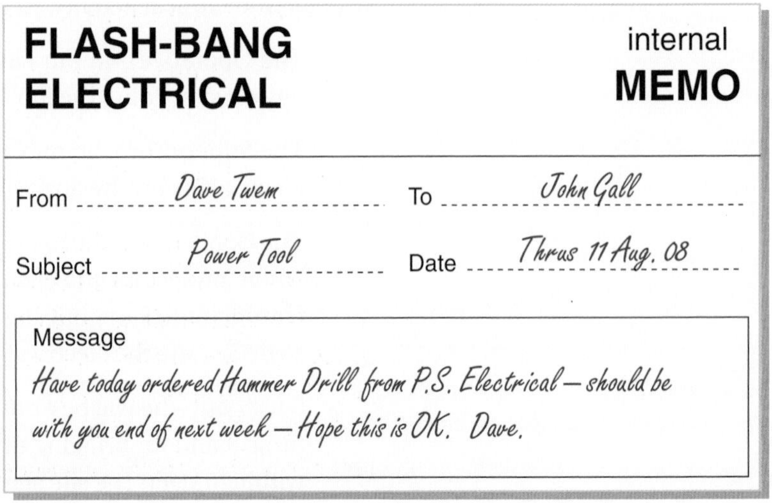

> **Definition**
>
> A *delivery note* is used to confirm that goods have been delivered by the supplier, who will then send out an invoice requesting payment.

FLASH-BANG ELECTRICAL   internal **MEMO**

From ..... *Dave Twem* .....   To ..... *John Gall* .....

Subject ..... *Power Tool* .....   Date ..... *Thrus 11 Aug. 08* .....

Message

Have today ordered Hammer Drill from P.S. Electrical — should be with you end of next week — Hope this is OK. Dave.

FIGURE 9.2
Typical standard memo form.

The person receiving the goods must carefully check that all the items stated on the delivery note have been delivered in good condition. Any missing or damaged items must be clearly indicated on the delivery note before signing, because, by signing the delivery note the person is saying 'yes, these items were delivered to me as my company's representative on that date and in good condition and I am now responsible for these goods'. Suppliers will replace materials damaged in transit provided that they are notified within a set time period, usually 3 days. The person receiving the goods should try to quickly determine their condition. Has the packaging been damaged, does the container 'sound' like it might contain broken items? It is best to check at the time of delivery if possible, or as soon as possible after delivery and within the notifiable period. Electrical goods delivered to site should be handled carefully and stored securely until they are installed. Copies of delivery notes are sent to head office so that payment can be made for the goods received.

Time sheets

A **time sheet** is a standard form completed by each employee to inform the employer of the actual time spent working on a particular contract or site. This helps the employer to bill the hours of work to an individual job. It is usually a weekly document and includes the number of hours worked, the name of the job and any travelling expenses claimed. Office personnel require time sheets such as that shown in Fig 9.3 so that wages can be made up.

Job sheets

A **job sheet** or job card such as that shown in Fig 9.4 carries information about a job which needs to be done, usually a small job. It gives the name and address of the customer, contact telephone numbers, often a job reference number and a brief description of the work to be carried out. A typical job sheet work description might be:

- Job 1 Upstairs lights not working.

- Job 2 Funny fishy smell from kettle socket in kitchen.

An electrician might typically have a 'jobbing day' where he picks up a number of job sheets from the office and carries out the work specified.

Job 1, for example, might be the result of a blown fuse which is easily rectified, but the electrician must search a little further for the fault which caused the fuse to blow in the first place. The actual fault might, for example, be a decayed flex on a pendant drop which has become shorted out, blowing the fuse. The pendant drop would be re-flexed or replaced, along with any others in poor condition. The installation would then be tested for correct operation and the customer given an account of what has been done to correct the fault. General information and assurances about the condition of the installation as a whole might be requested and given before setting off to job 2.

The kettle socket outlet at job 2 is probably getting warm and, therefore, giving off that 'fishy' bakelite smell because loose connections are causing

---

Definition

A *time sheet* is a standard form completed by each employee to inform the employer of the actual time spent working on a particular contract or site.

Definition

A *job sheet* or job card such as that shown in Fig 9.4 carries information about a job which needs to be done, usually a small job.

185

TIME SHEET

FLASH-BANG ELECTRICAL

Employee's name (Print) --

Week ending --

Day	Job number and/or Address	Start time	Finish time	Total hours	Travel time	Expenses
Monday						
Tuesday						
Wednesday						
Thursday						
Friday						
Saturday						
Sunday						

Employee's signature -------------------------------- Date --------------------

FIGURE 9.3
Typical time sheet.

the bakelite socket to burn locally. A visual inspection would confirm the diagnosis. A typical solution would be to replace the socket and repair any damage to the conductors inside the socket box. Check the kettle plug top for damage and loose connections. Make sure all connections are tight

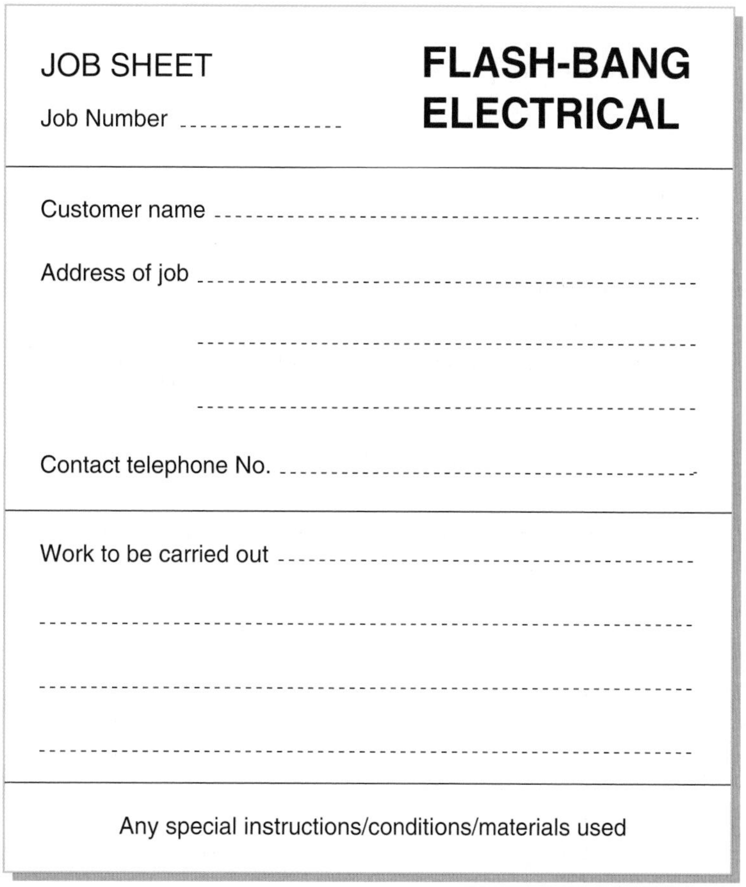

```
JOB SHEET                        FLASH-BANG
Job Number  _ _ _ _ _ _ _ _ _    ELECTRICAL

Customer name  _ _ _ _ _ _ _ _ _ _ _ _ _ _ _ _ _ _ _ _ _ _ _ _ _ _ _

Address of job  _ _ _ _ _ _ _ _ _ _ _ _ _ _ _ _ _ _ _ _ _ _ _ _ _ _

                _ _ _ _ _ _ _ _ _ _ _ _ _ _ _ _ _ _ _ _ _ _ _ _ _ _

                _ _ _ _ _ _ _ _ _ _ _ _ _ _ _ _ _ _ _ _ _ _ _ _ _ _

Contact telephone No.  _ _ _ _ _ _ _ _ _ _ _ _ _ _ _ _ _ _ _ _ _ _

Work to be carried out  _ _ _ _ _ _ _ _ _ _ _ _ _ _ _ _ _ _ _ _ _

                _ _ _ _ _ _ _ _ _ _ _ _ _ _ _ _ _ _ _ _ _ _ _ _ _ _

                _ _ _ _ _ _ _ _ _ _ _ _ _ _ _ _ _ _ _ _ _ _ _ _ _ _

                _ _ _ _ _ _ _ _ _ _ _ _ _ _ _ _ _ _ _ _ _ _ _ _ _ _

    Any special instructions/conditions/materials used
```

FIGURE 9.4
Typical job sheet.

before reassuring the customer that all is well; then, off to the next job or back to the office.

The time spent on each job and the materials used are sometimes recorded on the job sheet, but alternatively a daywork sheet can be used. This will depend upon what is normal practice for the particular electrical company. This information can then be used to 'bill' the customer for work carried out.

Daywork sheets

Definition

Daywork is one way of recording variations to a contract.

Daywork is one way of recording variations to a contract, that is, work done which is outside the scope of the original contract. If daywork is to be carried out, the site supervisor must first obtain a signature from the client's representative, for example, the architect, to authorize the extra work. A careful record must then be kept on the daywork sheets of all extra time and materials used so that the client can be billed for the extra work. A typical daywork sheet is shown in Fig. 9.5.

Reports

On large jobs, the foreman or supervisor is often required to keep a report of the relevant events which happen on the site – for example, how many

FLASH-BANG ELECTRICAL

DAYWORK SHEET

Client name _____

Job number/ref. _____

Date	Labour	Start time	Finish time	Total hours	Office use

Materials quantity	Description	Office use

Site supervisor or F.B. Electrical Representative responsible for carrying out work _____

Signature of person approving work and status e.g.

Client ☐   Architect ☐   Q.S. ☐   Main contractor ☐   Clerk of works ☐

Signature _____

FIGURE 9.5
Typical daywork sheet.

people from your company are working on site each day, what goods were delivered, whether there were any breakages or accidents, and records of site meetings attended. Some firms have two separate documents, a site diary to record daily events and a weekly report which is a summary of the week's events extracted from the site diary. The site diary remains on site and the weekly report is sent to head office to keep managers informed of the work's progress.

Personal communications and image

Remember that it is the customers who actually pay the wages of everyone employed in your company. You should always be polite and listen carefully to their wishes. They may be elderly or of a different religion or cultural background than you. In a domestic situation, the playing of loud music on a radio may not be approved of. Treat the property in which you are working with the utmost care. When working in houses, shops and offices use dust sheets to protect floor coverings and furnishings. Clean up periodically and make a special effort when the job is completed.

Dress appropriately: an unkempt or untidy appearance will encourage the customer to think that your work will be of poor quality.

The electrical installation in a building is often carried out alongside other trades. It makes good sense to help other trades where possible and to develop good working relationships with other employees. The customer will be most happy if the workers give an impression of working together as a team for the successful completion of the project.

Finally, remember that the customer will probably see more of the electrician and the electrical trainee than the managing director of your firm and, therefore, the image presented by you will be assumed to reflect the policy of the company. You are, therefore, your company's most important representative. Always give the impression of being capable and in command of the situation, because this gives customers confidence in the company's ability to meet their needs. However, if a problem does occur which is outside your previous experience and you do not feel confident to solve it successfully, then contact your supervisor for professional help and guidance. It is not unreasonable for a young member of the company's team to seek help and guidance from those employees with more experience. This approach would be preferred by most companies rather than having to meet the cost of an expensive blunder.

189

Check your Understanding

When you have completed the questions, check out the answers at the back of the book.
Note: more than one multiple choice answer may be correct.

1. A standard form completed by every employee to inform the employer of the time spent working on a particular site is called:
   a. job sheet
   b. time sheet
   c. delivery note
   d. daywork sheet.

2. A record that confirms that materials ordered have been delivered to site is called:
   a. job sheet
   b. time sheet
   c. delivery note
   d. daywork sheet.

3. A standard form containing information about work to be done usually distributed by a manager to an electrician is called:
   a. job sheet
   b. time sheet
   c. delivery note
   d. daywork sheet.

4. A standard form which records changes or extra work on a large project is called a:
   a. job sheet
   b. time sheet
   c. delivery note
   d. daywork sheet.

5. State three methods of receiving or sending technical information.

6. State four types of technical information and give a brief description of each one.

7. State three people who need access to technical information and briefly state why.

8. Use bullet points to describe the process of receiving goods on site on behalf of your employer.

9. Briefly state why we would use a daywork sheet to record additional work on a large contract. Why is it important to obtain a signature for this work from the client or his representative?

10. Briefly state why time sheets, fully and accurately completed, are important to:

    a. an employer

    b. an employee.

11. State the reasons why you should always present the right image to a client, customer or his representative.

12. Briefly describe what we mean by a schedule of work. Who would use a bar chart or schedule of work in your company and why?

CH 10

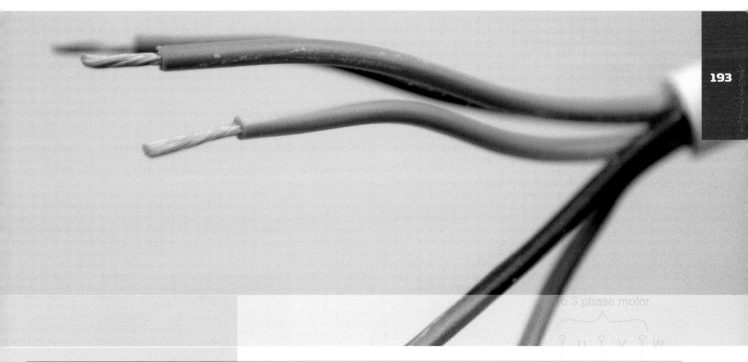

Alternating current theory and electrical machines

Unit 3 – Application of health and safety and electrical principles – Outcome 3

Underpinning knowledge: when you have completed this chapter you should be able to:

- state the effects of resistance, reactance and impedance in an a.c. circuit
- state how power factor correction may be achieved
- describe the concept of self and mutual inductance
- describe the energy stored in a magnetic field and magnetic hysteresis
- state the principle of operation of:
    - d.c. motors
    - a.c. motors
    - transformers
    - instrument transformers
    - fluorescent luminaires
    - a simple relay

Alternating current theory

Commercial quantities of electricity for industry, commerce and domestic use are generated as a.c. in large Power Stations and distributed around the United Kingdom on the National Grid to the end user. The d.c. electricity has many applications where portability or an emergency stand-by supply is important but for large quantities of power it has to be an a.c. supply because it is so easy to change the voltage levels using a transformer.

Rotating a simple loop of wire or coils of wire between the poles of a magnet such as that shown simplified in Fig. 10.1 will cut the north south lines of magnetic flux and induce an a.c. voltage in the loop or coils of wire as shown by the display on a cathode ray oscilloscope.

This is an a.c. supply, an alternating current supply. The basic principle of the a.c. supply generated in a Power Station is exactly the same as Fig. 10.1 except that powerful electromagnets are used and the power for rotation comes from a steam turbine.

In this section we will first of all consider the theoretical circuits of pure resistance, inductance and capacitance acting alone in an a.c. circuit before going on to consider the practical circuits of resistance, inductance and capacitance acting together. Let us first define some of our terms of reference.

194

Key Fact

Definitions

- Try to remember these a.c. circuit definitions
- Write them down if it helps.

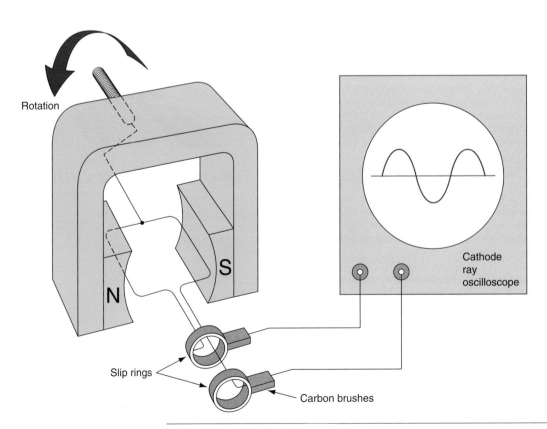

FIGURE 10.1

Simple a.c. generator or alternator.

Definition

In any circuit, *resistance* is defined as opposition to current flow.

Resistance

In any circuit, resistance is defined as opposition to current flow. From Ohm's law:

$$R = \frac{V_R}{I_R}\ (\Omega)$$

However, in an a.c. circuit, resistance is only part of the opposition to current flow. The inductance and capacitance of an a.c. circuit also cause an opposition to current flow, which we call *reactance*.

Definition

Inductive reactance (X_L) is the opposition to an a.c. current in an inductive circuit. It causes the current in the circuit to lag behind the applied voltage.

Inductive reactance (X_L) is the opposition to an a.c. current in an inductive circuit. It causes the current in the circuit to lag behind the applied voltage, as shown in Fig. 10.2. It is given by the formula:

$$X_L = 2\pi f L\ (\Omega)$$

where

π = 3.142 a constant

f = the frequency of the supply

L = the inductance of the circuit or by

$$X_L = \frac{V_L}{I_L}$$

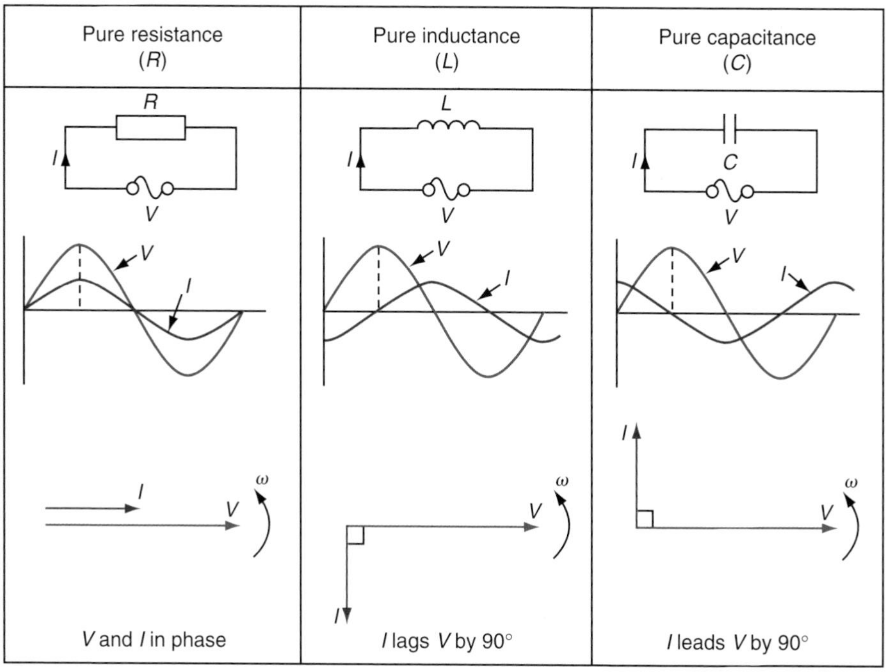
Pure resistance
(R)
Pure inductance
(L)
Pure capacitance
(C)
R
L
C
I
V
ω
V and I in phase
I lags V by 90°
I leads V by 90°

FIGURE 10.2
Voltage and current relationships in resistive, capacitive and inductive circuits.

Capacitive reactance (X_C) is the opposition to an a.c. current in a capacitive circuit. It causes the current in the circuit to lead ahead of the voltage, as shown in Fig. 10.2. It is given by the formula:

$$X_C = \frac{1}{2\pi fC} \ (\Omega)$$

where π and f are defined as before and C is the capacitance of the circuit. It can also be expressed as:

$$X_C = \frac{V_C}{I_C}$$

Example

Calculate the reactance of a 150 μF capacitor and a 0.05 H inductor if they were separately connected to the 50 Hz mains supply.

For capacitive reactance:

$$X_C = \frac{1}{2\pi fC}$$

where $f = 50$ Hz and $C = 150$ μF $= 150 \times 10^{-6}$ F.

$$\therefore X_C = \frac{1}{2 \times 3.142 \times 50\text{Hz} \times 150 \times 10^{-6}\text{F}} = 21.2\,\Omega$$

For inductive reactance:

$$X_L = 2\pi fL$$

where $f = 50$ Hz and $L = 0.05$ H.

$$\therefore X_L = 2 \times 3.142 \times 50\text{Hz} \times 0.05\text{H} = 15.7\,\Omega$$

Impedance

The total opposition to current flow in an a.c. circuit is called **impedance** and given the symbol Z. Thus impedance is the combined opposition to current flow of the resistance, inductive reactance and capacitive reactance of the circuit and can be calculated from the formula:

$$Z = \sqrt{R^2 + X^2} \ (\Omega)$$

or

$$Z = \frac{V_T}{I_T}$$

Example 1

Calculate the impedance when a 5Ω resistor is connected in series with a 12Ω inductive reactance.

$$Z = \sqrt{R^2 + X_L^2} \ (\Omega)$$

$$\therefore Z = \sqrt{5^2 + 12^2}$$

$$Z = \sqrt{25 + 144}$$

$$Z = \sqrt{169}$$

$$Z = 13\,\Omega$$

Example 2

Calculate the impedance when a 48 Ω resistor is connected in series with a 55 Ω capacitive reactance.

$$Z = \sqrt{R^2 + X_C^2} \ (\Omega)$$

$$\therefore Z = \sqrt{48^2 + 55^2}$$

$$Z = \sqrt{2304 + 3025}$$

$$Z = \sqrt{5329}$$

$$Z = 73\,\Omega$$

Resistance, inductance and capacitance in an a.c. circuit

When a resistor only is connected to an a.c. circuit the current and voltage waveforms remain together, starting and finishing at the same time. We say that the waveforms are *in phase*.

When a pure inductor is connected to an a.c. circuit the current lags behind the voltage waveform by an angle of 90°. We say that the current *lags* the voltage by 90°. When a pure capacitor is connected to an a.c. circuit the current *leads* the voltage by an angle of 90°. These various effects can be observed on an oscilloscope, but the circuit diagram, waveform diagram and phasor diagram for each circuit are shown in Fig. 10.2.

PHASOR DIAGRAMS

Phasor diagrams and a.c. circuits are an inseparable combination. Phasor diagrams allow us to produce a model or picture of the circuit under consideration which helps us to understand the circuit. A phasor is a straight line, having definite length and direction, which represents to scale the magnitude and direction of a quantity such as a current, voltage or impedance.

To find the combined effect of two quantities we combine their phasors by adding the beginning of the second phasor to the end of the first. The combined effect of the two quantities is shown by the resultant phasor, which is measured from the original zero position to the end of the last phasor.

Definition

A *phasor* is a straight line, having definite length and direction, which represents to scale the magnitude and direction of a quantity such as a current, voltage or impedance.

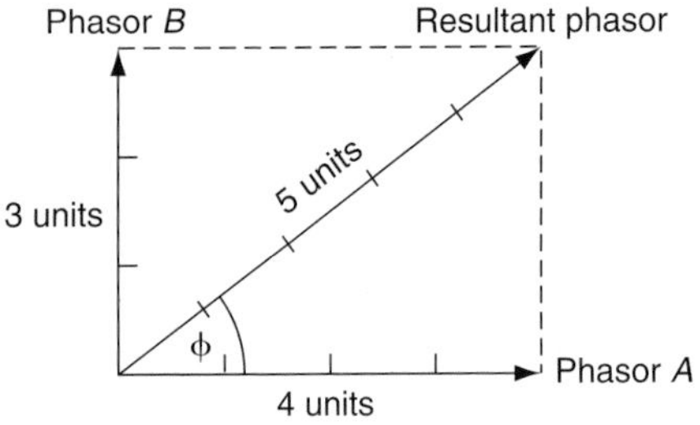

Phasor B   Resultant phasor

3 units

5 units

ϕ

4 units   → Phasor A

FIGURE 10.3

The phasor addition of currents A and B.

Example

Find by phasor addition the combined effect of currents A and B acting in a circuit. Current A has a value of 4 A, and current B a value of 3 A, leading A by 90°. We usually assume phasors to rotate anticlockwise and so the complete diagram will be as shown in Fig. 10.3. Choose a scale of, for example, 1 A = 1 cm and draw the phasors to scale, that is A = 4 cm and B = 3 cm, leading A by 90°.

The magnitude of the resultant phasor can be measured from the phasor diagram and is found to be 5 A acting at a phase angle ϕ of about 37° leading A. We therefore say that the combined effect of currents A and B is a current of 5 A at an angle of 37° leading A.

PHASE ANGLE ϕ

In an a.c. circuit containing resistance only, such as a heating circuit, the voltage and current are in phase, which means that they reach their peak and zero values together, as shown in Fig. 10.4(a).

In an a.c. circuit containing inductance, such as a motor or discharge lighting circuit, the current often reaches its maximum value after the voltage, which means that the current and voltage are out of phase with each other, as shown in Fig. 10.4(b). The phase difference, measured in degrees between the current and voltage, is called the phase angle of the circuit, and is denoted by the symbol ϕ, the lower-case Greek letter phi.

When circuits contain two or more separate elements, such as RL, RC or RLC, the phase angle between the total voltage and total current will be neither 0° nor 90° but will be determined by the relative values of resistance and reactance in the circuit. In Fig. 10.5 the phase angle between applied voltage and current is some angle ϕ.

Alternating current series circuits

In a circuit containing a resistor and inductor connected in series as shown in Fig. 10.5, the current I will flow through the resistor and the inductor causing the voltage V_R to be dropped across the resistor and V_L to be dropped across the inductor. The sum of these voltages will be equal to the total voltage V_T but because this is an a.c. circuit the voltages must be added by phasor addition. The result is shown in Fig. 10.5, where V_R is drawn to

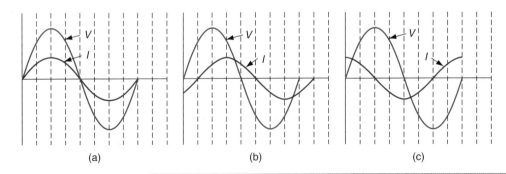

(a)   (b)   (c)

FIGURE 10.4

Phase relationship of a.c. waveform: (a) V and I in phase, phase angle $\phi = 0°$ and power factor $= \cos\phi = 1$; (b) V and I displaced by 45°, $\phi = 45°$ and p.f. $= 0.707$; and (c) V and I displaced by 90°, $\phi = 90°$ and p.f. $= 0$.

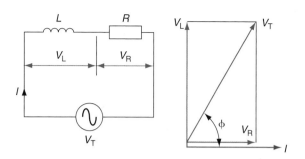

FIGURE 10.5

A series RL circuit and phasor diagram.

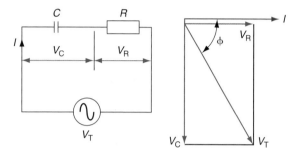

FIGURE 10.6

A series RC circuit and phasor diagram.

scale and in phase with the current and V_L is drawn to scale and leading the current by 90°. The phasor addition of these two voltages gives us the magnitude and direction of V_T, which leads the current by some angle ϕ.

In a circuit containing a resistor and capacitor connected in series as shown in Fig. 10.6, the current I will flow through the resistor and capacitor causing voltage drops V_R and V_C. The voltage V_R will be in phase with the current and V_C will lag the current by 90°. The phasor addition of these voltages is equal to the total voltage V_T which, as can be seen in Fig. 10.6, is lagging the current by some angle ϕ.

The impedance triangle

We have now established the general shape of the phasor diagram for a series a.c. circuit. Figures 10.5 and 10.6 show the voltage phasors, but we know that $V_R = IR$, $V_L = IX_L$, $V_C = IX_C$ and $V_T = IZ$, and therefore the phasor diagrams (a) and (b) of Fig. 10.7 must be equal. From Fig. 10.7(b), by the theorem of Pythagoras, we have:

$$(IZ)^2 = (IR)^2 + (IX)^2$$

$$I^2 Z^2 = I^2 R^2 + I^2 X^2$$

If we now divide throughout by I^2 we have:

$$Z^2 = R^2 + X^2$$
$$\text{or } Z = \sqrt{R^2 + X^2} \; \Omega$$

The phasor diagram can be simplified to the impedance triangle given in Fig. 10.7(c).

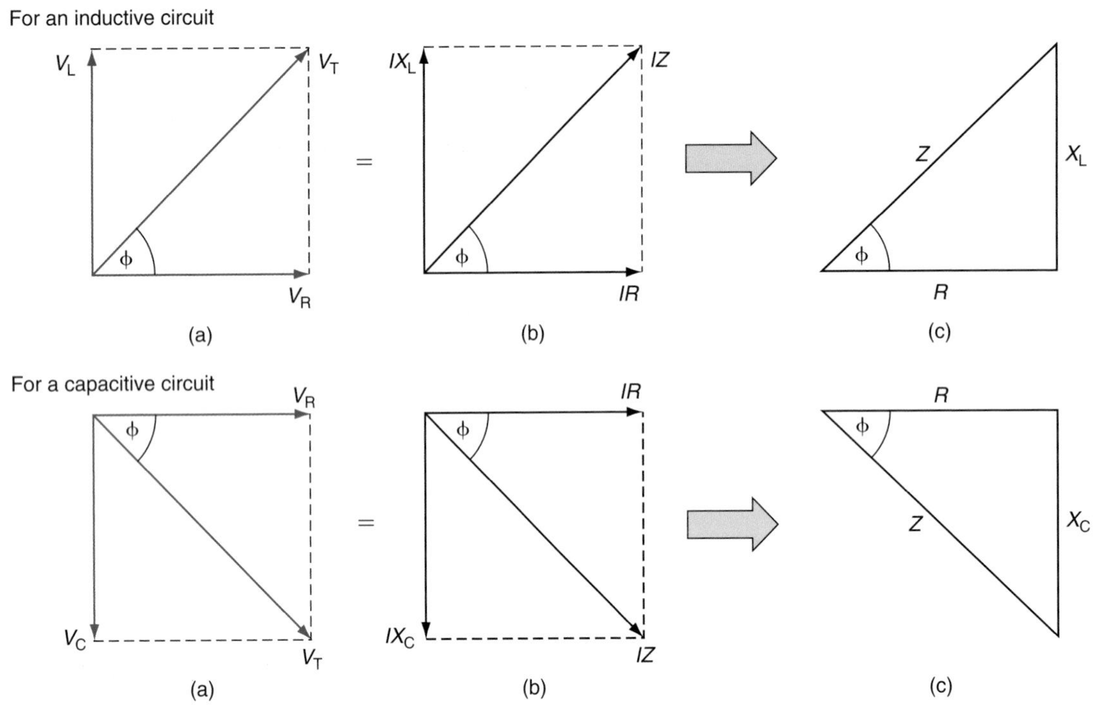

For an inductive circuit

(a)   (b)   (c)

For a capacitive circuit

(a)   (b)   (c)

FIGURE 10.7

Phasor diagram and impedance triangle.

Example 1

A coil of 0.15 H is connected in series with a 50 Ω resistor across a 100 V 50 Hz supply. Calculate (a) the reactance of the coil, (b) the impedance of the circuit and (c) the current.

For (a):

$$X_L = 2\pi fL \ (\Omega)$$
$$\therefore X_L = 2 \times 3.142 \times 50 \text{ Hz} \times 0.15 \text{ H} = 47.1 \Omega$$

For (b):

$$Z = \sqrt{R^2 + X^2} \ (\Omega)$$
$$\therefore Z = \sqrt{(50\,\Omega)^2 + (47.1\,\Omega)^2} = 68.69 \ \Omega$$

For (c):

$$I = \frac{V}{Z} \ (A)$$
$$\therefore I = \frac{100V}{68.69\,\Omega} = 1.46 \text{ A}$$

Example 2

A 60 μF capacitor is connected in series with a 100 Ω resistor across a 230 V 50 Hz supply. Calculate (a) the reactance of the capacitor, (b) the impedance of the circuit and (c) the current.

For (a):

$$X_C = \frac{1}{2\pi fC} \ (\Omega)$$

$$\therefore X_C = \frac{1}{2\pi \times 50\,\text{Hz} \times 60 \times 10^{-6}\,\text{F}} = 53.05\,\Omega$$

For (b):

$$Z = \sqrt{R^2 + X^2} \ (\Omega)$$

$$\therefore Z = \sqrt{(100\,\Omega)^2 + (53.05\,\Omega)^2} = 113.2\,\Omega$$

For (c):

$$I = \frac{V}{Z} \ (\text{A})$$

$$\therefore I = \frac{230\,\text{V}}{113.2\,\Omega} = 2.03\,\text{A}$$

Power and power factor

Definition

Power factor (p.f.) is defined as the cosine of the phase angle between the current and voltage.

Power factor (p.f.) is defined as the cosine of the phase angle between the current and voltage:

$$\text{p.f.} = \cos\phi$$

If the current lags the voltage as shown in Fig. 10.5, we say that the p.f. is lagging, and if the current leads the voltage as shown in Fig. 10.6, the p.f. is said to be leading. From the trigonometry of the impedance triangle shown in Fig. 10.7, p.f. is also equal to:

$$\text{p.f.} = \cos\phi = \frac{R}{Z} = \frac{V_R}{V_T}$$

The electrical power in a circuit is the product of the instantaneous values of the voltage and current. Figure 10.8 shows the voltage and current waveform for a pure inductor and pure capacitor. The power waveform is obtained from the product of V and I at every instant in the cycle. It can be seen that the power waveform reverses every quarter cycle, indicating that energy is alternately being fed into and taken out of the inductor and capacitor. When considered over one complete cycle, the positive and negative portions are equal, showing that the average power consumed by a pure inductor or capacitor is zero. This shows that inductors and capacitors store energy during one part of the voltage cycle and feed it back into the supply later in the cycle. Inductors store energy as a magnetic field and capacitors as an electric field.

In an electric circuit more power is taken from the supply than is fed back into it, since some power is dissipated by the resistance of the circuit, and therefore:

$$P = I^2 R \ (\text{W})$$

In any d.c. circuit the power consumed is given by the product of the voltage and current, because in a d.c. circuit voltage and current are in phase.

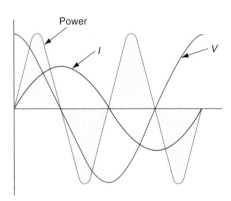

Power

I

V

Pure inductor

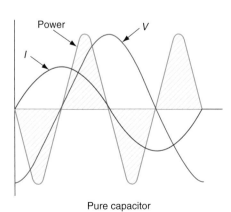

Power

V

I

Pure capacitor

FIGURE 10.8

Waveform for the a.c. power in purely inductive and purely capacitive circuits.

In an a.c. circuit the power consumed is given by the product of the current and that part of the voltage which is in phase with the current. The in-phase component of the voltage is given by $V \cos \phi$, and so power can also be given by the equation:

$$P = VI \cos \phi \, (W)$$

Example 1

A coil has a resistance of $30\,\Omega$ and a reactance of $40\,\Omega$ when connected to a 250V supply. Calculate (a) the impedance, (b) the current, (c) the p.f., and (d) the power.

For (a):

$$Z = \sqrt{R^2 + X^2} \, (\Omega)$$
$$\therefore Z = \sqrt{(30\,\Omega)^2 + (40\,\Omega)^2} = 50\,\Omega$$

For (b):

$$I = \frac{V}{Z} \, (A)$$
$$\therefore I = \frac{250\,V}{50\,\Omega} = 5\,A$$

For (c):

$$p.f. = \cos \phi = \frac{R}{Z}$$
$$\therefore p.f. = \frac{30\,\Omega}{50\,\Omega} = 0.6 \text{ lagging}$$

For (d):

$$P = VI \cos \phi \, (W)$$
$$\therefore P = 250\,V \times 5\,A \times 0.6 = 750\,W$$

Example 2

A capacitor of reactance $12\,\Omega$ is connected in series with a $9\,\Omega$ resistor across a 150V supply. Calculate (a) the impedance of the circuit, (b) the current, (c) the p.f., and (d) the power.

For (a):

$$Z = \sqrt{R^2 + X^2} \, (\Omega)$$
$$\therefore Z = \sqrt{(9\,\Omega)^2 + (12\,\Omega)^2} = 15\,\Omega$$

For (b):

$$I = \frac{V}{Z} \, (A)$$
$$\therefore I = \frac{150\,V}{15\,\Omega} = 10\,A$$

For (c):

$$p.f. = \cos \phi = \frac{R}{Z}$$
$$\therefore p.f. = \frac{9\,\Omega}{15\,\Omega} = 0.6 \text{ leading}$$

For (d):

$$P = VI \cos \phi \text{ (W)}$$
$$\therefore P = 150 \text{ V} \times 10 \text{ A} \times 0.6 = 900 \text{ W}$$

The power factor of most industrial loads is lagging because the machines and discharge lighting used in industry are mostly inductive. This causes an additional magnetizing current to be drawn from the supply, which does not produce power, but does need to be supplied, making supply cables larger.

Example 3

A 230V supply feeds three 1.84 kW loads with power factors of 1, 0.8 and 0.4. Calculate the current at each power factor.

The current is given by:

$$I = \frac{P}{V \cos \phi}$$

where $P = 1.84$ kW $= 1840$ W and $V = 230$V. If the p.f. is 1, then:

$$I = \frac{1840 \text{ W}}{230 \text{ V} \times 1} = 8 \text{ A}$$

For a p.f. of 0.8:

$$I = \frac{1840 \text{ W}}{230 \text{ V} \times 0.8} = 10 \text{ A}$$

For a p.f. of 0.4:

$$I = \frac{1840 \text{ W}}{230 \text{ V} \times 0.4} = 20 \text{ A}$$

It can be seen from these calculations that a 1.84 kW load supplied at a power factor of 0.4 would require a 20 A cable, while the same load at unity power factor could be supplied with an 8 A cable. There may also be the problem of higher voltage drops in the supply cables. As a result, the supply companies encourage installation engineers to improve their power factor to a value close to 1 and sometimes charge penalties if the power factor falls below 0.8.

POWER FACTOR CORRECTION

Most installations have a low or bad power factor because of the inductive nature of the load. A capacitor has the opposite effect of an inductor, and so it seems reasonable to add a capacitor to a load which is known to have a lower or bad power factor, for example, a motor.

Figure 10.9(a) shows an industrial load with a low power factor. If a capacitor is connected in parallel with the load, the capacitor current I_C leads the applied voltage by 90°. When this capacitor current is added to the load current as shown in Fig 10.9(b) the resultant load current has a much improved power factor. However, using a slightly bigger capacitor, the load current can be pushed up until it is 'in phase' with the voltage as can be seen in Fig. 10.9(c).

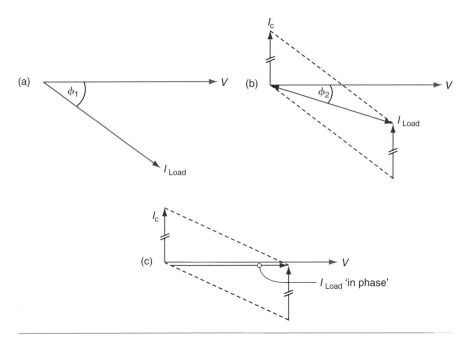

FIGURE 10.9
Power factor improvement using capacitors.

Capacitors may be connected across the main busbars of industrial loads in order to provide power factor improvement, but smaller capacitors may also be connected across an individual piece of equipment, as is the case for fluorescent light fittings.

Electrical machines

All electrical machines operate on the principles of magnetism. The basic rules of magnetism were laid down in Chapter 5 of this book. Here we will look at some of the laws of magnetism as they apply to electrical machines, such as generators, motors and transformers.

A current carrying conductor maintains a magnetic field around the conductor which is proportional to the current flowing. When this magnetic field interacts with another magnetic field, forces are exerted which describe the basic principles of electric motors.

Michael Faraday demonstrated on 29 August 1831 that electricity could be produced by magnetism. He stated that 'when a conductor cuts or is cut by a magnetic field an emf is induced in that conductor. The amount of induced emf is proportional to the rate or speed at which the magnetic field cuts the conductor'. This basic principle laid down the laws of present-day electricity generation where a strong magnetic field is rotated inside a coil of wire to generate electricity.

Self and mutual inductance

If a coil of wire is wound on to an iron core as shown in Fig. 10.10, a magnetic field will become established in the core when a current flows in the coil due to the switch being closed.

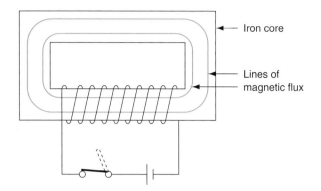

FIGURE 10.10
An inductive coil or choke.

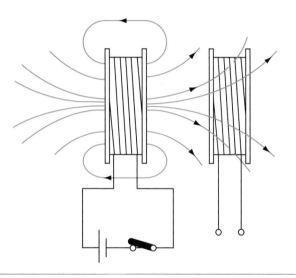

FIGURE 10.11
Mutual inductance between two coils.

When the switch is opened the current stops flowing and, therefore, the magnetic flux collapses. The collapsing magnetic flux induces an emf into the coil and this voltage appears across the switch contacts. The effect is known as *self-inductance*, or just *inductance*, and is one property of any coil. The unit of inductance is the henry (symbol H), to commemorate the work of the American physicist Joseph Henry (1797–1878), and a circuit is said to possess an inductance of 1 henry when an emf of 1 volt is induced in the circuit by a current changing at the rate of 1 ampere per second.

Fluorescent light fittings contain a choke or inductive coil in series with the tube and starter lamp. The starter lamp switches on and off very quickly, causing rapid current changes which induce a large voltage across the tube electrodes sufficient to strike an arc in the tube.

When two separate coils are placed close together, as they are in a transformer, a current in one coil produces a magnetic flux which links with the second coil. This induces a voltage in the second coil and is the basic principle of the transformer action which is described later in this chapter. The two coils in this case are said to possess **mutual inductance**, as shown by Fig. 10.11.

A mutual inductance of 1 henry exists between two coils when a uniformly varying current of 1 ampere per second in one coil produces an emf of 1 volt in the other coil.

The emf induced in a coil such as that shown on the right-hand side in Fig. 10.11 is dependent upon the rate of change of magnetic flux and the number of turns on the coil.

Energy stored in a magnetic field

When we open the switch of an inductive circuit such as an electric motor or fluorescent light circuit the magnetic flux collapses and produces an arc across the switch contacts. The arc is produced by the stored magnetic energy being discharged across the switch contacts.

Magnetic hysteresis

There are many different types of magnetic material and they all respond differently to being magnetized. Some materials magnetize easily, and some are difficult to magnetize. Some materials retain their magnetism, while others lose it. The result will look like the graphs shown in Fig. 10.12 and are called hysteresis loops.

Magnetic hysteresis loops describe the way in which different materials respond to being magnetized.

Materials from which permanent magnets are made should display a wide hysteresis loop, as shown by loop (b) in Fig. 10.12.

The core of an electromagnet is required to magnetize easily, and to lose its magnetism equally easily when switched off. Suitable materials will, therefore, display a narrow hysteresis loop, as shown by loop (a) in Fig. 10.12.

When an iron core is subjected to alternating magnetization, as in a transformer, the energy loss occurs at every cycle and so constitutes a continuous power loss, and, therefore, for applications such as transformers, a material with a narrow hysteresis loop is required.

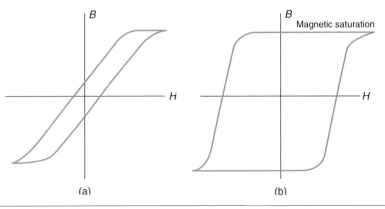

FIGURE 10.12

Magnetic hysteresis loops: (a) electromagnetic material and (b) permanent magnetic material.

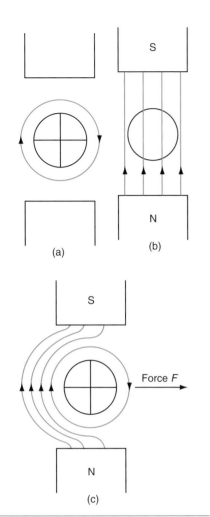

FIGURE 10.13

Force on a conductor in a magnetic field.

Direct current motors

All electric motors work on the principle that when a current carrying conductor is placed in a magnetic field it will experience a force. An electric motor uses this magnetic force to turn the shaft of the electric motor. Let us try to understand this action. If a current carrying conductor is placed into the field of a permanent magnet as shown in Fig. 10.13(c) a force F will be exerted on the conductor to push it out of the magnetic field.

To understand the force, let us consider each magnetic field acting alone. Figure 10.13(a) shows the magnetic field due to the current carrying conductor only. Figure 10.13(b) shows the magnetic field due to the permanent magnet in which is placed the conductor carrying no current. Figure 10.13(c) shows the effect of the combined magnetic fields which are distorted and, because lines of magnetic flux never cross, but behave like stretched elastic bands, always trying to find the shorter distance between a north and south pole, the force F is exerted on the conductor, pushing it out of the permanent magnetic field.

This is the basic motor principle, and the force F is dependent upon the strength of the magnetic field B, the magnitude of the current flowing in the conductor I and the length of conductor within the magnetic field l. The following equation expresses this relationship:

$$F = BlI \ (N)$$

where B is in tesla, l is in metres, I is in amperes and F is in newtons.

Example

A coil which is made up of a conductor some 15 m in length, lies at right angles to a magnetic field of strength 5 T. Calculate the force on the conductor when 15 A flows in the coil.

$$F = BlI \ (N)$$

$$F = 5\,\text{T} \times 15\,\text{m} \times 15\,\text{A} = 1125\,\text{N}$$

Practical d.c. motors

Practical motors are constructed as shown in Fig. 10.14. All d.c. motors contain a field winding wound on pole pieces attached to a steel yoke. The armature winding rotates between the poles and is connected to the commutator. Contact with the external circuit is made through carbon brushes rubbing on the commutator segments. Direct current motors are classified by the way in which the field and armature windings are connected, which may be in series or in parallel.

SERIES MOTOR

The field and armature windings are connected in series and consequently share the same current. The series motor has the characteristics of a high starting torque but a speed which varies with load. Figure 10.15 shows

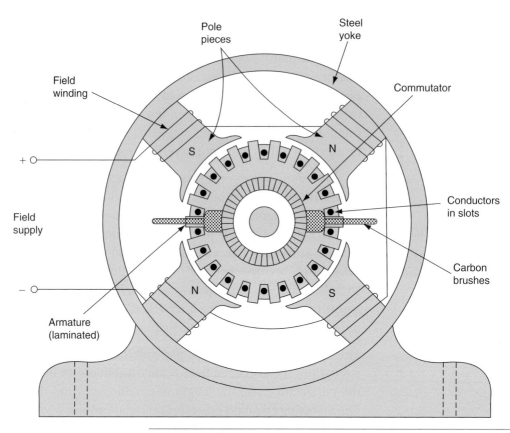

FIGURE 10.14

Showing d.c. machine construction.

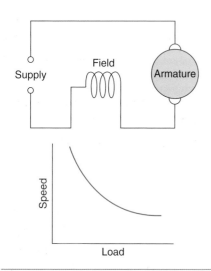

FIGURE 10.15

Series motor connections and characteristics.

series motor connections and characteristics. For this reason the motor is only suitable for direct coupling to a load, except in very small motors, such as vacuum cleaners and hand drills, and is ideally suited for applications where the machine must start on load, such as electric trains, cranes and hoists.

Reversal of rotation may be achieved by reversing the connections of either the field or armature windings but not both. This characteristic means that the machine will run on both a.c. or d.c. and is, therefore, sometimes referred to as a 'universal' motor.

Three-phase a.c. motors

If a three-phase supply is connected to three separate windings equally distributed around the stationary part or stator of an electrical machine, an alternating current circulates in the coils and establishes a magnetic flux. The magnetic field established by the three-phase currents travels around the stator, establishing a rotating magnetic flux, creating magnetic forces on the rotor which turns the shaft on the motor.

Three-phase induction motor

When a three-phase supply is connected to insulated coils set into slots in the inner surface of the stator or stationary part of an induction motor as

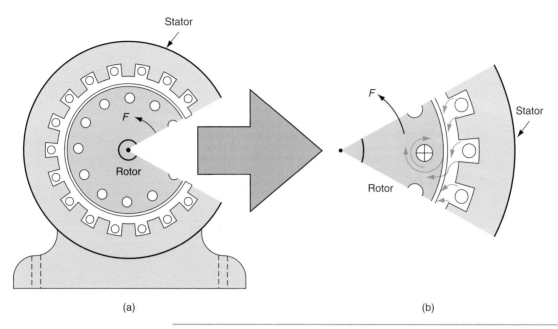

Stator

Rotor

F

(a)

F

Stator

Rotor

(b)

FIGURE 10.16

Segment taken out of an induction motor to show turning force: (a) construction of an induction motor and (b) production of torque by magnetic fields.

shown in Fig. 10.16(a), a rotating magnetic flux is produced. The rotating magnetic flux cuts the conductors of the rotor and induces an emf in the rotor conductors by Faraday's law, which states that when a conductor cuts or is cut by a magnetic field, an emf is induced in that conductor, the magnitude of which is proportional to the *rate* at which the conductor cuts or is cut by the magnetic flux. This induced emf causes rotor currents to flow and establish a magnetic flux which reacts with the stator flux and causes a force to be exerted on the rotor conductors, turning the rotor as shown in Fig. 10.16(b).

The turning force or torque experienced by the rotor is produced by inducing an emf into the rotor conductors due to the *relative* motion between the conductors and the rotating field. The torque produces rotation in the same direction as the rotating magnetic field.

Rotor construction

There are two types of induction motor rotor – the wound rotor and the cage rotor. The cage rotor consists of a laminated cylinder of silicon steel with copper or aluminium bars slotted in holes around the circumference and short circuited at each end of the cylinder as shown in Fig. 10.17. In small motors the rotor is cast in aluminium. Better starting and quieter running are achieved if the bars are slightly skewed. This type of rotor is extremely robust and since there are no external connections there is no need for slip rings or brushes. A machine fitted with a cage rotor does suffer from a low starting torque and the machine must be chosen which has a higher starting torque than the load, as shown by curve (b) in Fig. 10.18. A machine with the characteristic shown by curve (a) in Fig. 10.18 would not start since the load torque is greater than the machine starting torque.

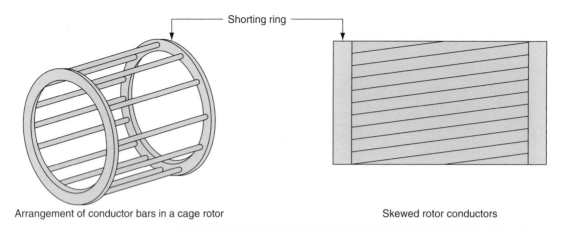

Arrangement of conductor bars in a cage rotor

Skewed rotor conductors

FIGURE 10.17
Construction of a cage rotor.

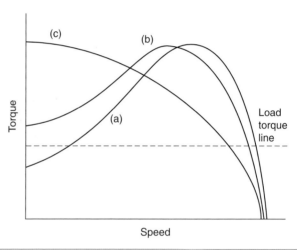

FIGURE 10.18
Various speed–torque characteristics for an induction motor.

Alternatively the load may be connected after the motor has been run up to full speed.

The wound rotor consists of a laminated cylinder of silicon steel with copper coils embedded in slots around the circumference. The windings may be connected in star or delta and the end connections brought out to slip rings mounted on the shaft. Connection by carbon brushes can then be made to an external resistance to improve starting.

The cage induction motor has a small starting torque and should be used with light loads or started with the load disconnected. The speed is almost constant. Its applications are for constant speed machines such as fans and pumps. Reversal of rotation is achieved by reversing any two of the stator winding connections.

Single-phase a.c. motors

A single-phase a.c. supply produces a pulsating magnetic field, not the rotating magnetic field produced by a three-phase supply. All a.c. motors

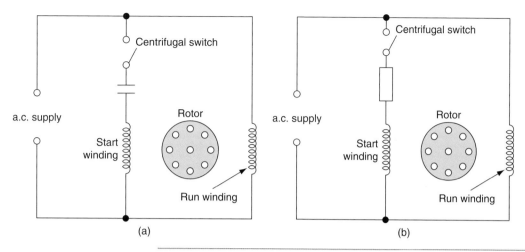

FIGURE 10.19
Circuit diagram of: (a) capacitor split-phase motors and (b) resistance split-phase motors.

require a rotating field to start. Therefore, single-phase a.c. motors have two windings which are electrically separated by about 90°. The two windings are known as the start and run windings. The magnetic fields produced by currents flowing through these out-of-phase windings create the rotating field and turning force required to start the motor. Once rotation is established, the pulsating field in the run winding is sufficient to maintain rotation and the start winding is disconnected by a centrifugal switch which operates when the motor has reached about 80% of the full load speed.

A cage rotor is used on single-phase a.c. motors, the turning force being produced in the way described previously for three-phase induction motors and shown in Fig. 10.16. Because both windings carry currents which are out of phase with each other, the motor is known as a 'split-phase' motor. The phase displacement between the currents in the windings is achieved in one of two ways:

- by connecting a capacitor in series with the start winding, as shown in Fig. 10.19(a), which gives a 90° phase difference between the currents in the start and run windings;

- by designing the start winding to have a high resistance and the run winding a high inductance, once again creating a 90° phase shift between the currents in each winding, as shown in Fig. 10.19(b).

When the motor is first switched on, the centrifugal switch is closed and the magnetic fields from the two coils produce the turning force required to run the rotor up to full speed. When the motor reaches about 80% of full speed, the centrifugal switch clicks open and the machine continues to run on the magnetic flux created by the run winding only.

Split-phase motors are constant speed machines with a low starting torque and are used on light loads such as fans, pumps, refrigerators and washing machines. Reversal of rotation may be achieved by reversing the connections to the start or run windings, but not both.

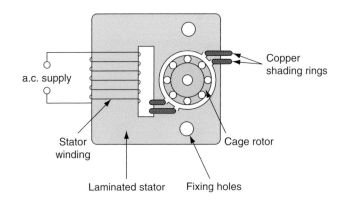

a.c. supply

Copper
shading rings

Stator
winding

Cage rotor

Laminated stator   Fixing holes

FIGURE 10.20
Shaded pole motor.

Shaded pole motors

The shaded pole motor is a simple, robust single-phase motor, which is suitable for very small machines with a rating of less than about 50W. Figure 10.20 shows a shaded pole motor. It has a cage rotor and the moving field is produced by enclosing one side of each stator pole in a solid copper or brass ring, called a shading ring, which displaces the magnetic field and creates an artificial phase shift.

Shaded pole motors are constant speed machines with a very low starting torque and are used on very light loads such as oven fans, record turntable motors and electric fan heaters. Reversal of rotation is theoretically possible by moving the shading rings to the opposite side of the stator pole face. However, in practice this is often not a simple process, but the motors are symmetrical and it is sometimes easier to reverse the rotor by removing the fixing bolts and reversing the whole motor.

There are more motors operating from single-phase supplies than all other types of motor added together. Most of them operate as very small motors in domestic and business machines where single-phase supplies are most common.

Definition

A *transformer* is an electrical machine which is used to change the value of an alternating voltage.

Transformers

A **transformer** is an electrical machine which is used to change the value of an alternating voltage. They vary in size from miniature units used in electronics to huge power transformers used in power stations. A transformer will only work when an alternating voltage is connected. It will not normally work from a d.c. supply such as a battery.

A transformer, as shown in Fig. 10.21, consists of two coils, called the primary and secondary coils, or windings, which are insulated from each other and wound on to the same steel or iron core.

An alternating voltage applied to the primary winding produces an alternating current, which sets up an alternating magnetic flux throughout the core. This magnetic flux induces an emf in the secondary winding, as

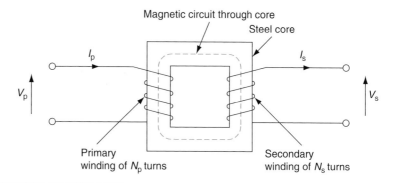

FIGURE 10.21
A simple transformer.

described by Faraday's law, which says that when a conductor is cut by a magnetic field, an emf is induced in that conductor. Since both windings are linked by the same magnetic flux, the induced emf per turn will be the same for both windings. Therefore, the emf in both windings is proportional to the number of turns. In symbols:

$$\frac{V_p}{N_p} = \frac{V_s}{N_s}$$

Where V_p = the primary voltage

V_s = the secondary voltage

N_p = the number of primary turns

N_s = the number of secondary turns

Moving the terms around, we have a general expression for a transformer:

$$\frac{V_p}{V_s} = \frac{N_p}{N_s}$$

Try This

Maths

Using the general equation for a transformer given above, follow this maths carefully, step by step, in the following example.

Example

A 230 V to 12 V emergency lighting transformer is constructed with 800 primary turns. Calculate the number of secondary turns required. Collecting the information given in the question into a usable form, we have:

$$V_p = 230 \text{ V}$$

$$V_s = 12 \text{ V}$$

$$N_p = 800$$

From the general equation:

$$\frac{V_p}{V_s} = \frac{N_p}{N_s}$$

the equation for the secondary turn is:

$$N_s = \frac{N_p V_s}{V_p}$$

$$\therefore N_s = \frac{800 \times 12V}{230V} = 42 \text{ turns}$$

42 turns are required on the secondary winding of this transformer to give a secondary voltage of 12V.

Types of transformer

Step down Transformers are used to reduce the output voltage, often for safety reasons. Figure 10.22 shows a step down transformer where the primary winding has twice as many turns as the secondary winding. The turns ratio is 2:1 and, therefore, the secondary voltage is halved.

Step up Transformers are used to increase the output voltage. The electricity generated in a power station is stepped up for distribution on the National Grid Network. Figure 10.23 shows a step up transformer where the primary winding has only half the number of turns as the secondary winding. The turns ratio is 1:2 and, therefore, the secondary voltage is doubled.

Instrument Transformers are used in industry and commerce so that large currents and voltage can be measured by small electrical instruments.

A Current Transformer (or CT) has the large load currents connected to the primary winding of the transformer and the ammeter connected to the secondary winding. The ammeter is calibrated to take account of the turns ratio of the transformer, so that the ammeter displays the actual current being taken by the load when the ammeter is actually only taking a small proportion of the load current.

Definition

Step down Transformers are used to reduce the output voltage, often for safety reasons.

Step up Transformers are used to increase the output voltage. The electricity generated in a power station is stepped up for distribution on the National Grid Network.

Key Fact

Transformers

- Transformers are very efficient (more than 80%) because they do not have any moving parts.

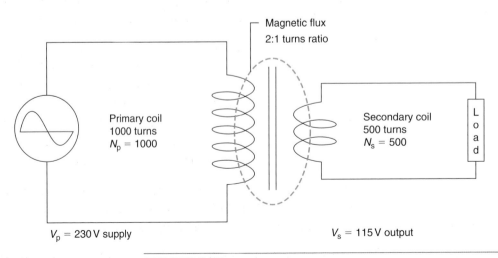

V_p = 230 V supply V_s = 115 V output

FIGURE 10.22
A step down transformer.

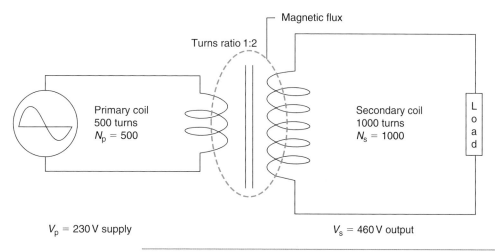

Magnetic flux

Turns ratio 1:2

Primary coil
500 turns
$N_p = 500$

Secondary coil
1000 turns
$N_s = 1000$

Load

$V_p = 230\,V$ supply

$V_s = 460\,V$ output

FIGURE 10.23
A step up transformer

A Voltage Transformer (or VT) has the main supply voltage connected to the primary winding of the transformer and the voltmeter connected to the secondary winding. The voltmeter is calibrated to take account of the turns ratio of the transformer, so that the voltmeter displays the actual supply voltage.

Separated Extra-low Voltage (SELV) Transformers If the primary winding and the secondary winding of a double wound transformer have a separate connection to earth, then the output of the transformer is effectively isolated from the input since the only connection between the primary and secondary windings is the magnetic flux in the transformer core. Such a transformer would give a very safe electrical supply which might be suitable for bathroom equipment such as shaver sockets and construction site 110V tools, providing that all other considerations are satisfied, such as water ingress, humidity, IP protection and robust construction.

Fluorescent luminaires

A **luminaire** is equipment which supports an electric lamp and distributes or filters the light created by the lamp. It is essentially the 'light fitting'.

A lamp is a device for converting electrical energy into light energy. There are many types of lamps. General lighting service (GLS) lamps and tungsten halogen lamps use a very hot wire filament to create the light and so they also become very hot in use. Fluorescent tubes operate on the 'discharge' principle; that is, the excitation of a gas within a glass tube. They are cooler in operation and very efficient in converting electricity into light. They form the basic principle of most energy efficient lamps.

Fluorescent lamps are linear arc tubes, internally coated with a fluorescent powder, containing a little low pressure mercury vapour and argon gas. The lamp construction is shown in Fig. 10.24.

Passing a current through the electrodes of the tube produces a cloud of electrons that ionize the mercury vapour and the argon in the tube,

Definition

A *luminaire* is equipment which supports an electric lamp and distributes or filters the light created by the lamp. A lamp is a device for converting electrical energy into light energy.

Key Fact

Energy efficiency

- fluorescent lamps and CFLs (compact fluorescent lamps) are energy efficient lamps
- they give more light out for every electrical, watt input.

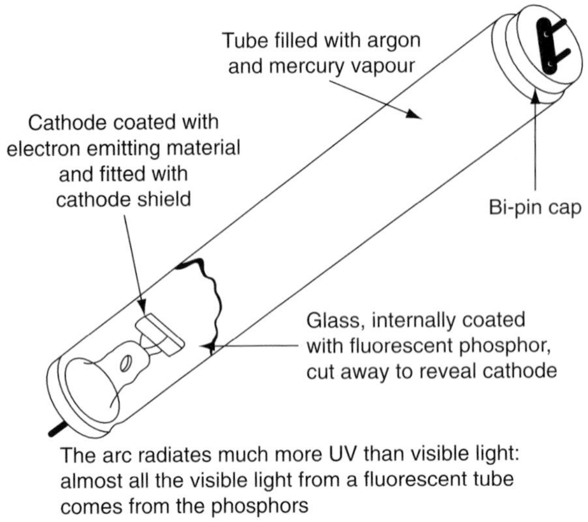

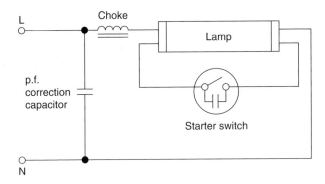

Tube filled with argon and mercury vapour

Cathode coated with electron emitting material and fitted with cathode shield

Bi-pin cap

Glass, internally coated with fluorescent phosphor, cut away to reveal cathode

The arc radiates much more UV than visible light: almost all the visible light from a fluorescent tube comes from the phosphors

FIGURE 10.24
Fluorescent lamp construction.

L

Choke

Lamp

p.f. correction capacitor

Starter switch

N

FIGURE 10.25
Fluorescent lamp circuit arrangement.

producing invisible ultraviolet light and some blue light. The fluorescent powder on the inside of the glass tube is very sensitive to ultraviolet rays and converts this radiation into visible light.

Fluorescent luminaires require a simple electrical circuit to initiate the ionization of the gas in the tube and a device to control the current once the arc is struck and the lamp is illuminated. Such a circuit is shown in Fig. 10.25.

A typical application for a fluorescent luminaire is in suspended ceiling lighting modules used in many commercial buildings.

The electrical relay

A **relay** is an electromagnetic switch operated by a solenoid. We looked at the action of a solenoid in Chapter 5 at Fig. 5.16. The solenoid in a relay operates a number of switch contacts as it moves under the electromagnetic

Definition

A *relay* is an electromagnetic switch operated by a solenoid.

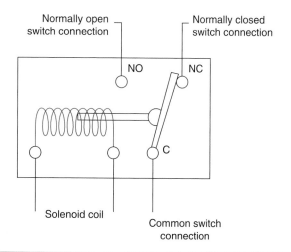

Normally open
switch connection

Normally closed
switch connection

NO   NC

Solenoid coil

C

Common switch
connection

FIGURE 10.26
A simple relay.

forces. Relays can be used to switch circuits on or off at a distance remotely. The energizing circuit, the solenoid, is completely separate to the switch contacts and, therefore, the relay can switch high voltage, high power circuits, from a low voltage switching circuit. This gives the relay many applications in motor control circuits, electronics and instrumentation systems. Figure 10.26 shows a simple relay.

Check Your Understanding

When you have completed the questions, check out the answers at the back of the book.
Note: more than one multiple choice answer may be correct.

1. The opposition to current flow in an a.c. resistive circuit is called:
   a. resistance
   b. inductance
   c. reactance
   d. impedance.

2. The opposition to current flow in an a.c. capacitive or inductive current is called:
   a. resistance
   b. inductance
   c. reactance
   d. impedance.

3. The total opposition to current flow in any a.c. circuit is called:
   a. resistance
   b. inductance
   c. reactance
   d. impedance.

4. A straight line having definite length and direction that represents to scale a quantity such as current voltage or impedance is called:
   a. a series a.c. circuit
   b. capacitive reactance
   c. a phasor, as in diagram
   d. the impedance triangle.

5. An a.c. series circuit has an inductive reactance of 4 Ω and a resistance of 3 Ω. The impedance of this circuit will be:
   a. 5 Ω
   b. 7 Ω
   c. 12 Ω
   d. 25 Ω.

6. An a.c. series circuit has a capacitive reactance of 12 Ω and a resistance of 9 Ω. The impedance of this current will be:
   a. 3 Ω
   b. 15 Ω
   c. 20 Ω
   d. 108 Ω.

7. The inductive reactance of a 100 mH coil when connected to a 50 Hz supply will be:

   a. 5 Ω

   b. 20 Ω

   c. 31.42 Ω

   d. 31.42 kΩ.

8. The capacitive reactance of a 100 μF capacitor when connected to a 50 Hz supply will be:

   a. 5 Ω

   b. 20 Ω

   c. 31.8 Ω

   d. 31.8 kΩ.

9. A circuit with *bad* power factor causes:

   a. fall in the supply voltage

   b. an increase in the supply voltage

   c. more current to be taken from the supply

   d. less current to be taken from the supply.

10. One application for a series d.c. motor is:

    a. an electric train

    b. a microwave oven

    c. a central heating pump

    d. an electric drill.

11. One application for an a.c. induction motor is:

    a. an electric train

    b. a microwave oven

    c. a central heating pump

    d. an electric drill.

12. One application for a shaded pole a.c. motor is:

    a. an electric train

    b. a microwave oven

    c. a central heating pump

    d. an electric drill.

13. A step down transformer has 1000 turns on the primary winding and 500 turns on the secondary winding. If the input voltage was 230 V the output voltage will be:

    a. 2 V

    b. 115 V

    c. 200 V

    d. 460 V.

14. An electromagnetic switch operated by a solenoid is one definition of:

    a. a transformer

    b. an a.c. motor

    c. a relay

    d. an inductive coil.

15. Use a sketch with notes of explanation to describe 'good' and 'bad' power factor.

16. State how power factor correction is achieved on:

    a. a fluorescent light fitting

    b. an electric motor.

17. Use a sketch to help you describe the meaning of the words:

    a. inductance

    b. mutual inductance.

18. Use a sketch with notes of explanation to describe how a force is applied to a conductor in a magnetic circuit and how this principle is applied to an electric motor.

19. Use a sketch with notes of explanation to show how a turning force is applied to the rotor and therefore, the drive shaft, of an electric motor.

20. Sketch the magnetic hysteresis loop of a magnetic material suitable for:

    a. a permanent magnet

    b. a transformer.

21. Give three applications for each of the following types of motor:

    a. a d.c. series motors

    b. an a.c. induction motor

    c. an a.c. split-phase motor

    d. an a.c. shaded pole motor.

Polyphase or three-phase electrical systems

221

Unit 3 – Application of health and safety and electrical principles – Outcome 4

Underpinning knowledge: when you have completed this chapter you should be able to:

- state the meaning of a polyphase system
- describe the production or generation of electricity by a polyphase system
- describe the transmission of electricity
- describe the distribution of electricity
- state the reasons for 'balancing' single-phase loads on a three-phase system
- distinguish between voltages and currents in a star and delta connected three-phase system

Generation, transmission and distribution of electricity

Generation

Figure 4.5 of Chapter 4 shows a simple a.c. generator or alternator producing an a.c. waveform. We generate electricity in large modern power stations using the same basic principle of operation. However, in place of a single loop of wire, the power station alternator has a three-phase winding and powerful electromagnets. The generated voltage is three identical sinusoidal waveforms each separated by 120° as shown in Fig. 11.1. The prime mover is not, of course, a simple crank handle, but a steam turbine. Hot water is heated until it becomes superheated steam, which drives the vanes of a steam turbine which is connected to the alternator. The heat required to produce the steam may come from burning coal or oil or from a nuclear reactor. Whatever the primary source of energy is, it is only being used to drive a turbine which is connected to an alternator, to generate electricity.

Transmission

Electricity is generated in the power station alternator at 25 kV. This electrical energy is fed into a transformer to be stepped up to a very high voltage for transmission on the National Grid Network at 400 kV, 275 kV or 132 kV. These very high voltages are necessary because, for a given power, the current is greatly reduced, which means smaller grid conductors and the transmission losses are reduced.

The National Grid Network consists of over 5,000 miles of overhead aluminium conductors suspended from steel pylons which link together all the power stations. Environmentalists say that these steel towers are ugly, but this method is about 16 times cheaper than the equivalent underground cable at these high voltages. Figure 11.2 shows a transmission line steel pylon.

222

Key Fact

Information

The individual conductors on the National Grid Network are only about 2.5 cm in diameter because they operate of very high voltage.

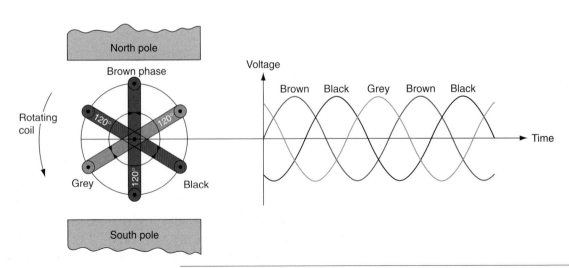

FIGURE 11.1

Generation of a three-phase voltage.

FIGURE 11.2
Transmission line steel pylon.

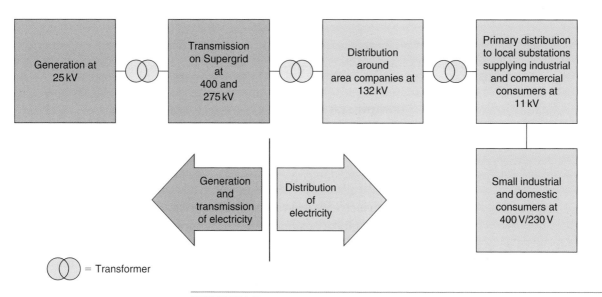

FIGURE 11.3
Generation, transmission and distribution of electrical energy.

Electricity is taken from the National Grid by appropriately located sub-stations which eventually transform the voltage down to 11 kV at a local sub-station. At the local sub-station the neutral conductor is formed for single-phase domestic supplies and three-phase supplies to shops, offices and garages. These supplies are usually underground radial supplies from the local sub-station, but in rural areas we still see transformers and over-head lines suspended on wooden poles. Figures 11.3 and 11.4 give an over-view of the system from power station to consumer.

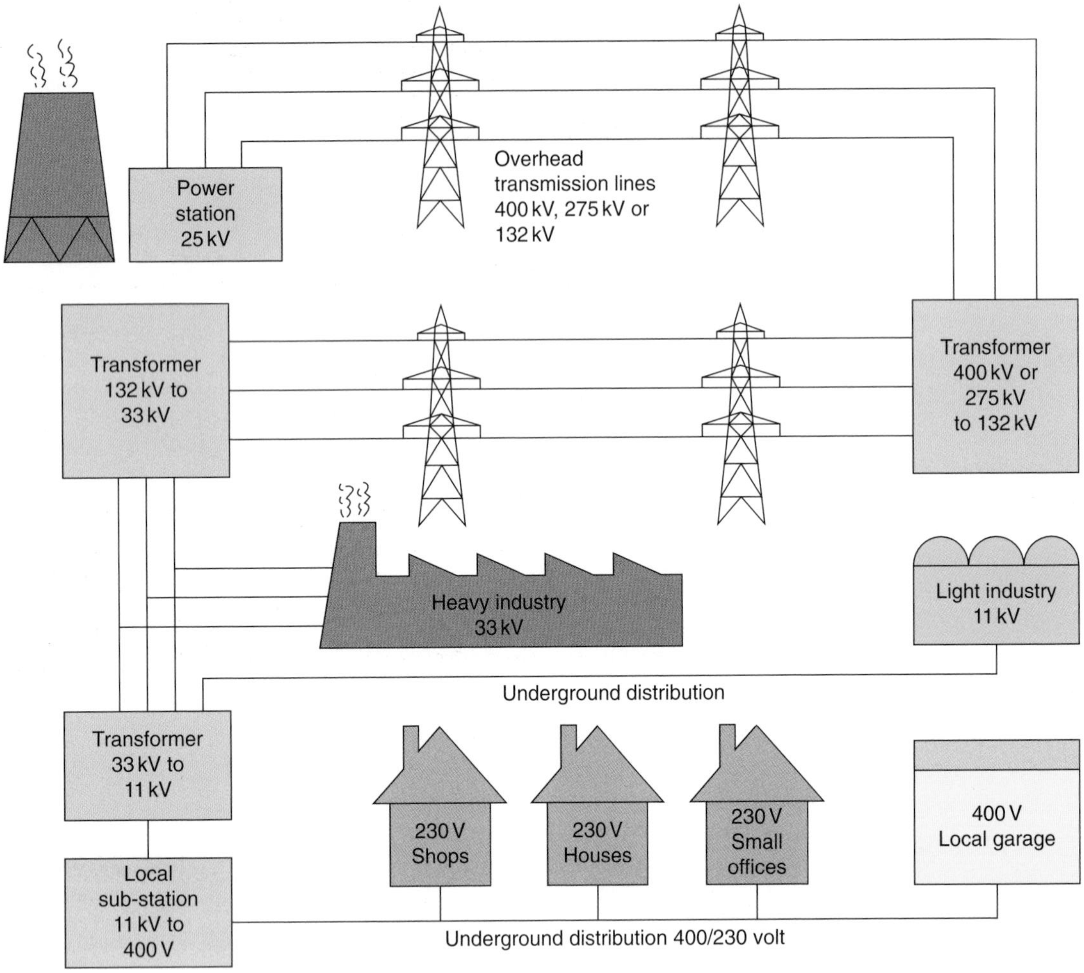

FIGURE 11.4

Simplified diagram of the distribution of electricity from power station to consumer.

Distribution to the consumer

The electricity leaves the local sub-station and arrives at the consumer's main's intake position. The final connections are usually by simple underground radial feeders at 400 V/230 V. Underground cable distribution is preferred within a city, town or village because people find the overhead distribution, which we see in rural and remote areas, unsightly. Also, at these lower distribution voltages, the cost of underground cables is not prohibitive. The 400 V/230 V is derived from the 11 kV/400 V sub-station transformer by connecting the secondary winding in star as shown in Fig. 11.5. The star point is earthed to an earth electrode sunk into the ground below the sub-station and from this point is taken the fourth conductor, and the neutral. Loads connected between phases are fed at 400 V and those fed between one phase and neutral at 230 V. A three-phase 400 V supply is used for supplying small industrial and commercial loads such as garages, schools and blocks of flats. A single-phase 230 V supply is usually provided for individual domestic consumers.

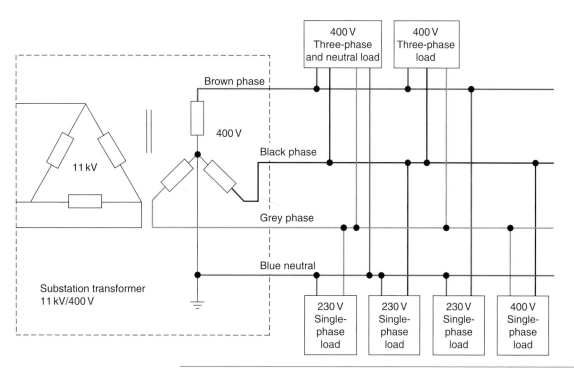

FIGURE 11.5
Three-phase four-wire distribution.

At the main's intake position, the supplier will provide a sealed HBC fuse and a sealed energy meter to measure the consumer's electricity consumption. It is after this point that we reach the consumer's installation.

Balancing single-phase loads

A three-phase load such as a motor has equally balanced phases since the resistance of each phase winding will be the same. Therefore, the current taken by each phase will be equal. When connecting single-phase loads to a three-phase supply, care should be taken to distribute the single-phase loads equally across the three phases so that each phase carries approximately the same current. Equally distributing the single-phase loads across the three-phase supply is known as 'balancing' the load. A lighting load of 18 luminaires would be 'balanced' if six luminaires were connected to each of the three phases.

Star and delta connections

The three-phase windings of an a.c. generator may be star connected or delta connected as shown in Fig. 11.6. The important relationship between phase and line currents and voltages is also shown. The square root of 3 ($\sqrt{3}$) is simply a constant for three-phase circuits, and has a value of 1.732. The delta connection is used for electrical power transmission because only three conductors are required. Delta connection is also used to

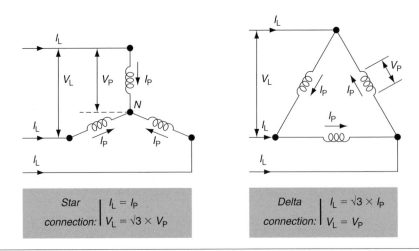

Star connection: $I_L = I_P$, $V_L = \sqrt{3} \times V_P$

Delta connection: $I_L = \sqrt{3} \times I_P$, $V_L = V_P$

FIGURE 11.6
Star and delta connections.

connect the windings of most three-phase motors because the phase windings are perfectly balanced and, therefore, do not require a neutral connection.

Making a star connection at the local sub-station has the advantage that two voltages become available – a line voltage of 400 V between any two phases, and a phase voltage of 230 V between line and neutral which is connected to the star point.

In any star-connected system currents flow along the lines (I_L), through the load and return by the neutral conductor connected to the star point. In a *balanced* three-phase system all currents have the same value and when they are added up by phasor addition, we find the resultant current is zero. Therefore, no current flows in the neutral and the star point is at zero volts. The star point of the distribution transformer is earthed because earth is also at zero potential. A star-connected system is also called a three-phase four-wire system and allows us to connect single-phase loads to a three-phase system.

Three-phase power

We know from our single-phase alternating current theory in Chapter 10 that power can be found from the following formula:

$$\text{Power} = VI \cos\phi \ (\text{W})$$

In any balanced three-phase system, the total power is equal to three times the power in any one-phase.

$$\therefore \text{Total three-phase power} = 3V_P I_P \cos\phi \ (\text{W}) \qquad (11.1)$$

Now for a star connection,

$$V_P = \frac{V_L}{\sqrt{3}} \quad \text{and} \quad I_L = I_P \tag{11.2}$$

Substituting Equation (11.2) into Equation (11.1), we have:

$$\text{Total three-phase power} = \sqrt{3} \, V_L I_L \cos\phi \text{ (W)}$$

Now consider a delta connection:

$$V_P = V_L \quad \text{and} \quad I_P = \frac{I_L}{\sqrt{3}} \tag{11.3}$$

Substituting Equation (11.3) into Equation (11.1) we have, for any balanced three-phase load,

$$\text{Total three-phase power} = \sqrt{3} \, V_L I_L \cos\phi \text{ (W)}$$

so, a general Equation for three-phase power is:

$$\text{Power} = \sqrt{3} \, V_L I_L \cos\phi$$

Example 1

A balanced star-connected three-phase load of $10\,\Omega$ per phase is supplied from a 400V, 50Hz mains supply at unity power factor. Calculate (a) the phase voltage, (b) the line current and (c) the total power consumed.

For a star connection,

$$V_L = \sqrt{3} \, V_P \text{ and } I_L = I_P$$

For (a),

$$V_P = \frac{V_L}{\sqrt{3}} \text{ (V)}$$

$$V_P = \frac{400\,\text{V}}{1.732} = 230.9\,\text{V}$$

For (b),

$$I_L = I_P = \frac{V_P}{R_P} \text{ (A)}$$

$$I_L = I_P = \frac{230.9\,\text{V}}{10\,\Omega} = 23.09\,\text{A}$$

For (c),

$$\text{Power} = \sqrt{3} \, V_L \, I_L \cos\phi \text{ (W)}$$

$$\therefore \text{Power} = 1.732 \times 400\,\text{V} \times 23.09\,\text{A} \times 1 = 16\,\text{kW}$$

Example 2

A 20 kW, 400 V balanced delta-connected load has a power factor of 0.8. Calculate (a) the line current and (b) the phase current.

We have that:

$$\text{Three-phase power} = \sqrt{3}\ V_L\ I_L\ \cos\phi\ \text{(W)}$$

For (a),

$$I_L = \frac{\text{Power}}{\sqrt{3}\ V_L\ \cos\phi}\ \text{(A)}$$

$$\therefore I_L = \frac{20,000\,\text{W}}{1.732 \times 400\,\text{V} \times 0.8}$$

$$I_L = 36.08\ \text{(A)}$$

For delta connection,

$$I_L = \sqrt{3}\ I_p\ \text{(A)}$$

Thus, for (b),

$$I_p = I_L / \sqrt{3}\ \text{(A)}$$

$$\therefore I_p = \frac{36.08\,\text{A}}{1.732} = 20.83\,\text{A}$$

Example 3

Three identical loads each having a resistance of $30\,\Omega$ and inductive reactance of $40\,\Omega$ are connected first in star and then in delta to a 400-V three-phase supply. Calculate the phase currents and line currents for each connection.

For each load,

$$Z = \sqrt{R^2 + X_L^2}\ (\Omega)$$

$$\therefore Z = \sqrt{30^2 + 40^2}$$

$$Z = \sqrt{2500} = 50\,\Omega$$

For star connection,

$$V_L = \sqrt{3}\ V_p \quad \text{and} \quad I_L = I_p$$

$$V_p = \frac{V_L}{\sqrt{3}}\,\text{(V)}$$

$$\therefore V_p = \frac{400\,\text{V}}{1.732} = 230.9\,\text{V}$$

$$I_p = \frac{V_p}{Z_p}\,\text{(A)}$$

$$\therefore I_p = \frac{230.9\,\text{V}}{50\,\Omega} = 4.62\,\text{A}$$

$$I_p = I_L$$

therefore phase and line currents are both equal to 4.62 A.

For delta connection,

$$V_L = V_P \quad \text{and} \quad I_L = \sqrt{3}\, I_P$$

$$V_L = V_P = 400\,V$$

$$I_P = V_P / Z_P \;(A)$$

$$\therefore I = \frac{400\,V}{50\,\Omega} = 8\,A$$

$$I_L = \sqrt{3}\, I_P \;(A)$$

$$\therefore I_L = 1.732 \times 8\,A = 13.86\,A$$

Check your Understanding

When you have completed the questions, check out the answers at the back of the book.
Note: more than one multiple choice answer may be correct.

1. Electricity is generated in large commercial power stations at:
    a. 230 V
    b. 400 V
    c. 25 kV
    d. 132 kV.

2. Transmission of electricity on the National Grid network takes place at:
    a. 230 V
    b. 400 V
    c. 25 kV
    d. 132 kV.

3. Electricity is transmitted at very high voltages because for a given power the:
    a. current is reduced
    b. current is increased
    c. losses are reduced
    d. losses are increased.

4. Electricity is distributed from the local sub-stations by underground cables at:
    a. 230 V
    b. 400 V
    c. 25 kV
    d. 132 kV.

5. A load connected to the three phases of a star-connected three-phase four-wire supply system from the local sub-station would have a voltage of:
    a. 230 V
    b. 400 V
    c. 25 kV
    d. 132 kV.

6. A load connected to phase and neutral of a star-connected three-phase four-wire supply system from the local sub-station would have a voltage of:

   a. 230 V

   b. 400 V

   c. 25 kV

   d. 132 kV.

7. The phase voltage of a star connected load is 100 V. The line voltage will be:

   a. 57.73 V

   b. 100 V

   c. 173.2 V

   d. 230 V.

8. The phase voltage of a delta connected load is 100 V. The line voltage will be:

   a. 57.73 V

   b. 100 V

   c. 173.2 V

   d. 230 V.

9. The phase current of a star connected load is 100 A. The line current will be:

   a. 57.73 A

   b. 100 A

   c. 173.2 A

   d. 230 A.

10. The phase current of a delta connected load is 100 A. The line current will be:

    a. 57.73 A

    b. 100 A

    c. 173.2 A

    d. 230 A.

11. State the meaning of a 'polyphase supply system'.

12. State the standard voltages used for generation, transmission and distribution in the United Kingdom.

13. Environmentalists often say that steel transmission towers are a 'blot on the landscape'. Why do we continue to use steel towers for transmission on the National Grid network?

14. Why is the distribution from local sub-stations to end users, for the most part, by underground cables in the United Kingdom?

15. State the reasons for balancing single-phase loads across a three-phase supply.

16. Briefly describe how and why we generate a three-phase supply compared to a single-phase supply.

17. What are the advantages of connecting a three-phase supply:
    a. in delta
    b. in star.

Overcurrent, short circuit and earth fault protection

233

Unit 3 – Application of health and safety and electrical principles – Outcome 5

Underpinning knowledge: when you have completed this chapter you should be able to:

- state the need for protective devices
- list the factors on which the selection of a protective device depends
- list the essential requirements for a device designed to protect against overcurrent
- state the action of a fuse under fault conditions
- state the need for 'discrimination' when a number of protective devices are fitted
- identify exposed and extraneous parts of a building
- state the path taken by an earth fault current
- state the need for RCDs and RCBOs
- state the meaning of 'Basic' and 'Fault' protection

Protecting electrical equipment, circuits and people

We know from the earlier chapters in this book that using electricity is one of the causes of accidents in the workplace. Using electricity is a hazard because it has the 'potential' and the possibility to cause harm. Therefore, the provision of protective devices in an electrical installation is fundamental to the whole concept of the safe use of electricity in buildings. The electrical installation as a whole must be protected against overload or short circuit and the people using the building must be protected against the risk of shock, fire or other risks arising from their own misuse of the installation or from a fault. The installation and maintenance of adequate and appropriate protective measures is a vital part of the safe use of electrical energy. I want to look at protection against an electric shock by both basic and fault protection, at protection by equipotential bonding and automatic disconnection of the supply, and protection against excess current.

Let us first define some of the words we will be using. Chapter 54 of the IEE Regulations describes the earthing arrangements for an electrical installation. It gives the following definitions:

Earth – the conductive mass of the earth. Whose electrical potential is taken as zero.

Earthing – the act of connecting the exposed conductive parts of an installation to the main protective earthing terminal of the installation.

Bonding conductor – a protective conductor providing equipotential bonding.

Bonding – the linking together of the exposed or extraneous metal parts of an electrical installation.

Circuit protective conductor (CPC) – a protective conductor connecting exposed conductive parts of equipment to the main earthing terminal. This is the green and yellow insulated conductor in twin and earth cable.

Exposed conductive parts – the metalwork of an electrical appliance or the trunking and conduit of an electrical system which can be touched because they are not normally live, but which may become live under fault conditions.

Extraneous conductive parts – the structural steelwork of a building and other service pipes such as gas, water, radiators and sinks. They do not form a part of the electrical installation but may introduce a potential, generally earth potential, to the electrical installation.

Shock protection – protection from electric shock is provided by basic protection and fault protection.

Basic protection – is provided by the insulation of live parts in accordance with Section 416 of the IEE Regulations.

Fault protection – is provided by protective equipotential bonding and automatic disconnection of the supply (by a fuse or miniature circuit breaker, MCB) in accordance with IEE Regulations 411.3 to 6.

Definitions

Earth – the conductive mass of the earth. Whose electrical potential is taken as zero.

Earthing – the act of connecting the exposed conductive parts of an installation to the main protective earthing terminal of the installation.

Bonding conductor – a protective conductor providing equipotential bonding.

Bonding – the linking together of the exposed or extraneous metal parts of an electrical installation.

Circuit protective conductor (CPC) – a protective conductor connecting exposed conductive parts of equipment to the main earthing terminal.

Exposed conductive parts – the metalwork of an electrical appliance or the trunking and conduit of an electrical system which can be touched because they are not normally live, but which may become live under fault conditions.

Extraneous conductive parts – the structural steelwork of a building and other service pipes such as gas, water, radiators and sinks.

Shock protection – protection from electric shock is provided by basic protection and fault protection.

Basic protection – is provided by the insulation of live parts in accordance with Section 416 of the IEE Regulations.

Fault protection – is provided by protective equipotential bonding and automatic disconnection of the supply (by a fuse or miniature circuit breaker, MCB) in accordance with IEE Regulations 411.3 to 6.

Definition

Protective equipotential bonding – this is equipotential bonding for the purpose of safety as shown in Fig. 6.7 of this book.

Protective equipotential bonding – this is equipotential bonding for the purpose of safety as shown in Fig. 6.7 of this book.

Try This

Definitions
- to help you to remember these definitions try writing them out in your own words
- just identify the essential words.

Basic protection and fault protection

The human body's movements are controlled by the nervous system. Very tiny electrical signals travel between the central nervous system and the muscles, stimulating operation of the muscles, which enable us to walk, talk and run and remember that the heart is also a muscle.

If the body becomes part of a more powerful external circuit, such as the electrical mains, and current flows through it, the body's normal electrical operations are disrupted. The shock current causes unnatural operation of the muscles and the result may be that the person is unable to release the live conductor causing the shock, or the person may be thrown across the room. The current which flows through the body is determined by the resistance of the human body and the surface resistance of the skin on the hands and feet.

This leads to the consideration of exceptional precautions where people with wet skin or wet surfaces are involved, and the need for special consideration in bathroom installations.

Two types of contact will result in a person receiving an electric shock. Direct contact with live parts which involves touching a terminal or line conductor that is actually live. The Regulations call this basic protection. Indirect contact results from contact with an exposed conductive part such as the metal structure of a piece of equipment that has become live as a result of a fault. The Regulations call this fault protection.

In installations operating at normal mains voltage, the primary method of protection against direct contact is by insulation. All live parts are enclosed in insulating material such as rubber or plastic, which prevents contact with those parts. The insulating material must, of course, be suitable for the circumstances in which they will be used and the stresses to which they will be subjected. The IEE Regulations call this basic protection (IEE Regulation 131.2.1).

Other methods of basic protection include the provision of barriers or enclosures which can only be opened by the use of a tool, or when the supply is first disconnected. Protection can also be provided by fixed obstacles such as a guardrail around an open switchboard or by placing live parts out of reach as with overhead lines.

Fault protection

Protection against indirect contact, called fault protection (IEE Regulation 131.2.2) is achieved by connecting exposed conductive parts of equipment to the main protective earthing terminal.

In Chapter 13 of the IEE Regulations we are told that where the metalwork of electrical equipment may become charged with electricity in such a manner as to cause danger, that metalwork will be connected with earth so as to discharge the electrical energy without danger. The application of protective equipotential bonding is one of the important principles for safety.

There are five methods of protection against contact with metalwork which has become unintentionally live, that is, indirect contact with exposed conductive parts recognized by the IEE Regulations. These are:

1. protective equipotential bonding coupled with automatic disconnection of the supply,

2. the use of Class II (double insulated) equipment,

3. the provision of a non-conducting location,

4. the use of earth free equipotential bonding,

5. electrical separation.

Methods 3 and 4 are limited to special situations under the effective supervision of trained personnel.

Method 5, electrical separation, is little used but does find an application in the domestic electric shaver supply unit which incorporates an isolating transformer.

Method 2, the use of Class II insulated equipment is limited to single pieces of equipment such as tools used on construction sites, because it relies upon effective supervision to ensure that no metallic equipment or extraneous earthed metalwork enters the area of the installation.

The method which is most universally used in the United Kingdom is, therefore, Method 1 – protective equipotential bonding coupled with automatic disconnection of the supply.

This method relies upon all exposed metalwork being electrically connected together to an effective earth connection. Not only must all the metalwork associated with the electrical installation be so connected, that is conduits, trunking, metal switches and the metalwork of electrical appliances, but Regulation 411.3.1.2 tells us to connect the extraneous metalwork of water service pipes, gas and other service pipes and ducting, central heating and air conditioning systems, exposed metallic structural parts of the building and lightning protective systems to the protective earthing terminal. In this way the possibility of a voltage appearing between two exposed metal parts is removed. Protective equipotential bonding is shown in Fig. 6.7 in Chapter 6.

The second element of this protection method is the provision of a means of automatic disconnection of the supply in the event of a fault occurring that causes the exposed metalwork to become live.

IEE Regulation 411.3.2 tells us that for final circuits not exceeding 32 A the maximum disconnection time shall not exceed 0.4 s.

The achievement of these disconnection times is dependent upon the type of protective device used, fuse or circuit breaker, the circuit conductors to the fault and the provision of adequate protective equipotential bonding. The resistance, or we call it the impedance, of the earth fault loop must be less than the values given in Appendix 2 of the *On Site Guide* and Tables 41.2 to 41.4 of the IEE Regulations. (Table 12.2 later in this chapter shows the maximum value of the earth fault loop impedance for circuits protected by MCBs to BS EN 60898.) We will look at this again later in this chapter under the heading 'Earth Fault Loop Impedance Z_S'. Chapter 54 of the IEE Regulations gives details of the earthing arrangements to be incorporated in the supply system to meet these Regulations and these are described in Chapter 14 of this book, under the heading 'Electricity supply systems'.

Residual current protection

The IEE Regulations recognize the particular problems created when electrical equipment such as lawnmowers, hedge-trimmers, drills and lights are used outside buildings. In these circumstances the availability of an adequate earth return path is a matter of chance. The Regulations, therefore, require that any socket outlet with a rated current not exceeding 20 A, for use by ordinary people and intended for general use shall have the additional protection of a residual current device (RCD) which has a rated operating current of not more than 30 milliamperes (mA) (IEE Regulation 411.3.3).

Try This

Definitions

- What do the Regulations mean by 'ordinary people'?
- Look back at the definitions in Chapter 2 and write it in the margin here.

An RCD is a type of circuit breaker that continuously compares the current in the line and neutral conductors of the circuit. The currents in a healthy circuit will be equal, but in a circuit that develops a fault, some current will flow to earth and the line and neutral currents will no longer balance. The RCD detects the imbalance and disconnects the circuit. Figure 12.8 later in this chapter shows an RCD's construction.

Isolation and switching

Part 4 of the IEE Regulations deals with the application of protective measures for safety and Chapter 53 with the regulations for switching devices

237

or switchgear required for protection, isolation and switching of a consumer's installation.

The consumer's main switchgear must be readily accessible to the consumer and be able to:

- isolate the complete installation from the supply,
- protect against overcurrent,
- cut off the current in the event of a serious fault occurring.

The Regulations identify four separate types of switching: switching for isolation, switching for mechanical maintenance, emergency switching and functional switching.

Isolation is defined as cutting off the electrical supply to a circuit or item of equipment in order to ensure the safety of those working on the equipment by making dead those parts which are live in normal service.

The purpose of isolation switching is to enable electrical work to be carried out safely on an isolated circuit or piece of equipment. *Isolation* is intended for use by electrically skilled or supervised persons.

An isolator is a mechanical device which is operated manually and used to open or close a circuit off load. An isolator switch must be provided close to the supply point so that all equipment can be made safe for maintenance. Isolators for motor circuits must isolate the motor and the control equipment, and isolators for discharge lighting luminaires must be an integral part of the luminaire so that it is isolated when the cover is removed or be provided with effective local isolation (Regulation 537.2.1.6). Devices which are suitable for isolation are isolation switches, fuse links, circuit breakers, plugs and socket outlets. They must isolate all live supply conductors and provision must be made to secure the isolation (IEE Regulation 537.2.2.4).

Isolation at the consumer's service position can be achieved by a double pole switch which opens or closes all conductors simultaneously. On three-phase supplies the switch need only break the live conductors with a solid link in the neutral, provided that the neutral link cannot be removed before opening the switch.

The switching for mechanical maintenance requirements is similar to those for isolation except that the control switch must be capable of switching the full load current of the circuit or piece of equipment.

The purpose of switching for mechanical maintenance is to enable non-electrical work to be carried out safely on the switched circuit or equipment.

Mechanical maintenance switching is intended for use by skilled but non-electrical persons. Switches for mechanical maintenance must be manually operated, not have exposed live parts when the appliance is opened, must be connected in the main electrical circuit and have a reliable on/off indication or visible contact gap (Regulations 537.3.2.2). Devices which are suitable for switching off for mechanical maintenance are switches, circuit breakers, plug and socket outlets.

Definition

Isolation is defined as cutting off the electrical supply to a circuit or item of equipment in order to ensure the safety of those working on the equipment by making dead those parts which are live in normal service.

Definition

The switching for mechanical maintenance requirements is similar to those for isolation except that the control switch must be capable of switching the full load current of the circuit or piece of equipment.

Definition

Emergency switching involves the rapid disconnection of the electrical supply by a single action to remove or prevent danger.

Emergency switching involves the rapid disconnection of the electrical supply by a single action to remove or prevent danger.

The purpose of emergency switching is to cut off the electrical energy *rapidly* to remove an unexpected hazard.

Emergency switching is for use by anyone. The device used for emergency switching must be immediately accessible and identifiable, and be capable of cutting off the full load current.

Electrical machines must be provided with a means of emergency switching, and a person operating an electrically driven machine must have access to an emergency switch so that the machine can be stopped in an emergency. The remote stop/start arrangement could meet this requirement for an electrically driven machine (Regulations 537.4.2.2). Devices which are suitable for emergency switching are switches, circuit breakers and contactors. Where contactors are operated by remote control they should *open* when the coil is de-energized, that is, fail safe. Push-buttons used for emergency switching must be coloured red and latch in the stop or off position. They should be installed where danger may arise and be clearly identified as emergency switches. Plugs and socket outlets cannot be considered appropriate for emergency disconnection of supplies.

Definition

Functional switching involves the switching on or off, or varying the supply, of electrically operated equipment in normal service.

Functional switching involves the switching on or off, or varying the supply, of electrically operated equipment in normal service.

The purpose of functional switching is to provide control of electrical circuits and equipment in normal service.

Functional switching is for the user of the electrical installation or equipment. The device must be capable of interrupting the total steady current of the circuit or appliance. When the device controls a discharge lighting circuit it must have a current rating capable of switching an inductive load. The Regulations acknowledge the growth in the number of electronic dimmer switches being used for the control and functional switching of lighting circuits. The functional switch must be capable of performing the most demanding duty it may be called upon to perform (IEE Regulations 537.5.2.1 and 2).

Overcurrent protection

Safety First

Isolation

Isolation and switching for protection is an important safety measure in the electrotechnical industry.

The consumer's mains equipment must provide protection against overcurrent, that is a current exceeding the rated value (Regulation 430.3). Fuses provide overcurrent protection when situated in the live conductors; they must not be connected in the neutral conductor. Circuit breakers may be used in place of fuses, in which case the circuit breaker may also provide the means of isolation, although a further means of isolation is usually provided so that maintenance can be carried out on the circuit breakers themselves.

When selecting a protective device we must give consideration to the following factors:

- the prospective fault current,
- the circuit load characteristics,

- the current carrying capacity of the cable,
- the disconnection time requirements for the circuit.

The essential requirements for a device designed to protect against overcurrent are:

- it must operate automatically under fault conditions,
- have a current rating matched to the circuit design current,
- have a disconnection time which is within the design parameters,
- have an adequate fault breaking capacity,
- be suitably located and identified.

We will look at these requirements below.

An overcurrent may be an overload current, or a short-circuit current. An **overload current** can be defined as a current which exceeds the rated value in an otherwise healthy circuit. Overload currents usually occur because the circuit is abused or because it has been badly designed or modified. A **short circuit** is an overcurrent resulting from a fault of negligible impedance connected between conductors. Short circuits usually occur as a result of an accident which could not have been predicted before the event.

An overload may result in currents of two or three times the rated current flowing in the circuit. Short-circuit currents may be hundreds of times greater than the rated current. In both cases the basic requirements for protection are that the fault currents should be interrupted quickly and the circuit isolated safely before the fault current causes a temperature rise or mechanical effects which might damage the insulation, connections, joints and terminations of the circuit conductors or their surroundings (IEE Regulation 130.3).

The selected protective device should have a current rating which is not less than the full load current of the circuit but which does not exceed the cable current rating. The cable is then fully protected against both overload and short-circuit faults (Regulation 435.1). Devices which provide overcurrent protection are:

- High breaking capacity (HBC) fuses to BS 88-6. These are for industrial applications having a maximum fault capacity of 80 kA.
- Cartridge fuses to BS 1361. These are used for a.c. circuits on industrial and domestic installations having a fault capacity of about 30 kA.
- Cartridge fuses to BS 1362. These are used in 13 A plug tops and have a maximum fault capacity of about 6 kA.
- Semi-enclosed fuses to BS 3036. These were previously called re-wirable fuses and are used mainly on domestic installations having a maximum fault capacity of about 4 kA.
- MCBs to BS EN 60898. These are miniature circuit breakers (MCBs) which may be used as an alternative to fuses for some installations.

Definitions

An *overload current* can be defined as a current which exceeds the rated value in an otherwise healthy circuit.

A *short circuit* is an overcurrent resulting from a fault of negligible impedance connected between conductors.

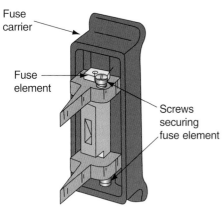

Fuse carrier

Fuse element

Screws securing fuse element

FIGURE 12.1
A semi-enclosed fuse.

The British Standard includes ratings up to 100 A and maximum fault capacities of 9 kA. They are graded according to their instantaneous tripping currents – that is, the current at which they will trip within 100 ms. This is less than the time taken to blink an eye.

By definition a fuse is the weakest link in the circuit. Under fault conditions it will melt when an overcurrent flows, protecting the circuit conductors from damage.

Semi-enclosed fuses (BS 3036)

The semi-enclosed fuse consists of a fuse wire, called the fuse element, secured between two screw terminals in a fuse carrier. The fuse element is connected in series with the load and the thickness of the element is sufficient to carry the normal rated circuit current. When a fault occurs an overcurrent flows and the fuse element becomes hot and melts or 'blows'.

This type of fuse is illustrated in Fig. 12.1. The fuse element should consist of a single strand of plan or tinned copper wire having a diameter appropriate to the current rating of the fuse. *This type of fuse was very popular in domestic installations, but less so these days because of their disadvantages.*

ADVANTAGE OF SEMI-ENCLOSED FUSES

- They are very cheap compared with other protective devices both to install and to replace.
- There are no mechanical moving parts.
- It is easy to identify a 'blown' fuse.

DISADVANTAGES OF SEMI-ENCLOSED FUSES

- The fuse element may be replaced with wire of the wrong size either deliberately or by accident.
- The fuse element weakens with age due to oxidization, which may result in a failure under normal operating conditions.
- The circuit cannot be restored quickly since the fuse element requires screw fixing.
- They have low breaking capacity since, in the event of a severe fault, the fault current may vaporize the fuse element and continue to flow in the form of an arc across the fuse terminals.
- They are not guaranteed to operate until up to twice the rated current is flowing.
- There is a danger from scattering hot metal if the fuse carrier is inserted into the base when the circuit is faulty.

Cartridge fuses (BS 1361)

The cartridge fuse breaks a faulty circuit in the same way as a semi-enclosed fuse, but its construction eliminates some of the disadvantages experienced with an open-fuse element. The fuse element is encased in a glass or ceramic tube and secured to end-caps which are firmly attached

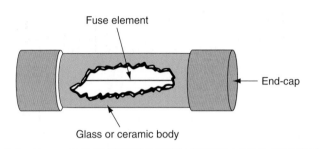

FIGURE 12.2
Cartridge fuse.

to the body of the fuse so that they do not blow off when the fuse operates. Cartridge fuse construction is illustrated in Fig. 12.2. With larger size cartridge fuses, lugs or tags are sometimes brazed on the end-caps to fix the fuse cartridge mechanically to the carrier. They may also be filled with quartz sand to absorb and extinguish the energy of the arc when the cartridge is brought into operation.

ADVANTAGES OF CARTRIDGE FUSES

- They have no mechanical moving parts.
- The declared rating is accurate.
- The element does not weaken with age.
- They have small physical size and no external arcing which permits their use in plug tops and small fuse carriers.
- Their operation is more rapid than semi-enclosed fuses. Operating time is inversely proportional to the fault current, so the bigger the fault current the quicker the fuse operates.
- They are easy to replace.

DISADVANTAGES OF CARTRIDGE FUSES

- They are more expensive to replace than fuse elements that can be re-wired.
- They can be replaced with an incorrect cartridge.
- The cartridge may be shorted out by wire or silver foil in extreme cases of bad practice.
- It is not possible to see if the fuse element is broken.

Miniature circuit breakers (BS EN 60898)

The disadvantage of all fuses is that when they have operated they must be replaced. An MCB overcomes this problem since it is an automatic switch which opens in the event of an excessive current flowing in the circuit and can be closed when the circuit returns to normal.

An MCB of the type shown in Fig. 12.3 incorporates a thermal and magnetic tripping device. The load current flows through the thermal and the

FIGURE 12.3
MCBs – B Breaker, fits Wylex Standard consumer unit (Courtesy of Wylex).

electromagnetic devices in normal operation but under overcurrent conditions they activate and trip the MCB.

The circuit can be restored when the fault is removed by pressing the ON toggle. This latches the various mechanisms within the MCB and 'makes' the switch contact. The toggle switch can also be used to disconnect the circuit for maintenance or isolation or to test the MCB for satisfactory operation.

ADVANTAGES OF MCBs

- They have factory set operating characteristics.
- Tripping characteristics and therefore circuit protection is set by the installer.
- The circuit protection is difficult to interfere with.
- The circuit is provided with discrimination.
- A faulty circuit may be quickly identified.
- A faulty circuit may be easily and quickly restored.
- The supply may be safely restored by an unskilled operator.

DISADVANTAGES OF MCBs

- They are relatively expensive but look at the advantages to see why they are so popular these days.
- They contain mechanical moving parts and therefore require regular testing to ensure satisfactory operation under fault conditions.

CHARACTERISTICS OF MCBs

MCB Type B to BS EN 60898 will trip instantly at between three and five times its rated current and is also suitable for domestic and commercial installations.

MCB Type C to BS EN 60898 will trip instantly at between five and ten times its rated current. It is more suitable for highly inductive commercial and industrial loads.

MCB Type D to BS EN 69898 will trip instantly at between 10 and 25 times its rated current. It is suitable for welding and X-ray machines where large inrush currents may occur.

Installing overcurrent protective devices

The general principle to be followed is that a protective device must be placed at a point where a reduction occurs in the current carrying capacity of the circuit conductors (IEE Regulations 433.2 and 434.2). A reduction may occur because of a change in the size or type of conductor or because of a change in the method of installation or a change in the environmental conditions. The only exceptions to this rule are where an overload protective device opening a circuit might cause a greater danger than the overload itself – for example, a circuit feeding an overhead electromagnet in a scrapyard.

Fault protection

The overcurrent protection device protecting circuits not exceeding 32 A shall have a disconnection time not exceeding 0.4 s (IEE Regulations 411.3.2.2).

The IEE Regulations permit us to assume that where an overload protective device is also intended to provide short-circuit protection, and has a rated breaking capacity greater than the prospective short-circuit current at the point of its installation, the conductors on the load side of the protective device are considered to be adequately protected against short-circuit currents without further proof. This is because the cable rating and the overload rating of the device are compatible. However, if this condition is not met or if there is some doubt, it must be verified that fault currents will be interrupted quickly before they can cause a dangerously high temperature rise in the circuit conductors. Regulation 434.5.2 provides an equation for calculating the maximum operating time of the protective device to prevent the permitted conductor temperature rise being exceeded as follows:

$$t = \frac{k^2 S^2}{I^2} \text{ (s)}$$

where

t = duration time in seconds

S = cross-sectional area of conductor in square millimetres

I = short-circuit rms current in amperes

k = a constant dependent upon the conductor metal and type of insulation (see Table 43 A of the IEE Regulations).

Example

A 10 mm PVC sheathed mineral insulated (MI) copper cable is short circuited when connected to a 400V supply. The impedance of the short-circuit path is 0.1 Ω. Calculate the maximum permissible disconnection time and show that a 50 A Type B MCB to BS EN 60898 will meet this requirement.

$$I = \frac{V}{Z} \text{ (A)} \qquad I = \frac{400 \text{ V}}{0.1 \text{ }\Omega} = 4000 \text{ A}$$

$$\therefore \text{ Fault current} = 4000 \text{ A}$$

For PVC Sheathed MI copper cables, Table 43.1 gives a value for k of 115. So,

$$t = \frac{k^2 S^2}{I^2} \text{ (s)}$$

$$\therefore t = \frac{115^2 \times 10^2 \text{ mm}^2}{4000 \text{ A}} = 82.66 \times 10^{-3} \text{ s}$$

The maximum time that a 4000 A fault current can be applied to this 10 mm² cable without dangerously raising the conductor temperature is 82.66 ms. Therefore, the protective device must disconnect the supply to the cable in less than 82.66 ms under short-circuit conditions. Manufacturers' information and Appendix 3 of the IEE Regulations give the operating times of protective devices at various short-circuit currents in the form of graphs. Let us come back to this problem in a few moments.

Time/current characteristics of protective devices

Disconnection times for various overcurrent devices are given in the form of a logarithmic graph. This means that each successive graduation of the axis represents a 10 times change over the previous graduation.

These logarithmic scales are shown in the graphs of Figs 12.4 and 12.5. From Fig. 12.4 it can be seen that the particular protective device represented by this characteristic will take 8 s to disconnect a fault current of 50 A and 0.08 s to clear a fault current of 1000 A.

Let us now go back to the problem and see if the Type B MCB will disconnect the supply in less than 82.66 ms.

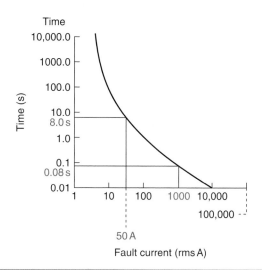

FIGURE 12.4

Time/current characteristic of an overcurrent protective device.

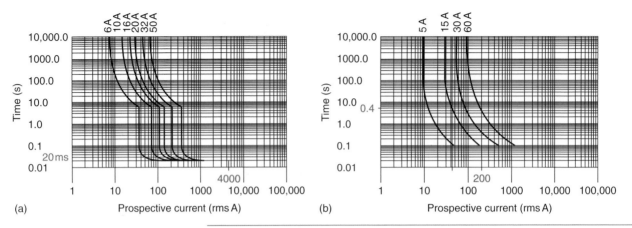

FIGURE 12.5

Time/current characteristics of (a) a Type B MCB to BS EN 60898 and (b) semi-enclosed fuse to BS 3036.

Figure 12.5(a) shows the time/current characteristics for a Type B MCB to BS EN 60898. This graph shows that a fault current of 4000 A will trip the protective device in 20 ms. Since this is quicker than 82.66 ms, the 50 A Type B MCB is suitable and will clear the fault current before the temperature of the cable is raised to a dangerous level.

Appendix 3 of the IEE Regulations gives the time/current characteristics and specific values of prospective short-circuit current for a number of protective devices.

These indicate the value of fault current which will cause the protective device to operate in the times indicated by IEE Regulation 411.

Figures 3.1, 3.2 and 3.3 in Appendix 3 of the IEE Regulations deal with fuses and Figs 3.4, 3.5 and 3.6 with MCBs.

It can be seen that the prospective fault current required to trip an MCB in the required time is a multiple of the current rating of the device. The multiple depends upon the characteristics of the particular devices. Thus:

- Type B MCB to BS EN 60898 has a multiple of 5
- Type C MCB to BS EN 60898 has a multiple of 10
- Type D MCB to BS EN 60898 has a multiple of 20.

Example

A 6 A Type B MCB to BS EN 60898 which is used to protect a domestic lighting circuit will trip within 0.4 s when 6 A times a multiple of 5, that is 30 A, flows under fault conditions.

Therefore if the earth fault loop impedance is low enough to allow at least 30 A to flow in the circuit under fault conditions, the protective device will operate within the time required by Regulation 411.

The characteristics shown in Appendix 3 of the IEE Regulations give the specific values of prospective short-circuit current for all standard sizes of protective device.

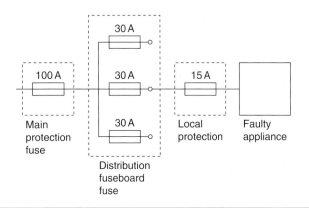

FIGURE 12.6

Effective discrimination achieved by graded protection.

Effective discrimination of protective devices

In the event of a fault occurring on an electrical installation only the protective device nearest to the fault should operate, leaving other healthy circuits unaffected. A circuit designed in this way would be considered to have effective discrimination. Effective discrimination can be achieved by graded protection since the speed of operation of the protective device increases as the rating decreases. This can be seen in Fig. 12.5(b). A fault current of 200 A will cause a 15 A semi-enclosed fuse to operate in about 0.1 s, a 30 A semi-enclosed fuse in about 0.4 s and a 60 A semi-enclosed fuse in about 5.0 s. If a circuit is arranged as shown in Fig. 12.6 and a fault occurs on the appliance, effective discrimination will be achieved because the 15 A fuse will operate more quickly than the other protective devices if they were all semi-enclosed types fuses with the characteristics shown in Fig. 12.5(b).

Security of supply, and therefore effective discrimination, is an important consideration for an electrical designer and is also a requirement of the IEE Regulations.

Earth fault loop impedance Z_S

In order that an overcurrent protective device can operate successfully, meeting the required disconnection times, of Regulations 411.3.2.2, that is, final circuits not exceeding 32 A shall have a disconnection time not exceeding 0.4 s. To achieve this, the earth fault loop impedance value measured in ohms must be less than those values given in Appendix 2 of the *On Site Guide* and Tables 41.2 and 41.3 of the IEE Regulations. The value of the earth fault loop impedance may be verified by means of an earth fault loop impedance test as described in Chapter 14 of this book. The formula is:

$$Z_S = Z_E + (R_1 + R_2)\ (\Omega)$$

Here Z_E is the impedance of the supply side of the earth fault loop. The actual value will depend upon many factors: the type of supply, the ground conditions, the distance from the transformer, etc. The value can be obtained from the area electricity companies, but typical values are 0.35 Ω for TN-C-S (protective multiple earthing, PME) supplies and 0.8 Ω for

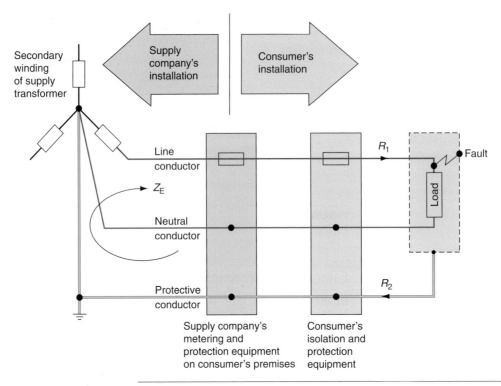

FIGURE 12.7
Earth fault loop path for a TN-S system.

TN-S (cable sheath earth) supplies. Also in the above formula, R_1 is the resistance of the line conductor and R_2 is the resistance of the earth conductor. The complete earth fault loop path is shown in Fig. 12.7.

Values of $R_1 + R_2$ have been calculated for copper and aluminium conductors and are given in Table 9A of the *On Site Guide* as shown in Table 12.1 of this book.

Example

A 20 A radial socket outlet circuit is wired in 2.5 mm² PVC cable incorporating a 1.5 mm² CPC. The cable length is 30 m installed in an ambient temperature of 20°C and the consumer's protection is by 20 A MCB Type B to BS EN 60898. The earth fault loop impedance of the supply is 0.5 Ω. Calculate the total earth fault loop impedance Z_S, and establish that the value is less than the maximum value permissible for this type of circuit.

We have:

$$Z_S = Z_E + (R_1 + R_2) \; (\Omega)$$

$$Z_E = 0.5\,\Omega \, (\text{value given in the question})$$

From the value given in Table 9A of the *On Site Guide* and reproduced in Table 12.1 a 2.5 mm phase conductor with a 1.5 mm protective conductor has an $(R_1 + R_2)$ value of $19.51 \times 10^{-3}\,\Omega/\text{m}$.

$$\text{For 30 m cable } (R_1 + R_2) = 19.51 \times 10^{-3}\,\Omega/\text{m} \times 30\,\text{m} = 0.585\,\Omega$$

However, under fault conditions, the temperature and therefore the cable resistance will increase. To take account of this, we must multiply the value of cable resistance by the factor

Table 12.1 Table 9A of the IEE *On Site* Guide: Value of Resistance/Metre for Copper and Aluminium Conductors and of $R_1 + R_2$ per Metre at 20°C in Milliohms/Metre

Cross-sectional area (mm²)		Resistance/metre or $(R_1 + R_2)$/metre (mΩ/m)	
Phase conductor	Protective conductor	Copper	Aluminium
1	–	18.10	
1	1	36.20	
1.5	–	12.10	
1.5	1	30.20	
1.5	1.5	24.20	
2.5	–	7.41	
2.5	1	25.51	
2.5	1.5	19.51	
2.5	2.5	14.82	
4	–	4.61	
4	1.5	16.71	
4	2.5	12.02	
4	4	9.22	
6	–	3.08	
6	2.5	10.49	
6	4	7.69	
6	6	6.16	
10	–	1.83	
10	4	6.44	
10	6	4.91	
10	10	3.66	
16	–	1.15	1.91
16	6s	4.23	–
16	10	2.98	–
16	16	2.30	3.82
25	–	0.727	1.20
25	10	2.557	–
25	16	1.877	–
25	25	1.454	2.40
35	–	0.524	0.87
35	16	1.674	2.78
35	25	1.251	2.07
35	35	1.048	1.74
50	–	0.387	0.64
50	25	1.114	1.84
50	35	0.911	1.51
50	50	0.774	1.28

Reproduced from the IEE *On Site Guide* by kind permission of the Institution of Electrical Engineers.

given in Table 9C of the *On Site Guide*. In this case the factor is 1.20 and therefore the cable resistance under fault conditions will be:

$$0.585\,\Omega \times 1.20 = 0.702\,\Omega$$

The total earth fault loop impedance is therefore:

$$Z_S = 0.5\,\Omega + 0.702\,\Omega = 1.202\,\Omega$$

Table 12.2 Maximum Earth Fault Loop Impedance Z_s MCB Type B Maximum Measured Earth Fault Loop Impedance (Ω) When Overcurrent Protective Device is MCB Type B to BS EN 60898

	MCB rating (A)							
	6	10	16	20	25	32	40	50
For 0.4 s disconnection Z_S (Ω)	7.67	4.6	2.87	2.3	1.84	1.44	1.15	0.92

The maximum permitted value given in Table 2A of the *On Site Guide* for a 20 A MCB protecting a socket outlet is 2.3 Ω as shown by Table 12.2. The circuit earth fault loop impedance is less than this value and therefore the protective device will operate within the required disconnection time of 0.4 s.

Protective conductor size

The CPC forms an integral part of the total earth fault loop impedance, so it is necessary to check that the cross-section of this conductor is adequate. If the cross-section of the CPC complies with Table 54.7 of the IEE Regulations, there is no need to carry out further checks. Where line and protective conductors are made from the same material, Table 54.7 tells us that:

- for line conductors equal to or less than 16 mm^2, the protective conductor should equal the line conductor;

- for line conductors greater than 16 mm^2 but less than 35 mm^2, the protective conductor should have a cross-sectional area of 16 mm^2;

- for line conductors greater than 35 mm^2, the protective conductor should be half the size of the line conductor.

However, where the conductor cross-section does not comply with this table, then the formula given in Regulation 543.1.3 must be used:

$$S = \frac{\sqrt{I^2 t}}{k} \ (\text{mm}^2)$$

where

S = cross-sectional area in mm^2

I = value of maximum fault current in amperes

t = operating time of the protective device

k = a factor for the particular protective conductor (see Tables 54.2 to 54.4 of the IEE Regulations).

Example 1

A 230 V ring main circuit of socket outlets is wired in 2.5 mm single PVC copper cables in a plastic conduit with a separate 1.5 mm CPC. An earth fault loop impedance test identifies Z_s

as 1.15 Ω. Verify that the 1.5 mm CPC meets the requirements of Regulation 543.1.3 when the protective device is a 30 A semi-enclosed fuse.

$$I = \text{Maximum fault current} = \frac{V}{Z_s}(A)$$

$$\therefore \frac{230}{1.15} = 200\ A$$

t = Maximum operating time of the protective device for a circuit not exceeding 32 A is 0.4 s from Regulation 411.3.2.2. From Fig. 12.5(b) you can see that the time taken to clear a fault of 200 A is about 0.4 s.

k = 115 (from Table 54.3).

$$S = \frac{\sqrt{I^2 t}}{k}(mm^2)$$

$$S = \frac{\sqrt{(200\ A)^2 \times 0.4\ s}}{115} = 1.10\ mm^2$$

A 1.5 mm² CPC is acceptable since this is the nearest standard-size conductor above the minimum cross-sectional area of 1.10 mm² found by calculation.

Example 2

A TN supply feeds a domestic immersion heater wired in 2.5 mm² PVC insulated copper cable and incorporates a 1.5 mm² CPC. The circuit is correctly protected with a 15 A semi-enclosed fuse to BS 3036. Establish by calculation that the CPC is of an adequate size to meet the requirements of Regulation 543.1.3. The characteristics of the protective device are given in IEE Regulation Table 3.2A.

For final circuits less than 32 A the maximum operating time of the protective device is 0.4 s. From Table 3.2A it can be seen that a current of about 90 A will trip the 15 A fuse in 0.4 s. The small insert Table on the top right of Table 3.2 A of the IEE Regulations gives the value of the prospective fault current required to operate the device within the various disconnection times given.

So, in this case the table states that 90 A will trip a 15 A semi-enclosed fuse in 0.4 s.

$$\therefore I = 90\ A$$
$$t = 0.4\ s$$
$$k = 115\ \text{(from Table 54.3)}$$
$$S = \frac{\sqrt{I^2 t}}{k}(mm^2)\ \text{(from Regulation 543.1.3)}$$
$$S = \frac{\sqrt{(90\ A)^2 \times 0.4\ s}}{115} = 0.49\ mm^2$$

The CPC of the cable is greater than 0.49 mm² and is therefore suitable. If the protective conductor is a separate conductor, that is, it does not form part of a cable as in this example and is not enclosed in a wiring system as in Example 1, the cross-section of the protective conductor must be not less than 2.5 mm² where mechanical protection is provided or 4.0 mm² where mechanical protection is **not** provided in order to comply with Regulation 544.2.3.

Additional protection: RCDs

When it is required to provide the very best protection from electric shock and fire risk, earth fault protection devices are incorporated in the installation. The object of the Regulations concerning these devices, 411.3.2.1 to 411.3.2.6, is to remove an earth fault current very quickly, less than 0.4 s for all final circuits not exceeding 32 A and limit the voltage which might appear on any exposed metal parts under fault conditions to not more than 50 V. They will continue to provide adequate protection throughout the life of the installation even if the earthing conditions deteriorate. This is in direct contrast to the protection provided by overcurrent devices, which require a low resistance earth loop impedance path.

The Regulations recognize RCDs as 'additional protection' in the event of failure of the provision for basic protection, fault protection or carelessness by the users of the installation (IEE Regulation 415.1.1).

The basic circuit for a single-phase RCD is shown in Fig. 12.8. The load current is fed through two equal and opposing coils wound on to a common transformer core. The phase and neutral currents in a healthy circuit produce equal and opposing fluxes in the transformer core, which induces no voltage in the tripping coil. However, if more current flows in the line conductor than in the neutral conductor as a result of a fault between live and earth, an out-of-balance flux will result in an emf being induced in the trip coil which will open the double pole switch and isolate the load. Modern RCDs have tripping sensitivities between 10 and 30 mA, and therefore a faulty circuit can be isolated before the lower lethal limit to human beings (about 50 mA) is reached.

Consumer units can now be supplied which incorporate an RCD, split boards, where half of the final circuits are RCD protected so that any equipment supplied by the consumer unit to socket outlet circuits or out of doors circuits which are outside the zone created by the protective equipotential bonding, such as a garage or greenhouse, can have the special protection required by IEE Regulations 411.3.3 and 415.1.1.

RCBOs (residual current circuit breaker with overload protection) combines RCD protection and MCB protection into one unit.

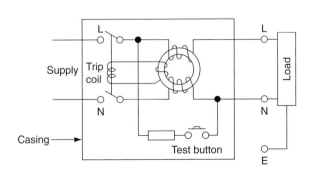

FIGURE 12.8
Construction of a RCD.

In a split board consumer unit, about half of the total number of final circuits are protected by the RCD. A fault on any one final circuit will trip out all of the RCD protected circuits which may cause inconvenience.

The RCBO gives the combined protection of an MCB plus RCD for each final circuit so protected and in the event of a fault occurring only the faulty circuit is interrupted.

Finally, it should perhaps be said that a foolproof method of giving protection to people or animals who simultaneously touch both live and neutral has yet to be devised. The ultimate safety of an installation depends upon the skill and experience of the electrical contractor and the good sense of the user.

Check your Understanding

When you have completed the questions, check out the answers at the back of the book.

Note: more than one multiple choice answer may be correct.

1. 'The conductive mass of the earth' is one definition of:
    a. earth
    b. earthing
    c. bonding conductor
    d. circuit protective conductor.

2. A protective conductor connecting exposed conductive parts of equipment to the main earthing terminal is one definition of:
    a. earth
    b. earthing
    c. bonding conductor
    d. circuit protective conductor.

3. A protective conductor connecting exposed and extraneous parts together is one definition of:
    a. earth
    b. earthing
    c. bonding conductor
    d. circuit protective conductor.

4. The act of connecting the exposed conductive parts of the installation to the main earthing terminal is called:
    a. earth
    b. earthing
    c. bonding conductor
    d. circuit protective conductor.

5. The act of linking together the exposed and extraneous metal parts is called:
    a. earthing
    b. bonding
    c. basic protection
    d. fault protection.

6. The protection provided by insulating live parts is called:

   a. extraneous conductive parts

   b. basic protection

   c. exposed conductive parts

   d. fault protection.

7. The metalwork of the electrical installation is called:

   a. extraneous conductive parts

   b. basic protection

   c. exposed conductive parts

   d. fault protection.

8. The metalwork of the building and other service pipes is called:

   a. extraneous conductive parts

   b. basic protection

   c. exposed conductive parts

   d. fault protection.

9. The protection provided by equipotential bonding and automatic disconnection of the supply is called:

   a. extraneous conductive parts

   b. basic protection

   c. exposed conductive parts

   d. fault protection.

10. Cutting off the electrical supply in order to ensure the safety of those working on the equipment is one definition of:

    a. basic protection

    b. fault protection

    c. equipotential bonding

    d. isolation switching.

11. A current which exceeds the rated current in an otherwise healthy circuit is one definition of:

    a. fault protection

    b. an overcurrent

    c. an overload current

    d. a short-circuit current.

12. The weakest link in the circuit designed to melt when an overcurrent flows is one definition of:

    a. fault protection

    b. a circuit protective conductor

    c. a fuse

    d. a consumer unit.

13. According to IEE Regulation 411.3.2.2 all final circuits not exceeding 32 A in a building supplied with a 230 V TN supply shall have a maximum disconnection time not exceeding:

    a. 0.2 s

    b. 0.4 s

    c. 5.0 s

    d. unlimited.

14. To ensure the effective operation of the overcurrent protective devices, the earth fault loop path must have:

    a. a 230 V supply

    b. a very low resistance

    c. fuses or MCBs in the live conductor

    d. a very high resistance.

15. Briefly explain why an electrical installation needs protective devices.

16. List the four factors on which the selection of a protective device depends.

17. List the five essential requirements for a device designed to protect against overcurrent.

18. Briefly describe the action of a fuse under fault conditions.

19. State the meaning of 'discrimination' as applied to circuit protective devices.

20. Use a sketch to show how 'discrimination' can be applied to a piece of equipment connected to a final circuit.

21. List typical 'exposed parts' of an installation.

22. List typical 'extraneous parts' of a building.

23. Use a sketch to show the path taken by an earth fault current.

24. Use bullet points and a simple sketch to briefly describe the operation of an RCD.

25. State the need for RCDs in an electrical installation.
    a. supplying socket outlets with a rated current not exceeding 20 A and
    b. for use by mobile equipment out of doors as required by IEE Regulation 411.3.3.

26. Briefly describe an application for RCBOs.

27. State the meaning of fault protection.

28. State the meaning of basic protection.

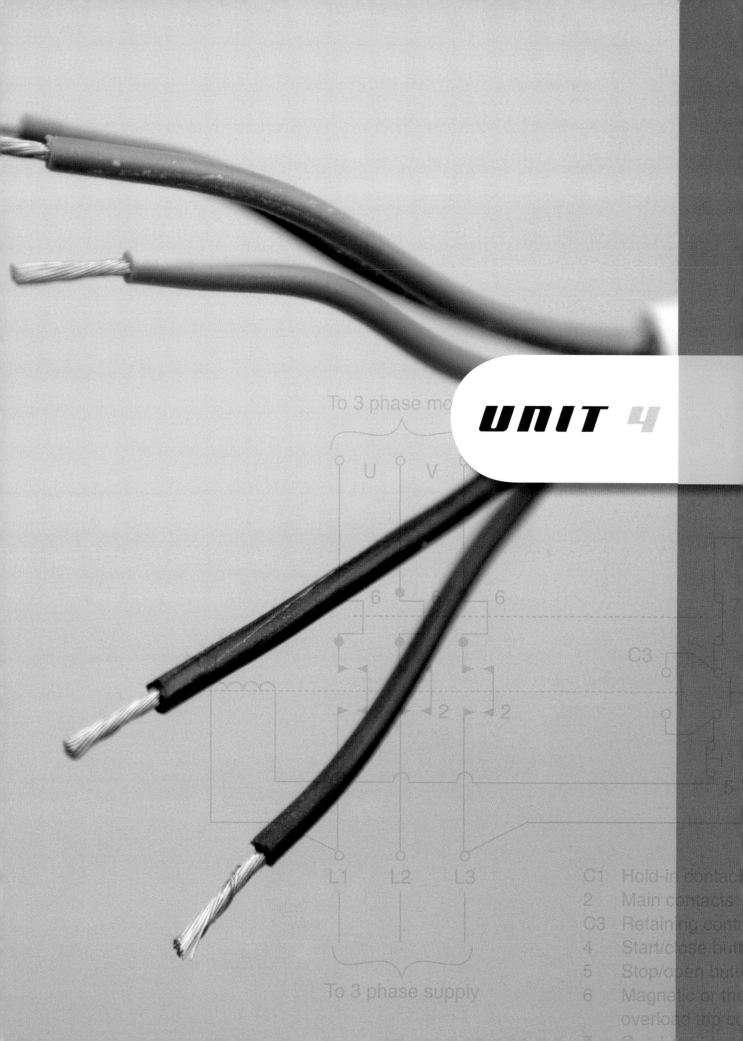

UNIT 4

To 3 phase motor

U V W

6 6 6

C3

2 2

L1 L2 L3

C1 Hold-in contact
2 Main contacts
C3 Retaining cont
4 Start/close butt
5 Stop/open butt
To 3 phase supply
6 Magnetic or the
overload trip

Statutory regulations and codes of practice

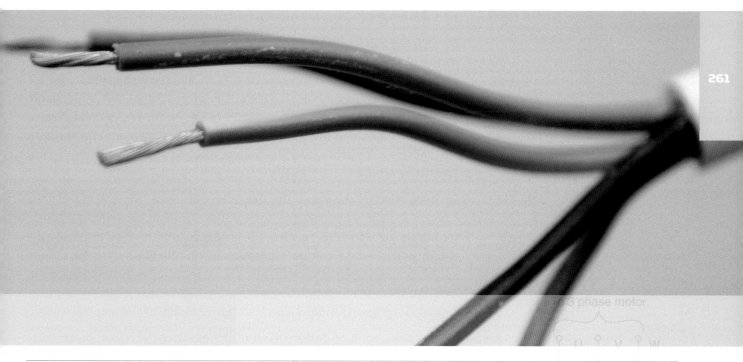

Unit 4 – Installation (buildings and structures) – Outcome 1

Underpinning knowledge: when you have completed this chapter you should be able to:

- list the statutory regulations and codes of practice which impact upon all aspects of electrotechnical systems

- define the meaning of 'duty holder'

- define 'absolute' and 'reasonably practicable' qualifying terms

- state the meaning of a BS EN number

- identify Index of Protection codes

- identify how the Regulations and Codes of Practice impact upon hazardous installations and lightning protection systems

Regulations and responsibilities

Electricity generation as we know it today began when Michael Faraday conducted the famous ring experiment in 1831. This experiment, together with many other experiments of the time, made it possible for Lord Kelvin and Sebastian de Ferranti to patent in 1882 the designs for an electrical machine called the Ferranti–Thompson dynamo, which enabled the generation of electricity on a commercial scale.

In 1887 the London Electric Supply Corporation was formed with Ferranti as chief engineer. This was one of many privately owned electricity generating stations supplying the electrical needs of the United Kingdom. As the demand for electricity grew, more privately owned generating stations were built until eventually the government realized that electricity was a national asset which would benefit from nationalization.

In 1926 the Electricity Supply Act placed the responsibility for generation in the hands of the Central Electricity Board. In England and Wales, the Central Electricity Generating Board (CEGB) had the responsibility for the generation and transmission of electricity on the Supergrid. In Scotland, generation was the joint responsibility of the North of Scotland Hydro-Electricity Board and the South of Scotland Electricity Board. In Northern Ireland electricity generation was the responsibility of the Northern Ireland Electricity Service.

In 1988 Cecil Parkinson, the Secretary of State for Energy in the Conservative government, proposed the denationalization of the electricity supply industry; this became law in March 1991, thereby returning the responsibility for generation, transmission and distribution to the private sector. It was anticipated that this action, together with new legislation over the security of supplies, would lead to a guaranteed quality of provision, with increased competition leading eventually to cheaper electricity.

During the period of development of the electricity services, particularly in the early days, poor design and installation led to many buildings being damaged by fire and the electrocution of human beings and livestock. It was the insurance companies which originally drew up a set of rules and guidelines of good practice in the interest of reducing the number of claims made upon them. The first rules were made by the American Board of Fire Underwriters and were quickly followed by the Phoenix Rules of 1882. In the same year the first edition of the Rules and Regulations for the Prevention of Fire Risk arising from Electrical Lighting was issued by the Institution of Electrical Engineers.

The current edition of these regulations is called the Requirements for Electrical Installations, IEE Wiring Regulations (BS 7671: 2008), and since June 2008 we have been using the 17th edition. All the rules have been revised, updated and amended at regular intervals to take account of modern developments, and the 17th edition brought the UK Regulations into harmony with those of the rest of Europe. The electrotechnical industry is now controlled by many rules, regulations and standards.

In Chapter 1 of this book we looked at a number of Statutory and Non-Statutory Regulations as they apply to the electrotechnical industry. In this section we will look at our responsibilities in our working environment.

The Health and Safety at Work Act 1974

The Health and Safety at Work Act provides a legal framework for stimulating and encouraging high standards of health and safety for everyone at work and the public at large from risks arising from work activities. The Act was a result of recommendations made by a Royal Commission in 1970 which looked at the health and safety of employees at work, and concluded that the main cause of accidents was apathy on the part of employer and employee. The new Act places the responsibility for safety at work on both the employer and the employee.

The employer has a duty to care for the health and safety of employees (Section 2 of the Act). To do this he must ensure that:

- the working conditions and standard of hygiene are appropriate;

- the plant, tools and equipment are properly maintained;

- the necessary safety equipment – such as personal protective equipment, dust and fume extractors and machine guards – is available and properly used;

- the workers are trained to use equipment and plant safely.

Employees have a duty to care for their own health and safety and that of others who may be affected by their actions (Section 7 of the Act). To do this they must:

- take reasonable care to avoid injury to themselves or others as a result of their work activity;

- co-operate with their employer, helping him or her to comply with the requirements of the Act;

- not interfere with or misuse anything provided to protect their health and safety.

Failure to comply with the Health and Safety at Work Act is a criminal offence and any infringement of the law can result in heavy fines, a prison sentence or both.

Enforcement

Laws and rules must be enforced if they are to be effective. The system of control under the Health and Safety at Work Act comes from the Health and Safety Executive (HSE) which is charged with enforcing the law. The HSE is divided into a number of specialist inspectorates or sections which operate from local offices throughout the United Kingdom. From the local offices the inspectors visit individual places of work.

263

The HSE inspectors have been given wide-ranging powers to assist them in the enforcement of the law. They can:

1. enter premises unannounced and carry out investigations, take measurements or photographs;

2. take statements from individuals;

3. check the records and documents required by legislation;

4. give information and advice to an employee or employer about safety in the workplace;

5. demand the dismantling or destruction of any equipment, material or substance likely to cause immediate serious injury;

6. issue an improvement notice which will require an employer to put right, within a specified period of time, a minor infringement of the legislation;

7. issue a prohibition notice which will require an employer to stop immediately any activity likely to result in serious injury, and which will be enforced until the situation is corrected;

8. prosecute all persons who fail to comply with their safety duties, including employers, employees, designers, manufacturers, suppliers and the self-employed.

Safety documentation

Under the Health and Safety at Work Act, the employer is responsible for ensuring that adequate instruction and information is given to employees to make them safety-conscious. Part 1, Section 3 of the Act instructs all employers to prepare a written health and safety policy statement and to bring this to the notice of all employees.

To promote adequate health and safety measures, the employer must consult with the employees' safety representatives. All actions of the safety representatives should be documented and recorded as evidence that the company takes seriously its health and safety policy.

The Act provides for criminal proceedings to be taken against those who do not satisfy the requirements of the Regulations.

Under the general protective umbrella of the Health and Safety at Work Act, other pieces of legislation also affect those working in the electrotechnical industry.

The Electricity at Work Regulations 1989 (EWR)

This legislation came into force in 1990 and replaced earlier regulations such as the Electricity (Factories Act) Special Regulations 1944. The Regulations are made under the Health and Safety at Work Act 1974, and enforced by the Health and Safety Executive. The purpose of the

Regulations is to 'require precautions to be taken against the risk of death or personal injury from electricity in work activities'.

Section 4 of the EWR tells us that 'all systems must be constructed so as to prevent danger ..., and be properly maintained.... Every work activity shall be carried out in a manner which does not give rise to danger.... In the case of work of an electrical nature, it is preferable that the conductors be made dead before work commences'.

The EWR do not tell us specifically how to carry out our work activities and ensure compliance, but if proceedings were brought against an individual for breaking the EWR, the only acceptable defence would be 'to prove that all reasonable steps were taken and all diligence exercised to avoid the offence' (Regulation 29).

An electrical contractor could reasonably be expected to have 'exercised all diligence' if the installation was wired according to the IEE Wiring Regulations (see below) and this is confirmed in the 17th edition of Regulations 114.

Duty of care

The Health and Safety at Work Act and the Electricity at Work Regulations make numerous references to employer and employees having a '**duty of care**' for the health and safety of others in the work environment. In this context the Electricity at Work Regulations refer to a person as a '**duty holder**'. This phrase recognizes the level of responsibility which electricians are expected to take on as a part of their job in order to control electrical safety in the work environment.

Everyone has a duty of care, but not everyone is a duty holder. The Regulations recognize the amount of control that an individual might exercise over the whole electrical installation. The person who exercises 'control over the whole systems, equipment and conductors' and is the Electrical Company's representative on-site, is the *duty holder*. He might be a supervisor or manager, but he will have a duty of care on behalf of his employer for the electrical, health, safety and environmental issues on that site.

Duties referred to in the Regulations may have the qualifying terms '**reasonably practicable**' or '**absolute**'. If the requirement of the regulation is absolute, then that regulation must be met regardless of cost or any other consideration. If the regulation is to be met 'so far as is reasonably practicable', then risks, cost, time, trouble and difficulty can be considered.

Often there is a cost effective way to reduce a particular risk and prevent an accident occurring. For example, placing a fire-guard in front of the fire at home when there are young children in the family is a reasonably practicable way of reducing the risk of a child being burned.

If a regulation is not qualified with 'so far as is reasonably practicable', then it must be assumed that the regulation is absolute. In the context of the

Electricity at Work Regulations, where the risk is very often death by electrocution, the level of duty to prevent danger more often approaches that of an absolute duty of care.

The IEE Wiring Regulations 17th edition (BS 7671: 2008)

The Institution of Electrical Engineers Requirements for Electrical Installations (the IEE Regulations) are non-statutory regulations. They relate principally to the design, selection, erection, inspection and testing of electrical installations, whether permanent or temporary, in and about buildings generally and to agricultural and horticultural premises, construction sites and caravans and their sites. Paragraph 7 of the introduction to the EWR says: 'the IEE Wiring Regulations is a code of practice which is widely recognized and accepted in the United Kingdom and compliance with them is likely to achieve compliance with all relevant aspects of the Electricity At Work Regulations'. BS 7671: 2008 says, the IEE Regulations are non-statutory. However, they may be used in a Court of Law to claim compliance with a statutory requirement (IEE Regulation 114). The IEE Wiring Regulations only apply to installations operating at a voltage up to 1000 V a.c.

That is electrical installations in:

- domestic dwellings
- commercial buildings
- industrial situations
- agricultural and horticultural situations
- caravans and caravan parks
- construction sites and other temporary situations.

They do not apply to electrical installations in mines and quarries where special regulations apply because of the adverse conditions experienced there.

The current edition of the IEE Wiring Regulations is the 17th edition which became law in 2008. The main reason for incorporating the IEE Wiring Regulations into British Standard BS 7671 was to create harmonization with European Standards.

Specific sections within the IEE Regulations have an impact on all aspects of electrical safety in buildings, for example:

- selection and erection of equipment – Part 5
- isolation and switching – Chapter 53
- inspection and testing – Part 6
- protection against fire and thermal effects – Chapter 42

- protection against electric shock – Chapter 41

- protection against overcurrent – Chapter 43

- special installations or locations – Part 7.

To assist electricians in their understanding of the Regulations a number of guidance notes have been published. The guidance notes, to which I will frequently make reference in this book, are those contained in the *On Site Guide*. Seven other guidance note booklets are also currently available. These are:

- *Selection and Erection*

- *Isolation and Switching*

- *Inspection and Testing*

- *Protection against Fire*

- *Protection against Electric Shock*

- *Protection against Overcurrent*

- *Special Locations*

- *Earthing and Bonding*

The IEE *On Site Guide* is prepared by the Institution of Electrical Engineers to simplify some aspects of the IEE Regulations BS 7671: 2008.

The Guide is intended to be used by skilled persons (electricians) carrying out limited applications of BS 7671 in:

(a) domestic installations generally, including off-peak supplies and supplies to associated garages, outbuildings and the like;

(b) industrial and commercial single- and three-phase installations where the local distribution fuse boards or consumer unit is located at or near the supplier's cut-out.

These guidance notes are intended to be read in conjunction with the Regulations.

The IEE Wiring Regulations are the electrician's bible and provide an authoritative framework for all the work activities which we undertake as electricians.

British and European Standards

Goods manufactured to the exacting specifications laid down by the British Standards Institution (BSI) are suitable for the purpose for which they were made. There seems to be a British Standard for practically everything made today, and compliance with the relevant British Standard is, in most cases, voluntary. However, when specifying or installing equipment, the electrical

BSI kite mark

BSI safety mark

FIGURE 13.1
BSI kite and safety marks.

FIGURE 13.2
European Commission safety mark.

designer or contractor needs to be sure that the materials are suitable for their purpose and offer a degree of safety, and should only use equipment which carries the appropriate British Standards number.

British Standard numbers begin as BS followed by a number and a date of issue, for example: BS 7671: 2008 is the latest IEE Wiring Regulations published in the year 2008.

Some British Standard numbers are written as BS EN followed by a number. This means that the British Standard is in harmony or agreement with the European Standards. These standards apply to all countries in the EEC, for example: BS EN 60898: 1991 is the British and European Standard for miniature and moulded case circuit breakers which were issued in 1991.

The BSI has created two important marks of safety, the BSI kite mark and the BSI safety mark, which are shown in Fig. 13.1.

The BSI kite mark is an assurance that the product carrying the label has been produced under a system of supervision, control and testing, and can only be used by manufacturers who have been granted a licence under the scheme. It does not necessarily cover safety unless the appropriate British Standard specifies a safety requirement.

The BSI safety mark is a guarantee of the product's electrical, mechanical and thermal safety. It does not guarantee the product's performance.

The CE mark (Fig. 13.2) is not a quality mark, but an indication given by the manufacturer or importer that the product or system meets the legal safety requirements of the European Commission and can therefore be presumed safe to use. The mark is applied by the manufacturer after carrying out the appropriate tests to ensure compliance with the relevant safety standards. The CE mark gives the manufacturer the right to sell the product in all the countries of the European Economic Area. All electrical products used by electrical contractors after 1 January 1997 must bear the CE mark.

Index of Protection (IP) BS EN 60529

IEE Regulation 612.4.5 tells us that where barriers and enclosures have been installed to prevent direct contact with live parts, they must afford a degree of protection not less than IP2X and IP4X, but what does this mean?

The Index of Protection is a code which gives us a means of specifying the suitability of equipment for the environmental conditions in which it will be used. The tests to be carried out for the various degrees of protection are given in the British and European Standard BS EN 60529.

The code is written as IP (Index of Protection) followed by two numbers XX. The first number gives the degree of protection against the penetration of solid objects into the enclosure. The second number gives the degree of protection against water penetration. For example, a piece of equipment classified as IP45 will have barriers installed which prevent a 1-mm diameter rigid steel bar from making contact with live parts and be protected against the ingress of water from jets of water applied from any direction.

Where a degree of protection is not specified, the number is replaced by an 'X' which simply means that the degree of protection is not specified although some protection may be afforded. The 'X' is used instead of '0' since '0' would indicate that no protection was given. The index of protection codes is shown in Fig. 13.3.

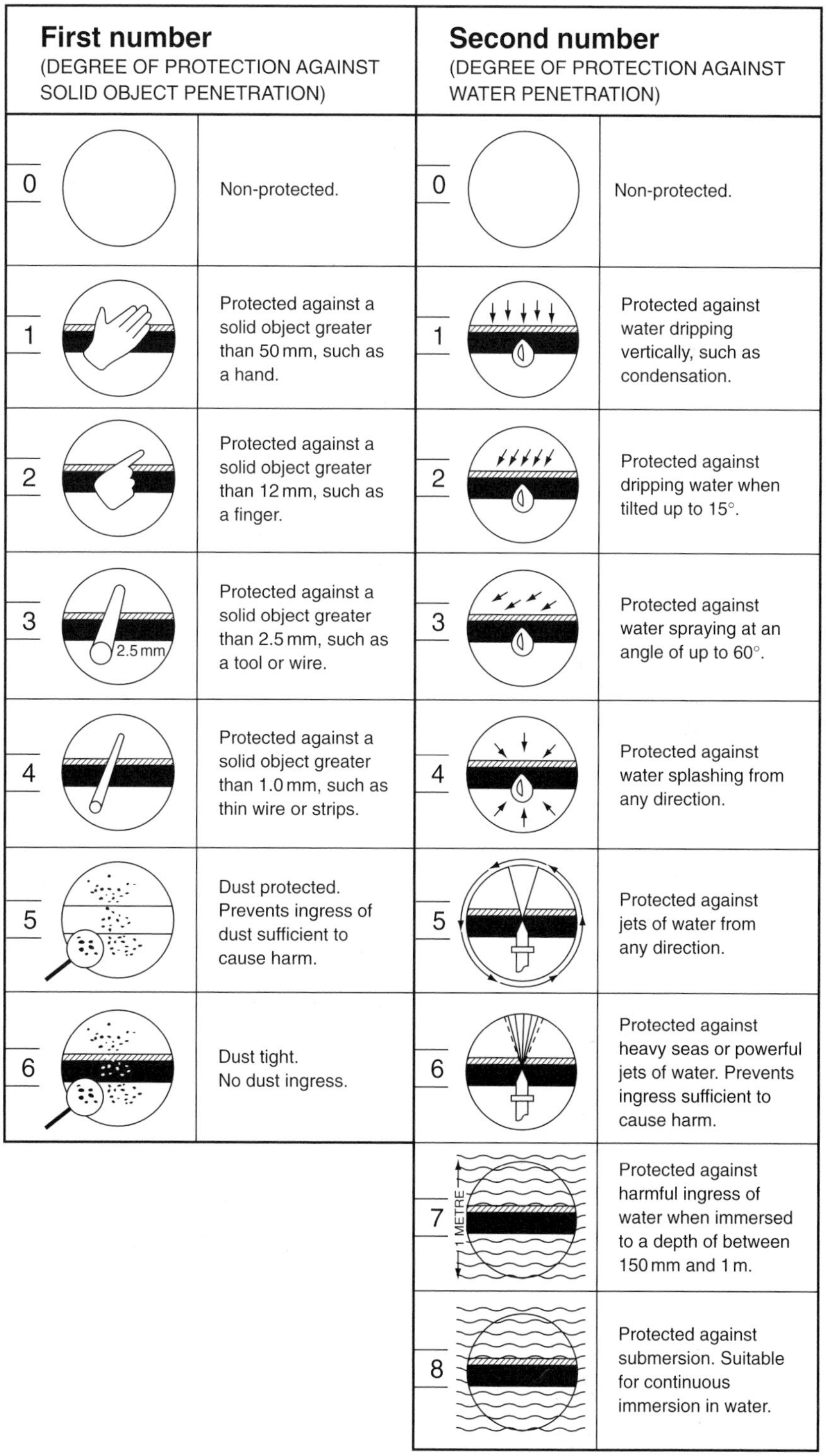

First number (DEGREE OF PROTECTION AGAINST SOLID OBJECT PENETRATION)		**Second number** (DEGREE OF PROTECTION AGAINST WATER PENETRATION)	
0	Non-protected.	0	Non-protected.
1	Protected against a solid object greater than 50 mm, such as a hand.	1	Protected against water dripping vertically, such as condensation.
2	Protected against a solid object greater than 12 mm, such as a finger.	2	Protected against dripping water when tilted up to 15°.
3	Protected against a solid object greater than 2.5 mm, such as a tool or wire.	3	Protected against water spraying at an angle of up to 60°.
4	Protected against a solid object greater than 1.0 mm, such as thin wire or strips.	4	Protected against water splashing from any direction.
5	Dust protected. Prevents ingress of dust sufficient to cause harm.	5	Protected against jets of water from any direction.
6	Dust tight. No dust ingress.	6	Protected against heavy seas or powerful jets of water. Prevents ingress sufficient to cause harm.
		7	Protected against harmful ingress of water when immersed to a depth of between 150 mm and 1 m.
		8	Protected against submersion. Suitable for continuous immersion in water.

FIGURE 13.3
Index of protection codes.

Hazardous area installations

The British Standards concerned with hazardous areas were first published in the 1920s and were concerned with the connection of electrical apparatus in the mining industry. Since those early days many national and international standards, as well as codes of practice, have been published to inform the manufacture, installation and maintenance of electrical equipment in all hazardous areas.

The relevant British Standards for Electrical Apparatus for Potentially Explosive Atmospheres are BS 5345, BS EN 60079 and BS EN 50014: 1998.

They define a **hazardous area** as 'any place in which an explosive atmosphere may occur in such quantity as to require special precautions to protect the safety of workers'. Clearly these regulations affect the petroleum industry, but they also apply to petrol filling stations.

Most flammable liquids only form an explosive mixture between certain concentration limits. Above and below this level of concentration the mix will not explode. The lowest temperature at which sufficient vapour is given off from a flammable substance to form an explosive gas–air mixture is called the **flashpoint**. A liquid which is safe at normal temperatures will require special consideration if heated to flashpoint. An area in which an explosive gas–air mixture is present is called a **hazardous area**, as defined by the British Standards, and any electrical apparatus or equipment within a hazardous area must be classified as flameproof.

Flameproof electrical equipment is constructed so that it can withstand an internal explosion of the gas for which it is certified, and prevent any spark or flame resulting from that explosion leaking out and igniting the surrounding atmosphere. This is achieved by manufacturing flameproof equipment to a robust standard of construction. All access and connection points have wide machined flanges which damp the flame in its passage across the flange. Flanged surfaces are firmly bolted together with many recessed bolts, as shown in Fig. 13.4. Wiring systems within

Definition

They define a *hazardous area* as 'any place in which an explosive atmosphere may occur in such quantity as to require special precautions to protect the safety of workers'.

Definitions

The lowest temperature at which sufficient vapour is given off from a flammable substance to form an explosive gas–air mixture is called the *flashpoint*.

An area in which an explosive gas–air mixture is present is called a *hazardous area*, as defined by the British Standards, and any electrical apparatus or equipment within a hazardous area must be classified as flameproof.

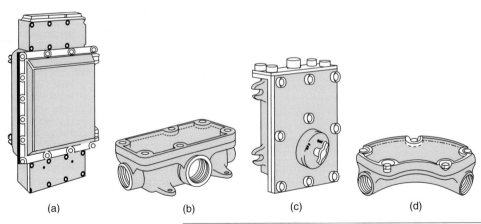

(a)   (b)   (c)   (d)

FIGURE 13.4

Flameproof fittings: (a) flameproof distribution board; (b) flameproof rectangular junction box; (c) double-pole switch; (d) flameproof inspection bend.

a hazardous area must be to flameproof fittings using an appropriate method, such as:

- PVC cables encased in solid drawn heavy-gauge screwed steel conduit terminated at approved enclosures having wide flanges and bolted covers.

- Mineral insulated cables terminated into accessories with approved flameproof glands. These have a longer gland thread than normal MICC glands of the type shown in Fig. 6.3 in Chapter 6 of this book. Where the cable is laid underground, it must be protected by a PVC sheath and laid at a depth of not less than 500 mm.

- PVC armoured cables terminated into accessories with approved flameproof glands or any other wiring system which is approved by the British Standard. All certified flameproof enclosures will be marked Ex, indicating that they are suitable for potentially explosive situations, or EEx, where equipment is certified to the harmonized European Standard. All the equipment used in a flameproof installation must carry the appropriate markings, as shown in Fig. 13.5, if the integrity of the wiring system is to be maintained.

Flammable and explosive installations are to be found in the petroleum and chemical industries, which are classified as group II industries. Mining is classified as group I and receives special consideration from the Mining Regulations because of the extreme hazards of working underground. Petrol filling pumps must be wired and controlled by flameproof equipment to meet the requirements of the Petroleum Regulation Act 1928 and 1936 and any local licensing laws concerning the keeping and dispensing of petroleum spirit.

Hazardous area classification

The British Standard divides the risk associated with inflammable gases and vapours into three classes or zones.

- Zone 0 is the most hazardous, and is defined as a zone or area in which an explosive gas–air mixture is *continuously present* or present for long periods. ('Long periods' is usually taken to mean that the gas–air mixture will be present for longer than 1000 h per year.)

FIGURE 13.5
Flameproof equipment markings.

- Zone 1 is an area in which an explosive gas–air mixture is *likely to occur* in normal operation. (This is usually taken to mean that the gas–air mixture will be present for up to 1000 h per year.)

- Zone 2 is an area in which an explosive gas–air mixture is *not likely* to occur in normal operation and if it does occur it will exist for a very short time. (This is usually taken to mean that the gas–air mixture will be present for less than 10 h per year.)

If an area is not classified as zone 0, 1 or 2, then it is deemed to be non-hazardous, so that normal industrial electrical equipment may be used.

The electrical equipment used in zone 2 will contain a minimum amount of protection. For example, normal sockets and switches cannot be installed in a zone 2 area, but oil-filled radiators may be installed if they are directly connected and controlled from outside the area. Electrical equipment in this area should be marked Ex'o' for oil-immersed or Ex'p' for powder-filled.

In Zone 1 all electrical equipment must be flameproof, as shown in Fig. 13.4, and marked Ex'd' to indicate a flameproof enclosure.

Ordinary electrical equipment cannot be installed in Zone 0, even when it is flameproof protected. However, many chemical and oil-processing plants are entirely dependent upon instrumentation and data transmission for their safe operation. Therefore, very low-power instrumentation and data-transmission circuits can be used in special circumstances, but the equipment must be *intrinsically safe*, and used in conjunction with a 'safety barrier' installed outside the hazardous area. Intrinsically safe equipment must be marked Ex'ia' or Ex's', specially certified for use in zone 0.

Intrinsic safety

By definition, an **intrinsically safe circuit** is one in which no spark or thermal effect is capable of causing ignition of a given explosive atmosphere. The intrinsic safety of the equipment in a hazardous area is assured by incorporating a Zener diode safety barrier into the control circuit such as that shown in Fig. 13.6. In normal operation, the voltage across a

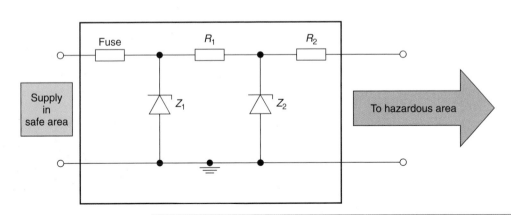

FIGURE 13.6
Zener safety barrier.

Zener diode is too low for it to conduct, but if a fault occurs, the voltage across Z_1 and Z_2 will rise, switching them on and blowing the protective fuse. Z_2 is included in the circuit as a 'back-up' in case the first Zener diode fails.

An intrinsically safe system, suitable for use in Zone 0, is one in which *all* the equipment, apparatus and interconnecting wires and circuits are intrinsically safe.

Lightning protection systems (BS EN 62305: 2008)

Many buildings of historical interest or tall buildings such as churches are fitted with a lightning protection system to protect the building and those who use it from the harmful effects of a lightning strike upon the building.

The average lightning discharge current is in the order of 20,000 A and it is this discharge of energy through a structure that causes the damage.

Lighting protective systems offer the lightning strike an alternative low resistance path to the general mass of earth. Such systems have three main parts

- an air terminal, a sharp copper spike;

- a system of down conductors made of robust copper strip;

- an earth termination, providing a solid and reliable connection with the general mass of earth so that the energy can be harmlessly dispersed into the ground.

The building receives a zone of protection from a lightning strike in the shape of a cone, starting at the highest point of the air terminal of the lightning protective system as shown in Fig. 13.7.

FIGURE 13.7

The zone of protection for a structure is given by a cone with its apex at the highest point of the air termination.

Lightning protective systems are discussed in detail in Advanced Electrical Installation Work under the sub-heading 'Protection of structures against lightning'.

The IEE Regulations make specific reference to many British Standard specifications and British Standard codes of practice in the 17th edition of the Regulations.

The Electricity Supply Regulations forbid electricity supply authorities to connect electric lines and apparatus to the supply system unless their insulation is capable of withstanding the tests prescribed by the appropriate British Standards.

It is clearly in the interests of the electrical contractor to be aware of, and to comply with, any regulations which are relevant to the particular installation. IEE Regulation 134.1.1 states that good workmanship by competent persons and the use of proper materials are essential for compliance with the Regulations.

In order to try to ensure that all electrical installation work is carried out to a minimum standard, the National Inspection Council for Electrical Installation Contracting (NICEIC) was established in 1956. NICEIC is supported by all sections of the electrical industry and its aims are to provide consumers with protection against faulty, unsafe or otherwise defective electrical installations. It maintains an approved roll of members who regularly have their premises, equipment and installations inspected by NICEIC engineers. Through this inspectorate the council is able to ensure a minimum standard of workmanship among its members. Electricians employed by an NICEIC-approved contractor are also, by association with their employer, accepted as being competent to carry out electrical installation work to an approved standard.

Safety First

Site Safety

- the Health and Safety at Work Act;
- makes you responsible for demonstrating;
- safe working practices;
- following site safety rules;
- wearing appropriate PPE.

Safe working environment

In Chapter 1, we looked at some of the laws and regulations that affect our working environment. We looked at Safety Signs and PPE and how to recognize and use different types of fire extinguishers. The structure of companies within the electrotechnical industry and the ways in which they communicate information by drawings, symbols and standard forms was also discussed in Chapter 9.

We began to look at safe electrical isolation procedures in Chapter 1 and then discussed this topic further in Chapter 8, together with safe manual handling techniques and safe procedures for working above ground.

In Chapter 8, under the heading 'Risks', 'Precautions and procedures', we looked at the common causes of accidents at work and how to control the risks associated with various hazards.

If your career in the electrotechnical industry is to be a long, happy and safe one, you must always behave responsibly and sensibly in order to maintain a safe working environment. Before starting work, make a safety assessment. What is going to be hazardous, will you require PPE, do you need any special access equipment. Carry out safe isolation procedures before beginning any work. You do not necessarily have to do these things formally, such as carrying out a risk assessment as described in Chapter 8, but just get into the habit of always working safely and being aware of the potential hazards around you when you are working. Finally, when the job is finished, clean up and dispose of all waste material responsibly as described in Chapter 8 under the heading 'Disposing of waste'.

Check your Understanding

When you have completed these questions, check out the answers at the back of the book.
Note: more than one multiple choice answer may be correct.

1. Identify the Statutory Regulations from the following list:
   a. Health and Safety at Work Act 1974
   b. the BSI Kite Safety Mark
   c. the Electricity at Work Regulations 1989
   d. the IEE Regulations BS 7671: 2008.

2. Identify the Non-Statutory Regulations from the following list which impact upon electrotechnical systems:
   a. Health and Safety at Work Act 1974
   b. Electricity at Work Regulations 1989
   c. the IEE Regulations BS 7671: 2008
   d. the BSI Kite Safety Mark.

3. Identify the types of electrical installations to which the IEE Regulations apply from the list below:
   a. domestic homes
   b. construction sites
   c. mines and quarries
   d. caravan parks.

4. 'Any place in which an explosive gas–air mixture is present' is one definition of:
   a. Index of protection
   b. Flash point
   c. Intrinsically safe
   d. Hazardous area.

5. 'The lowest temperature at which sufficient vapour is given off from a flammable substance to form an explosive gas' is one definition of:
   a. Index of protection
   b. Flash point
   c. Intrinsically safe
   d. Hazardous area.

6. A circuit in which no spark or thermal effect is capable of causing ignition of a given explosive atmosphere' us called:

    a. Index of protection

    b. Flash point

    c. Intrinsically safe

    d. Hazardous area.

7. The person who is the company representative responsible for electrical safety in the work environment is called the:

    a. Supervisor

    b. Duty of care

    c. Duty holder

    d. Competent person.

8. Everyone in the workplace has this responsibility for safety:

    a. First aid

    b. Slips, trips and falls

    c. Maintenance of tools and equipment

    d. Duty of care.

9. If an Electricity at Work Regulation must be met regardless of cost we say it is:

    a. Important

    b. Practically impossible

    c. Reasonably practicable

    d. Absolute.

10. If an Electricity at Work Regulation is to be met *but* time, trouble and difficulty can be taken into account we say it is:

    a. Important

    b. Practically impossible

    c. Reasonably practicable

    d. Absolute.

11. List at least two Statutory Regulations which have an impact upon all electrotechnical activities and state your reasons why.

12. List at least one code of practice which has an impact upon all electrotechnical activities and state the reason why.

13. List seven sections within the IEE Regulations which have an impact upon electrical safety in buildings.

14. List six types of building or situation to which the IEE Regulations alone will apply.

15. List two types of installation or situation to which the IEE Regulations alone will not apply because the installation or situation is considered to be too dangerous.

16. BS EN 60702: 2002 is the British Standard number for MICC cables and glands. What do these letters and numbers mean?

17. A piece of electrical equipment is classified as having IP 24 protection. What does this mean?

18. Describe the type of equipment and cable that would be suitable for wiring a petrol pump on a garage forecourt.

19. Briefly describe (a) how a lightning protective system gives protection to a building and its occupants and (b) state the three main parts of such a system.

20. Briefly describe how you would protect a tall thin construction, like a church spire, from a lightning strike.

Electrical installations and wiring systems

Unit 4 – Installation (Buildings and Structures) – Outcomes 2–4

Underpinning knowledge: when you have completed this chapter you should be able to:

- state the supply system earthing arrangements
- state the different types of electrical installation, components and functions
- state the different types of wiring enclosure and the factors that determine the choice of wiring system
- state the special arrangements required for:
  - locations containing a bath or shower
  - construction and demolition sites
  - agricultural and horticultural premises
  - caravans and camping parks
- state the devices for fixing equipment to surfaces
- carry out calculations to determine cable size
- state the requirements for testing electrical installations

Before beginning this final chapter on electrical installation wiring systems, the City and Guilds syllabus tells us to revise some of the work done in previous units because they are relevant to installing electrical equipment and systems in buildings. I will tell you where to find the appropriate revision material in this book.

- *Revision material 1*

Identify the sources of technical information.

This can be found in Chapter 3.

- *Revision material 2*

Using technical information.

This can be found in Chapter 9.

These are two very small chapters and you should refresh your memory of these topics. This will probably be a good homework project before you start this final chapter of the Level 2 Certificate in Electrotechnical Technology, looking at the different types of electrical installations and wiring systems.

Electricity supply systems

The British government agreed on 1 January 1995 that the electricity supplies in the United Kingdom would be harmonized with those of the rest of Europe. Thus the voltages used previously in low-voltage supply systems of 415 and 240V have become 400V for three-phase supplies and 230V for single-phase supplies. The Electricity Supply Regulations 1988 have also been amended to permit a range of variation from the new declared nominal voltage. From January 1995 the permitted tolerance is the nominal voltage +10% or −6%. Previously it was ±6%. This gives a voltage range of 216 to 253V for a nominal voltage of 230V and 376 to 440V for a nominal voltage of 400V (IEE Regulation Appendix 2 para 14).

It is further proposed that the tolerance levels will be adjusted to ±10% of the declared nominal voltage. All European Union countries will adjust their voltages to comply with a nominal voltage of 230V single phase and 400V three phase.

The supply to a domestic, commercial or small industrial consumer's installation is usually protected at the incoming service cable position with a 100A high breaking capacity (HBC) fuse. The maximum, that is, worst case value of external earth fault loop impedance outside of the consumer's domestic installation is:

0.8 Ω for cable sheath earth supplies (TN-S system)
0.35 Ω for protective multiple earthing (PME) supplies (TN-C-S system)
21.0 Ω excluding the consumer's earth electrode for no earth supplies (TT system).

The maximum, that is, worst case value of prospective short-circuit current is 16 kA at the supply terminals (Ref. On Site Guide and Part P Chapter 3).

Other items of equipment at this position are the energy meter and the consumer's distribution unit, providing the protection for the final circuits and the earthing arrangements for the installation.

An efficient and effective earthing system is essential to allow protective devices to operate. The limiting values of earth fault loop impedance are given in Tables 41.2 to 41.4 of the IEE Regulations, and Chapter 54 and Wiring Systems of Part 2 gives details of the earthing arrangements to be incorporated in the supply system to meet the requirements of the Regulations. Five systems are described in the definitions but only the TN-S, TN-C-S and TT systems are suitable for public supplies.

A system consists of an electrical installation connected to a supply. Systems are classified by a capital letter designation.

THE SUPPLY EARTHING

The supply earthing arrangements are indicated by the first letter, where T means one or more points of the supply are directly connected to earth and I means the supply is not earthed or one point is earthed through a fault-limiting impedance.

THE INSTALLATION EARTHING

The installation earthing arrangements are indicated by the second letter, where T means the exposed conductive parts are connected directly to earth and N means the exposed conductive parts are connected directly to the earthed point of the source of the electrical supply.

THE EARTHED SUPPLY CONDUCTOR

The earthed supply conductor arrangements are indicated by the third letter, where S means a separate neutral and protective conductor and C means that the neutral and protective conductors are combined in a single conductor.

CABLE SHEATH EARTH SUPPLY (TN-S SYSTEM)

This is one of the most common types of supply system to be found in the United Kingdom where the electricity companies' supply is provided by underground cables. The neutral and protective conductor are separate throughout the system. The protective earth conductor (PE) is the metal sheath and armour of the underground cable, and this is connected to the consumer's main earthing terminal. All extraneous conductive parts of the installation, gas pipes, water pipes and any lightning protective system are connected to the protective conductor via the main earthing terminal of the installation. The arrangement is shown in Fig. 14.1 and Fig. 2.3 of the IEE Regulations.

PROTECTIVE MULTIPLE EARTHING SUPPLIES (TN-C-S SYSTEM)

This type of underground supply is becoming increasingly popular to supply new installations in the United Kingdom. It is more commonly referred to as protective multiple earthing (PME). The supply cable uses a combined

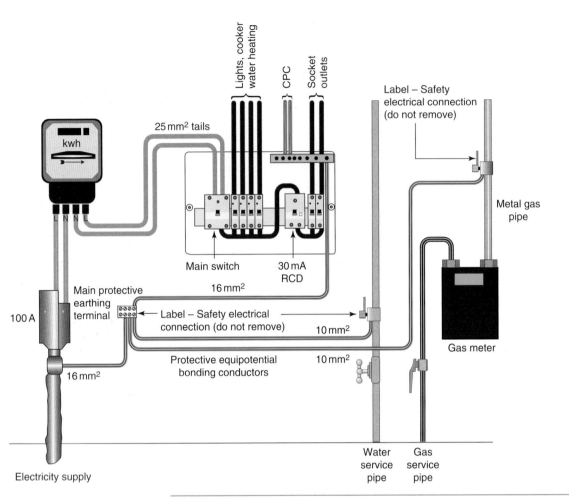

25 mm² tails

Lights, cooker water heating

CPC

Socket outlets

Label – Safety electrical connection (do not remove)

Metal gas pipe

kwh

Main switch

30 mA RCD

16 mm²

100 A

Main protective earthing terminal

Label – Safety electrical connection (do not remove)

10 mm²

Gas meter

Protective equipotential bonding conductors

10 mm²

16 mm²

Electricity supply

Water service pipe

Gas service pipe

FIGURE 14.1

Cable Sheath Earth Supplies (TN-S system): showing earthing and bonding arrangements.

protective earth and neutral conductor (PEN conductor). At the supply intake point a consumer's main earthing terminal is formed by connecting the earthing terminal to the neutral conductor. All extraneous conductive parts of the installation, gas pipes, water pipes and any lightning protective system are then connected to the main earthing terminals. Thus phase to earth faults are effectively converted into phase to neutral faults. The arrangement is shown in Fig. 14.2 and Fig. 2.4 of the IEE Regulations.

NO EARTH PROVIDED SUPPLIES (TT SYSTEM)

This is the type of supply more often found when the installation is fed from overhead cables. The supply authorities do not provide an earth terminal and the installation's circuit protective conductors (CPCs) must be connected to earth via an earth electrode provided by the consumer. An effective earth connection is sometimes difficult to obtain and in most cases a residual current device (RCD) is provided when this type of supply is used. The arrangement is shown in Fig. 14.3.

Figures 14.1, 14.2 and 14.3 show the layout of a typical domestic service position for these three supply systems. There are two other systems of supply, the TN-C and IT systems but they do not comply with the supply

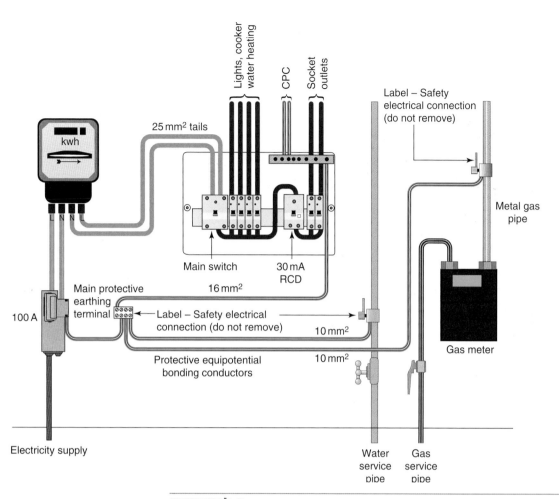

FIGURE 14.2

Protective Multiple Earthing Supply (TN-C-S system): showing earthing and bonding arrangements

regulations and therefore cannot be used for public supplies. Their use is restricted to private generating plants. For this reason I shall not include them here but they can be seen in Part 2 of the IEE Regulations.

Wiring and lighting circuits

Table 1A in Appendix 1 of the *On Site Guide* deals with the assumed current demand of points, and states that for lighting outlets we should assume a current equivalent to a minimum of 100W per lampholder. This means that for a domestic lighting circuit rated at 5A, a maximum of 11 lighting outlets could be connected to each circuit. In practice, it is usual to divide the fixed lighting outlets into a convenient number of circuits of seven or eight outlets each. In this way the whole installation is not plunged into darkness if one lighting circuit fuses and complies with Regulation 314.1 which tells us to 'divide into circuits to minimize inconvenience and avoid danger'.

Lighting circuits are usually wired in 1.0 or 1.5 mm cable using either a loop-in or joint-box method of installation. The loop-in method is universally employed with conduit installations or when access from above

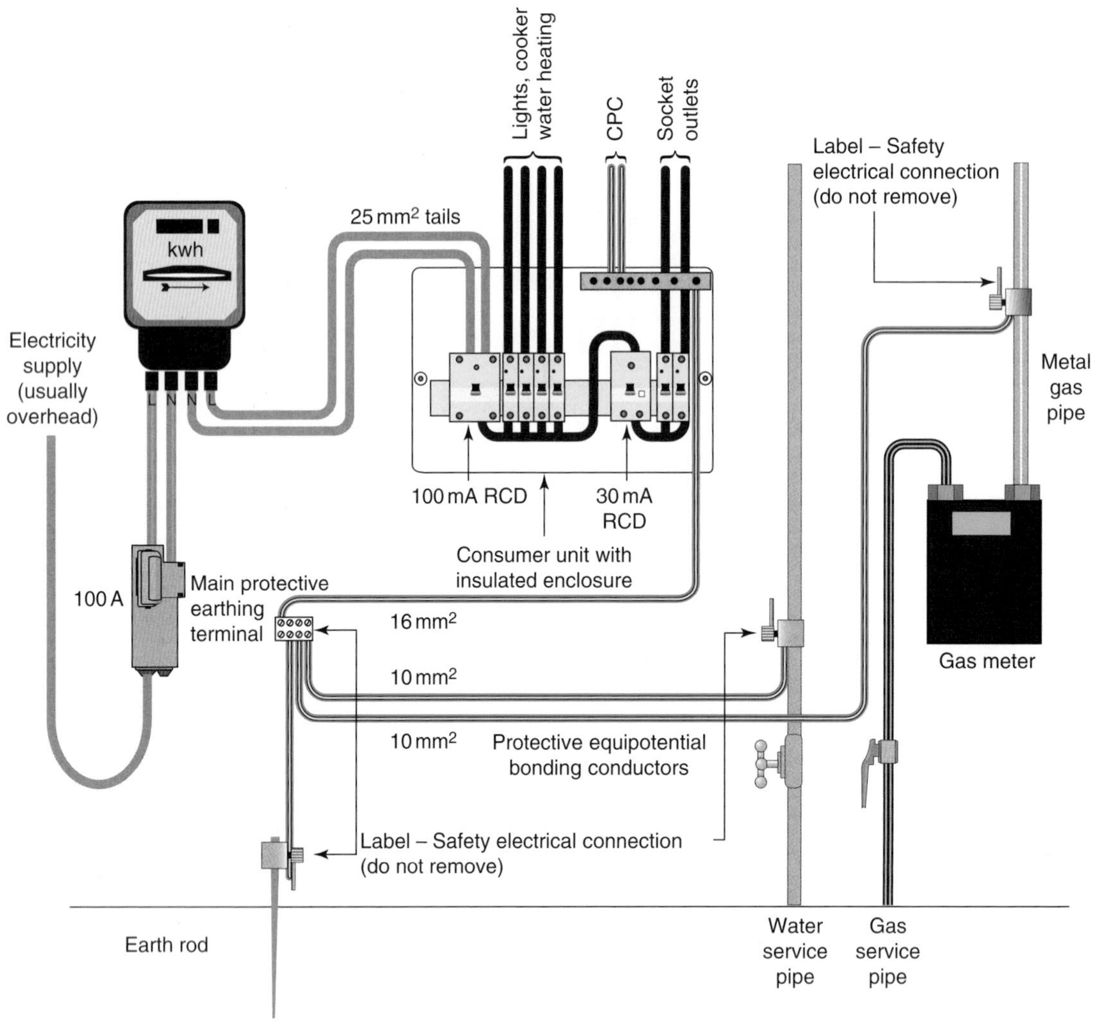

FIGURE 14.3

No Supply Earth Provided (TT systems): showing earthing and bonding arrangements

or below is prohibited after installation, as is the case with some industrial installations or blocks of flats. In this method the only joints are at the switches or lighting points, the live conductors being looped from switch to switch and the neutrals from one lighting point to another.

The use of junction boxes with fixed brass terminals is the method often adopted in domestic installations, since the joint boxes can be made accessible but are out of site in the loft area and under floorboards. However, every connection must remain accessible for inspection, testing and maintenance (Regulation 526.3).

The live conductors must be broken at the switch position in order to comply with the polarity regulations 612.7. A ceiling rose may only be connected to installations operating at 250V maximum and must only accommodate one flexible cord unless it is specially designed to take more than one (Regulations 559.6.1.2 and 559.6.1.3). Lampholders suspended from flexible cords must be capable of suspending the mass of the luminaire fixed to the lampholder (Regulation 559.6.1.5).

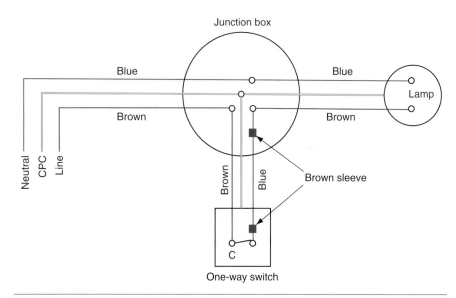

FIGURE 14.4

Wiring diagram of one-way switch control.

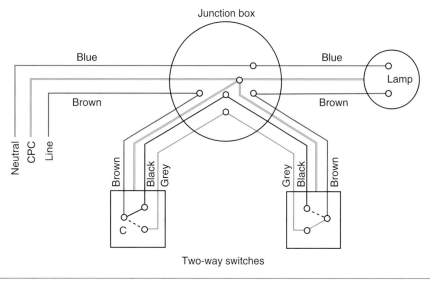

FIGURE 14.5

Wiring diagram of two-way switch control.

The method of fixing must be capable of carrying a mass of not less than 5 kg. Suspended ceilings are considered to be 'stable' or firmly fixed and may, therefore, support luminaires.

A luminaire, that is, a light fitting or small spot light must be fixed at an adequate distance from combustible material, or as recommended by the manufacturer, or be enclosed in non-flammable material (Regulations 422.3.1, 422.3.8, 422.4.4 and 559.5.1).

The type of circuit used will depend upon the installation conditions and the customer's requirements. One light controlled by one switch is called one-way switch control (see Fig. 14.4). A room with two access doors might benefit from a two-way switch control (see Fig. 14.5) so that the lights may be switched on or off at either position. A long staircase with more than two switches controlling the same lights would require intermediate switching (see Fig. 14.6).

One-way, two-way or intermediate switches can be obtained as plate switches for wall mounting or ceiling mounted cord switches. Cord switches can provide a convenient method of control in bedrooms or bathrooms and for independently controlling an office luminaire.

To convert an existing one-way switch control into a two-way switch control, a three-core and earth cable is run from the existing switch position to the proposed second switch position. The existing one-way switch is replaced by a two-way switch and connected as shown in Fig. 14.7.

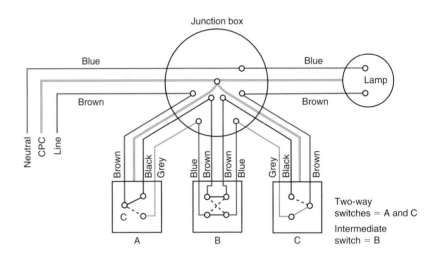

FIGURE 14.6
Wiring diagram of intermediate switch control.

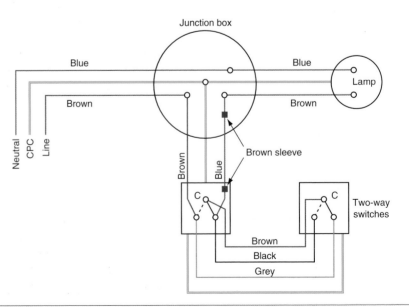

FIGURE 14.7
Wiring diagram of one-way converted to two-way switch control.

Building regulations for switches and sockets

Socket outlets must be mounted at a height above the floor or work surface so as to minimize the risk of mechanical damage (IEE Regulation 553.1.6).

Part M of the Building Regulations requires switches and socket outlets in dwellings to be installed so that all persons, including those whose reach is limited, can easily reach them. The recommendation is that they should be installed in habitable rooms at a height of between 450 and 1200 mm from the finished floor level. This is shown in Fig. 14.8. The guidance given applies to all new dwellings but not to rewires. However, these recommendations will undoubtedly 'influence' decisions taken when rewiring dwellings.

Building regulations for energy-efficient lighting

Part P of the Building Regulations relates to Electrical Safety in Dwellings. All new installations must comply with the Part P Regulations and any other relevant parts of the Building Regulations. Approved document L1, Conservation of Fuel and Power is relevant to us as electricians because it says that reasonable provision shall be made to provide lighting systems with energy-efficient lamps and sufficient controls so that electrical energy can be used efficiently. Part L describes two methods of compliance with these Regulations for both internal and external lighting. It says:

- A reasonable number of internal lighting points should be wired that will only take energy-efficient lamps such as fluorescent tubes and compact fluorescent lamps, CFLs.

- External lighting fixed to the building, including lighting in porches but not lighting in garages or carports, should provide reasonable provision for energy-efficient lamps such as fluorescent tubes and CFLs. These lamps should automatically extinguish in daylight and when not required at night, by being controlled by Passive infra-red (PIR) detectors.

The traditional light bulb, called a GLS (general lighting service) lamp is hopelessly bad in energy efficiency terms, producing only 14 lumens of light output for every electrical watt input. Fluorescent tubes and CFLs

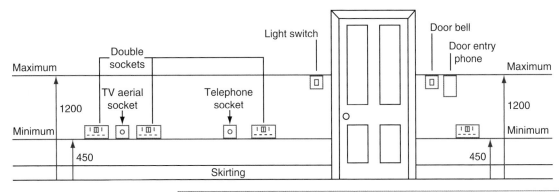

FIGURE 14.8

Fixing positions of switches and socket outlets.

produce more than 40 lumens of light output for every watt input. The Government calculates that if every British household was to replace three 60 or 100W light bulbs with CFLs, the energy saving would be greater than the power used by the entire street lighting network.

Mr. Hilary Benn, the Environment Secretary, has announced that the traditional GLS light bulbs of 150, 100, 60 and 40W will begin to be phased out by 2010. Thus, households will have to use more energy-efficient lamps in the future.

Let us now look at four different types of lamp.

GLS lamps

GLS lamps produce light as a result of the heating effect of an electrical current. Most of the electricity goes to producing heat and a little to producing light. A fine tungsten wire is first coiled and coiled again to form the incandescent filament of the GLS lamp. The coiled coil arrangement reduces filament cooling and increases the light output by allowing the filament to operate at a higher temperature. The light output covers the visible spectrum, giving a warm white to yellow light with a colour rendering quality classified as fairly good. The efficacy of the GLS lamp is 14 lumens per watt over its intended lifespan of 1000 h.

The filament lamp in its simplest form is a purely functional light source which is unchallenged on the domestic market despite the manufacture of more efficient lamps. One factor which may have contributed to its popularity is that lamp designers have been able to modify the glass envelope of the lamp to give a very pleasing decorative appearance, as shown by Fig. 14.9.

FIGURE 14.9
Some decorative GLS lamp shapes.

FIGURE 14.10
Tungsten Halogen Dichroic Reflector Lamp.

Tungsten Halogen Dichroic Reflector Miniature Spot Lamps

Tungsten Halogen Dichroic Reflector Miniature Spot Lamps such as the one shown in Fig. 14.10 are extremely popular in the lighting schemes of the new millennium. Their small size and bright white illumination makes them very popular in both commercial and domestic installations. They are available as a 12 V bi-pin package in 20, 35 and 50W and as a 230 V bayonet type cap (called a GU10 or GZ10 cap) in 20, 35 and 50 W. At 20 lumens of light output over its intended lifespan of 2000h they are more energy efficient than GLS lamps. However, only lamps offering more than 40 lumens of light output are considered energy efficient by the Government's criteria.

Discharge lamps

Discharge lamps do not produce light by means of an incandescent filament but by the excitation of a gas or metallic vapour contained within a glass envelope. A voltage applied to two terminals or electrodes sealed into the end of a glass tube containing a gas or metallic vapour will excite the contents and produce light directly. Fluorescent tubes and CFLs operate on this principle.

FLUORESCENT TUBE

A **fluorescent lamp** is a linear arc tube, internally coated with a fluorescent powder, containing a low-pressure mercury vapour discharge. The lamp construction is shown in Fig. 10.23 and the control circuit in Fig 10.24.

COMPACT FLUORESCENT LAMPS

CFLs are miniature fluorescent lamps designed to replace ordinary GLS lamps. They are available in a variety of shapes and sizes so that they can be fitted into existing light fittings. Figure 14.11 shows three typical shapes. The 'stick' type give most of their light output radially while the flat 'double D' type give most of their light output above and below.

Energy-efficient lamps use electricity much more efficiently than an equivalent GLS lamp. For example, a 20W energy-efficient lamp will give the same light output as a 100 W GLS lamp. An 11W energy-efficient lamp is

> **Definition**
>
> A *fluorescent lamp* is a linear arc tube, internally coated with a fluorescent powder, containing a low-pressure mercury vapour discharge.

> **Definition**
>
> *CFLs* are miniature fluorescent lamps designed to replace ordinary GLS lamps.

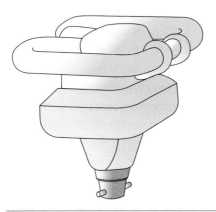

FIGURE 14.11
Energy-efficient lamps.

equivalent to a 60W GLS lamp. Energy-efficient lamps also have a lifespan of about eight times longer than a GLS lamp and so, they do use energy very efficiently.

However, energy-efficient lamps are expensive to purchase and they do take a few minutes to attain full brilliance after switching on. They cannot always be controlled by a dimmer switch and are unsuitable for incorporating in an automatic presence detector because they are usually not switched on long enough to be worthwhile, but energy-efficient lamps are excellent for outside security lighting which is left on for several hours each night.

The electrical contractor, in discussion with a customer, must balance the advantages and disadvantages of energy-efficient lamps compared to other sources of illumination for each individual installation.

Wiring and socket outlet circuits

Where portable equipment is to be used it should be connected by a plug top to a conveniently accessible socket outlet, taking into account the length of the flexible cord normally fitted to portable equipment (IEE Regulation 553.1.7). The length of flexible cord is usually no longer than 2 m and so, pressing the plug top into a socket outlet connects the appliance to the source of supply. **Socket outlets** therefore provide an easy and convenient method of connecting portable electrical appliances to a source of supply.

Socket outlets can be obtained in 15, 13, 5 and 2 A ratings (see IEE Table 55.1), but the 13 A flat pin shuttered type complying with BS 1363 is the most popular for domestic installations in the United Kingdom. Each 13 A plug top contains a cartridge fuse to give maximum potential protection to the flexible cord and the appliances which it serves.

Socket outlets may be wired on a ring or radial circuit and in order that every appliance can be fed from an adjacent and convenient socket outlet, the number of sockets is unlimited provided that the floor area covered by the circuit does not exceed that given in Appendix 15 of the *IEE Regulations*.

Radial circuits

In a radial circuit each socket outlet is fed from the previous one. Live is connected to live, neutral to neutral and earth to earth at each socket outlet. The fuse and cable sizes are given in Appendix 15 of the *IEE Regulations* but circuits may also be expressed with a block diagram, as shown in Fig. 14.12. The number of permitted socket outlets is unlimited but each radial circuit must not exceed the floor area stated and the known or estimated load.

Where two or more circuits are installed in the same premises, the socket outlets and permanently connected equipment should be reasonably shared out among the circuits, so that the total load is balanced.

When designing ring or radial circuits special consideration should be given to the loading in kitchens which may require separate circuits. This is because the maximum demand of current-using equipment in kitchens may exceed the rating of the circuit cable and protection devices.

Ring and radial circuits may be used for domestic or other premises where the maximum demand of the current-using equipment is estimated not to exceed the rating of the protective devices for the chosen circuit.

Ring circuits

Ring circuits are very similar to radial circuits in that each socket outlet is fed from the previous one, but in ring circuits the last socket is wired back to the source of supply. Each ring final circuit conductor must be looped into every socket outlet or joint box which forms the ring and must be electrically continuous throughout its length. The number of permitted socket outlets is unlimited but each ring circuit must not cover more than 100 m² of floor area.

The circuit details are given in Appendix 15 of the *IEE Regulations* but may also be expressed by the block diagram given in Fig. 14.13.

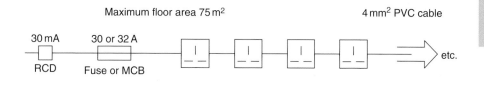

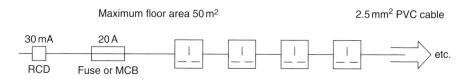

FIGURE 14.12
Block diagram of radial circuits.

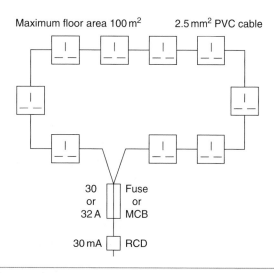

FIGURE 14.13
Block diagram of ring circuits.

SPURS TO RING CIRCUITS

A spur is defined in Part 2 of the Regulations as a branch cable from a ring final circuit.

NON-FUSED SPURS

The total number of non-fused spurs must not exceed the total number of socket outlets and pieces of stationary equipment connected directly in the circuit. The cable used for non-fused spurs must not be less than that of the ring circuit. The requirements concerning spurs are given in Appendix 15 of the *IEE Regulations* but the various circuit arrangements may be expressed by the block diagrams of Fig. 14.14.

A non-fused spur may only feed one single or one twin socket outlet.

Non-fused spurs may be connected into the ring circuit at the terminals of socket outlets or at joint boxes or at the origin of the circuit.

FUSED SPURS

The total number of fused spurs is unlimited. A fused spur is connected to the circuit through a fused connection unit, the rating of which should be suitable for the conductor forming the spur but should not exceed 13 A. The requirements for fused spurs are also given in Appendix 15 but the various circuit arrangements may be expressed by the block diagrams of Fig. 14.15.

The general arrangement shown in Fig. 14.16 shows 11 socket outlets connected to the ring, three non-fused spur connections and two fused spur connections.

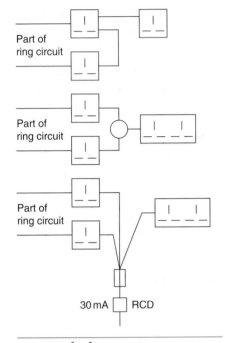

FIGURE 14.14

Connection of non-fused spurs.

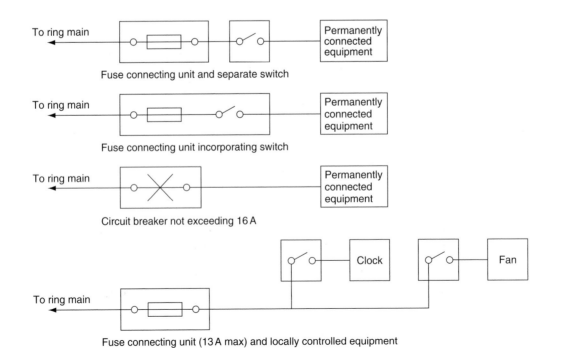

FIGURE 14.15

Connection of fused spurs.

Socket outlet numbers

The Regulations allow us to install an unlimited number of socket outlets, the restriction being that each circuit must not exceed a given floor area as shown in Figs. 14.12 and 14.13.

These days most households have lots of domestic appliances and electronic equipment, so how many sockets should be installed? Ultimately this is a matter for the customer and the electrical designer but most consumer organizations, the house builders NHBC and the Royal Society for the Prevention of Accidents (ROSPA) make the following general recommendations:

- The hard wiring for a single socket outlet is the same as the hard wiring for a double socket outlet. So, always install a double switched socket outlet unless there is a reason not to.

- Kitchens will require between six and ten double sockets, fitted both above and below the work surface for specific appliances.

- Utility room – two double sockets.

- Sitting rooms will require between six and ten double sockets with one double socket situated next to any telephone outlet to power telecommunication equipment and two double sockets adjacent to the TV aerial outlet for TV, video and DVD supplies.

- Double bedrooms – four to six double sockets.

- Single bedrooms – four to six double sockets.

- Hallways – two double sockets with one situated next to any telephone outlet.

- Home Office – six double sockets.

- Garage – two double sockets.

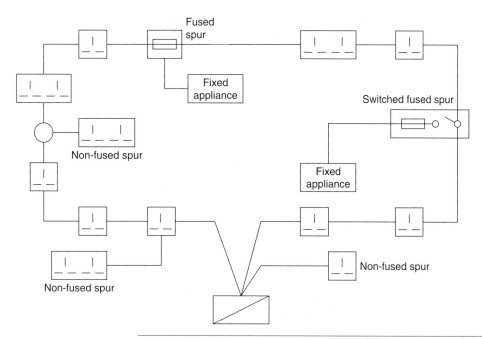

FIGURE 14.16

Typical ring circuit with spurs.

Additional protection for socket outlets

Additional protection by 30 mA RCD is required in addition to overcurrent protection for all socket outlet circuits to be used by ordinary persons and intended for general use.

This additional protection is provided in case basic protection or fault protection fails or if the user of the installation is careless (IEE Regulation 415.1.1).

An **ordinary person** is one who is neither an electrically skilled or instructed person.

Water heating circuits

A small, single-point over-sink type water heater may be considered as a permanently connected appliance and so may be connected to a ring circuit through a fused connection unit. A water heater of the immersion type is usually rated at a maximum of 3 kW, and could be considered as a permanently connected appliance, fed from a fused connection unit. However, many immersion heating systems are connected into storage vessels of about 150 litres in domestic installations, and the *On Site Guide* states that immersion heaters fitted to vessels in excess of 15 litres should be supplied by their own circuit.

Therefore, immersion heaters must be wired on a separate radial circuit when they are connected to water vessels which hold more than 15 litres . Figure. 14.17 shows the wiring arrangements for an immersion heater. Every switch must be a double-pole (DP) switch and out of reach of anyone using a fixed bath or shower when the immersion heater is fitted to a vessel in a bathroom.

Supplementary equipotential bonding to pipework *will only be required* as an addition to fault protection (IEE Regulation 415.2) if the immersion heater vessel is in a bathroom that *does not have*:

(i) all circuits protected by a 30 mA RCD **and**

(ii) protective equipotential bonding (IEE Regulation 701.415.2) as shown in Fig. 6.7.

Definition

An *ordinary person* is one who is neither an electrically skilled or instructed person.

294

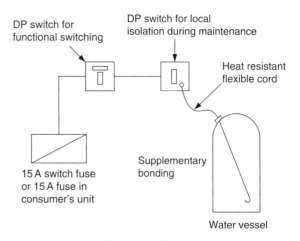

FIGURE 14.17
Immersion heater wiring.

See the section on supplementary equipotential bonding later in this chapter and Fig. 14.64.

Electric space heating circuits

Electrical heating systems can be broadly divided into two categories: unrestricted local heating and off-peak heating.

Unrestricted local heating may be provided by portable electric radiators which plug into the socket outlets of the installation. Fixed heaters that are wall mounted or inset must be connected through a fused connection and incorporate a local switch, either on the heater itself or as a part of the fuse connecting unit, as shown in Fig. 14.15. Heating appliances where the heating element can be touched must have a DP switch which disconnects all conductors. This requirement includes radiators which have an element inside a silica-glass sheath.

Off-peak heating systems may provide central heating from storage radiators, ducted warm air or underfloor heating elements. All three systems use the thermal storage principle, whereby a large mass of heat-retaining material is heated during the off-peak period and allowed to emit the stored heat throughout the day. The final circuits of all off-peak heating installations must be fed from a separate supply controlled by an electricity board time clock.

When calculating the size of cable required to supply a single-storage radiator, it is good practice to assume a current demand equal to 3.4 kW at each point. This will allow the radiator to be changed at a future time with the minimum disturbance to the installation. Each radiator must have a 20 A DP means of isolation adjacent to the heater and the final connection should be via a flex outlet. See Fig. 14.18 for wiring arrangements.

Ducted warm air systems have a centrally sited thermal storage heater with a high storage capacity. The unit is charged during the off-peak period, and a fan drives the stored heat in the form of warm air through large air ducts to outlet grilles in the various rooms. The wiring arrangements for this type of heating are shown in Fig. 14.19.

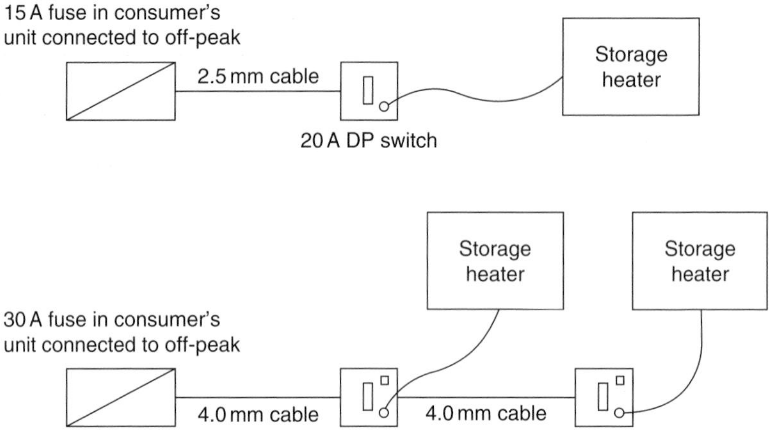

15 A fuse in consumer's
unit connected to off-peak

2.5 mm cable

20 A DP switch

Storage heater

30 A fuse in consumer's
unit connected to off-peak

4.0 mm cable

4.0 mm cable

Storage heater

Storage heater

FIGURE 14.18

Possible wiring arrangements for storage heaters.

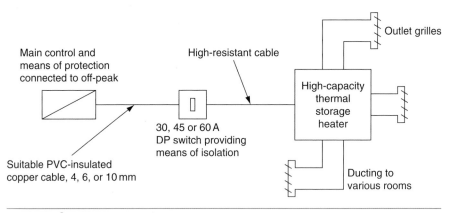

Main control and
means of protection
connected to off-peak

High-resistant cable

Outlet grilles

High-capacity thermal storage heater

30, 45 or 60 A
DP switch providing
means of isolation

Suitable PVC-insulated
copper cable, 4, 6, or 10 mm

Ducting to
various rooms

FIGURE 14.19

Ducted warm air heating system.

The single-storage heater is heated by an electric element embedded in bricks and rated between 6 and 15 kW depending upon its thermal capacity. A radiator of this capacity must be supplied on its own circuit, in cable capable of carrying the maximum current demand and protected by a fuse or miniature circuit breaker (MCB) of 30, 45 or 60 A as appropriate. At the heater position, a DP switch must be installed to terminate the fixed heater wiring. The flexible cables used for the final connection to the heaters must be of the heat-resistant type.

Floor warming installations use the thermal storage properties of concrete. Special cables are embedded in the concrete floor screed during construction. When current is passed through the cables they become heated, the concrete absorbs this heat and radiates it into the room. The wiring arrangements are shown in Fig. 14.20. Once heated, the concrete will give off heat for a long time after the supply is switched off and is, therefore, suitable for connection to an off-peak supply.

Underfloor heating cables installed in bathrooms or shower rooms must incorporate an earthed metallic sheath or be covered by an earthed

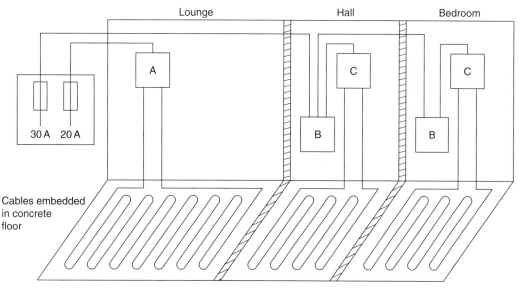

A = Thermostat incorporating DP switch fed by 2.5 mm PVC/copper
B = DP switch fuse fed by 4.0 mm PVC/copper
C = Thermostat fed by 2.5 mm PVC/copper

FIGURE 14.20
Floor warming installations.

metallic grid connected to the protective conductor of the supply circuit (Regulation 701.753).

Cooker circuit

A cooker with a rating above 3 kW must be supplied on its own circuit but since it is unlikely that in normal use every heating element will be switched on at the same time, a diversity factor may be applied in calculating the cable size, as detailed in the *On Site Guide.*

Consider, as an example, a cooker with the following elements fed from a cooker control unit incorporating a 13 A socket:

$$4 \times 2 \text{ kW fast boiling rings} = 8000 \text{ W}$$
$$1 \times 2 \text{ kW grill} = 2000 \text{ W}$$
$$1 \times 2 \text{ kW oven} = 2000 \text{ W}$$
$$\text{Total loading} = 12\,000 \text{ W}$$

When connected to 230 V

$$\text{Current rating} = \frac{12\,000}{230} = 52.17 \text{ A}$$

Applying the diversity factor of Table 1A

$$\text{Total current rating} = 52.17 \text{ A}$$
$$\text{First 10 amperes} = 10 \text{ A}$$
$$30\% \text{ of } 42.17 \text{A} = 12.65 \text{ A}$$
$$\text{Socket outlet} = 5 \text{ A}$$
$$\text{Assessed current demand} = 10 + 12.65 + 5 = 27.65 \text{ A}$$

Therefore, a cable capable of carrying 27.65 A may be used safely rather than a 52.17 A cable.

A cooking appliance must be controlled by a switch separate from the cooker but in a readily accessible position. Where two cooking appliances are installed in one room, such as split-level cookers, one switch may be used to control both appliances provided that neither appliance is more than 2 m from the switch (*On Site Guide*, Appendix 8).

Conductor size calculations

The size of a cable to be used for an installation depends upon:

- the current rating of the cable under defined installation conditions and

- the maximum permitted drop in voltage as defined by Regulation 525.

The factors which influence the current rating are:

1. *Design current*: cable must carry the full load current.

2. *Type of cable*: PVC, MICC, copper conductors or aluminium conductors.

3. *Installed conditions*: clipped to a surface or installed with other cables in a trunking.

4. *Surrounding temperature*: cable resistance increases as temperature increases and insulation may melt if the temperature is too high.

5. *Type of protection*: for how long will the cable have to carry a fault current?

Regulation 525 states that the drop in voltage from the supply terminals to the fixed current-using equipment must not exceed 3% for lighting circuits and 5% for other uses of the mains voltage. That is a maximum of 6.9V for lighting and 11.5V for other uses on a 230V installation. The volt drop for a particular cable may be found from:

$$VD = \text{Factor} \times \text{Design current} \times \text{Length of run}$$

The factor is given in the tables of Appendix 4 of the IEE Regulations and Appendix 6 of the *On Site Guide*. They are also given in Table 14.3.

The cable rating, denoted I_t, may be determined as follows:

$$I_t = \frac{\text{Current rating of protective device}}{\text{Any applicable correction factors}}$$

Table 14.1 Ambient Temperature Correction Factors. Adapted from the IEE *On Site Guide* by Kind Permission of the Institution of Electrical Engineers

Table 4B1 of IEE Regulations
Correction factors for ambient temperature where protection is against short-circuit and overload

Type of insulation	Operating temperature	Ambient temperature (°C)								
		25	30	35	40	45	50	55	60	65
Thermoplastic (general purpose PVC)	70°C	1.03	1.0	0.94	0.87	0.79	0.71	0.61	0.50	NA

The cable rating must be chosen to comply with Regulation 433.1. The correction factors which may need applying are given below as:

Ca the ambient or surrounding temperature correction factor, which is given in Tables 4B1 and 4B2 of Appendix 4 of the IEE Regulations. They are also shown in Table 14.1.

Cg the grouping correction factor given in Tables 4C1 to 4C5 of the IEE Regulations and Table 6C of the *On Site Guide*.

Cc the 0.725 correction factor to be applied when semi-enclosed fuses protect the circuit as described in item 5.1.1 of the preface to Appendix 4 of the IEE Regulations.

Ci the correction factor to be used when cables are enclosed in thermal insulation. Regulation 523.6.6 gives us three possible correction values:

- Where one side of the cable is in contact with thermal insulation we must read the current rating from the column in the table which relates to reference method A (see Table 14.2).

- Where the cable is *totally* surrounded over a length greater than 0.5 m we must apply a factor of 0.5.

- Where the cable is *totally* surrounded over a short length, the appropriate factor given in Table 52.2 of the IEE Regulations or Table 6B of the *On Site Guide* should be applied.

Note: A cable should preferably *not* be installed in thermal insulations.

Having calculated the cable rating, I_t the smallest cable should be chosen from the appropriate table which will carry that current. This cable must also meet the voltage drop Regulation 525 and this should be calculated as described earlier. When the calculated value is less than 3% for lighting and 5% for other uses of the mains voltage the cable may be considered suitable. If the calculated value is greater than this value, the next larger cable size must be tested until **a cable is found which meets both the current rating and voltage drop criteria.**

Table 14.2 Current Carrying Capacity of Cables. Adapted from the IEE *On Site Guide* by Kind Permission of the Institution of Electrical Engineers

Multicore cables having thermoplastic (PVC) or thermosetting insulation, non-armoured, (COPPER CONDUCTORS)
Table 4D1A of IEE Regulations
Ambient temperature: 30°C. Conductor operating temperature: 70°C
Current-carrying capacity (Amperes): BS 6004, BS 7629

Conductor cross-sectional area	Reference Method A (enclosed in conduit in an insulated wall, etc.)		Reference Method B (enclosed in conduit on a wall or ceiling, or in trunking)		Reference Method C (clipped direct)		Reference Method F (on a perforated cable tray) or (free air)	
	Two cables single-phase a.c. or d.c.	Three or four cables three-phase a.c.	One two-core cable, single-phase a.c. or d.c.	One three-core cable or one four-core cable, three-phase a.c.	One two-core cable, single-phase a.c. or d.c.	One three-core cable or one four-core cable, three-phase a.c.	One two-core cable, single-phase a.c. or d.c.	One three-core cable or one four-core cable, three-phase a.c.
1	2	3	4	5	6	7	8	9
mm²	A	A	A	A	A	A	A	A
1	11	10	13	11.5	15.5	13.5	17	14.5
1.5	14.5	13	16.5	15	20	17.5	22	18.5
2.5	20	17.5	23	20	27	24	30	25
4	26	23	30	27	37	32	40	34
6	34	29	38	34	47	41	51	43
10	46	39	52	46	65	57	70	60
16	61	52	69	62	87	76	94	80
25	80	68	90	80	114	96	119	101
35	99	83	111	99	141	119	148	126
50	119	99	133	118	182	144	180	153
70	151	125	168	149	234	184	232	196
95	182	150	201	179	284	223	282	238

Example

A house extension has a total load of 6 kW installed some 18 m away from the mains consumer unit for lighting. A PVC insulated and sheathed twin and earth cable will provide a sub-main to this load and be clipped to the side of the ceiling joists over much of its length in a roof space which is anticipated to reach 35°C in the summer and where insulation is installed up to the top of the joists. Calculate the minimum cable size if the circuit is to be protected by a type B MCB to BS EN 60878. Assume a TN-S supply, that is, a supply having a separate neutral and protective conductor throughout.

Let us solve this question using only the tables given in the *On Site Guide*. The tables in the Regulations will give the same values, but this will simplify the problem because we can. Refer to Tables 14.1, 14.2 and 14.3 which give the relevant *On Site Guide* Tables.

$$\text{Design current, } I_b = \frac{\text{Power}}{\text{Volts}} = \frac{6000\,\text{W}}{230\,\text{V}} = 26.09\,\text{A}$$

Table 14.3 Voltage Drop in Cables Factor. Adapted from the IEE *On site Guide* by Kind Permission of the Institution of Electrical Engineers

Table 4D1B of IEE Regulations

Voltage drop (per ampere per metre)			Conductor operating temperature: 70°C
Conductor cross-sectional area (mm²) 1	Two-core cable, d.c. (mV/A/m) 2	Two-core cable, single-phase a.c. (mV/A/m) 3	Three- or four-core cable, three-phase (mV/A/m) 4
1	44	44	38
1.5	29	29	25
2.5	18	18	15
4	11	11	9.5
6	7.3	7.3	6.4
10	4.4	4.4	3.8
16	2.8	2.8	2.4
		r	r
25	1.75	1.80	1.50
35	1.25	1.30	1.10
50	0.93	0.95	0.81
70	0.63	0.65	0.56
95	0.46	0.49	0.42

Nominal current setting of the protection for this load $I_n = 32$ A.

The cable rating, I_t is given by:

$$I_t = \frac{\text{Current rating of protective device } (I_n)}{\text{The product of the correction factors}}$$

the correction factors to be included in this calculation are:

Ca ambient temperature; from the table shown in Table 14.1 the correction factor for 35°C is 0.94.

Cg grouping factors need not be applied.

Cc since protection is by MCB no factor need be applied.

Ci thermal insulation demands that we assume installed method A (see Table 14.2).

The design current is 26.09 A and we will therefore choose a 32 A MCB for the nominal current setting of the protective device, I_n.

$$\text{Cable rating, } I_t = \frac{32}{0.94} = 34.04 \text{ A}$$

From column 2 of the table shown in Table 14.2, a 10 mm cable, having a rating of 46 A, is required to carry this current.

Key Facts

Volt drop

Maximum permissible volt drop on 230V supplies:

- 3% for lighting = 6.9V
- 5% for other uses = 11.5V

IEE Regulation 525 and Appendix 12.

Now test for volt drop: from the table shown in Table 14.3 the volt drop per ampere per metre for a 10mm cable is 4.4mV. So the volt drop for this cable length and load is equal to:

$$4.4 \times 10^{-3} \text{ V/Am} \times 26.09 \text{ A} \times 18 \text{ m} = 2.06 \text{ V}$$

Since this is less than the maximum permissible value for a lighting circuit of 6.9 V, a 10mm cable satisfies the current and drop in voltage requirements when the circuit is protected by an MCB. This cable is run in a loft that gets hot in summer and has thermal insulation touching one side of the cable. We must, therefore, use installed reference method A of Table 4.2. If we were able to route the cable under the floor, clipped direct or in conduit or trunking on a wall, we may be able to use a 6mm cable for this load. You can see how the current carrying capacity of a cable varies with the installed method by looking at Table 4.2. Compare the values in column 2 with those in column 6. When the cable is clipped direct on to a wall or surface the current rating is higher because the cable is cooler. If the alternative route was longer, you would need to test for volt drop before choosing the cable. These are some of the decisions which the electrical contractor must make when designing an installation which meets the requirements of the customer and the IEE Regulations.

If you are unsure of the standard fuse and MCB rating of protective devices, you can refer to Fig. 3.4 Appendix 3 of the IEE Regulations.

Cable size for standard domestic circuits

Appendix 4 of the IEE Regulations (BS: 7671) and Appendix 6 of the IEE *On Site Guide* contain tables for determining the current carrying capacities of conductors which we looked at in the last section. However, for standard domestic circuits, Table 14.4 gives a guide to cable size.

In this table, I am assuming a standard 230V domestic installation, having a sheathed earth or PME supply terminated in a 100A HBC fuse at the mains position. Final circuits are fed from a consumer unit, having Type B, MCB protection and wired in PVC insulated and sheathed cables with copper conductors having a grey thermoplastic PVC outer sheath or a white

Table 14.4 Cable Size for Standard Domestic Circuits

Type of final circuit	Cable size (twin and earth)	MCB rating, Type B (A)	Maximum floor area covered by circuit (m²)	Maximum length of cable run (m)
Fixed lighting	1.0	6	–	40
Fixed lighting	1.5	6	–	60
Immersion heater	2.5	16	–	30
Storage radiator	2.5	16	–	30
Cooker (oven only)	2.5	16	–	30
13A socket outlets (Radial circuit)	2.5	20	50	30
13A socket outlets (Ring circuit)	2.5	32	100	90
13A socket outlets (Ring circuit)	4.0	32	75	35
Cooker (oven and hob)	6.0	32	–	40
Shower (up to 7.5kw)	6.0	32	–	40
Shower (up to 9.6kw)	10	40	–	40

thermosetting cable with LSF (low smoke and fume properties). I am also assuming that the surrounding temperature throughout the length of the circuit does not exceed 30°C and the cables are run singly and clipped to a surface.

Wiring systems and enclosures

The final choice of a wiring system must rest with those designing the installation and those ordering the work, but whatever system is employed, good workmanship by competent persons and the use of proper materials is essential for compliance with the Regulations (IEE Regulation 134.1.1). The necessary skills can be acquired by an electrical trainee who has the correct attitude and dedication to his craft.

PVC insulated and sheathed cable installations

PVC insulated and sheathed wiring systems are used extensively for lighting and socket installations in domestic dwellings. Mechanical damage to the cable caused by impact, abrasion, penetration, compression or tension must be minimized during installation (Regulation 522.6.1). The cables are generally fixed using plastic clips incorporating a masonry nail, which means the cables can be fixed to wood, plaster or brick with almost equal ease. Cables should be run horizontally or vertically, not diagonally, down a wall. All kinks should be removed so that the cable is run straight and neatly between clips fixed at equal distances providing adequate support for the cable so that it does not become damaged by its own weight (Regulation 522.8.4 and Table 4A of the *On Site Guide*). Table 4A of the IEE *On Site Guide* is shown in Table 14.5. Where cables are bent, the radius of the bend should not cause the conductors to be damaged (Regulation 522.8.3 and Table 4E of the *On Site Guide*).

Terminations or joints in the cable may be made in ceiling roses, junction boxes, or behind sockets or switches, provided that they are enclosed in a non-ignitable material, are properly insulated and are mechanically and electrically secure (IEE Regulation 526). All joints must be accessible for inspection testing and maintenance when the installation is completed (IEE Regulation 526.3).

Where PVC insulated and sheathed cables are concealed in walls, floors or partitions, they must be provided with a box incorporating an earth terminal at each outlet position. PVC cables do not react chemically with plaster, as do some cables, and consequently PVC cables may be buried under plaster. Further protection by channel or conduit is only necessary if mechanical protection from nails or screws is required or to protect them from the plasterer's trowel. However, Regulation 522.6.6 now tells us that where PVC cables are to be embedded in a wall or partition at a depth of less than 50 mm they should be run along one of the permitted routes shown in Fig. 14.22. Figure 14.21 shows a typical PVC installation. To identify the most probable cable routes, Regulation 522.6.6 tells us that outside a zone formed by a 150 mm border all around a wall edge, cables can only be

Table 14.5 Spacing of Cable Supports. Adapted from the IEE *On Site Guide* by Kind Permission of the Institution of Electrical Engineers

Spacings of supports for cables in accessible positions

Overall diameter of cable* (mm)	Maximum spacings of clips							
	Non-armoured thermosetting, thermoplastic or lead sheathed cables				Armoured cables		Mineral insulated copper sheathed or aluminium sheathed cables	
	Generally		In caravans					
	Horizontal** (mm)	Vertical** (mm)	Horizontal** (mm)	Vertical** (mm)	Horizontal** (mm)	Vertical** (mm)	Horizontal** (mm)	Vertical** (mm)
1	2	3	4	5	6	7	8	9
Not exceeding 9	250	400			–	–	600	800
Exceeding 9 and not exceeding 15	300	400	250 (for all sizes)	400 (for all sizes)	350	450	900	1200
Exceeding 15 and not exceeding 20	350	450			400	550	1500	2000
Exceeding 20 and not exceeding 40	400	550			450	600	–	–

Note: For the spacing of supports for cables having an overall diameter exceeding 40 mm, and for single-core cables having conductors of cross-sectional area 300 mm^2 and larger, the manufacturer's recommendations should be observed.
*For flat cables taken as the dimension of the major axis.
**The spacings stated for horizontal runs may be applied also to runs at an angle of more than 30 from the vertical. For runs at an angle of 30° or less from the vertical, the vertical spacings are applicable.

run horizontally or vertically to a point or accessory unless they are contained in a substantial earthed enclosure, such as a conduit, which can withstand nail penetration, as shown in Fig. 14.22.

Where the accessory or cable is fixed to a wall which is less than 100 mm thick, protection must also be extended to the reverse side of the wall if a position can be determined.

Where none of this protection can be complied with and the installation is to be used by ordinary people, then the cable must be given additional protection with a 30 mA RCD (IEE Regulation 522.6.7).

Where cables pass through walls, floors and ceilings the hole should be made good with incombustible material such as mortar or plaster to prevent the spread of fire (Regulations 527.1.2 and 527.2.1). Cables passing through metal boxes should be bushed with a rubber grommet to prevent abrasion of the cable. Holes drilled in floor joists through which cables are run should be 50 mm below the top or 50 mm above the bottom of the joist to prevent damage to the cable by nail penetration (Regulation 522.6.5), as shown in Fig. 14.23. PVC cables should not be installed when the surrounding temperature is below 0°C or when the cable temperature

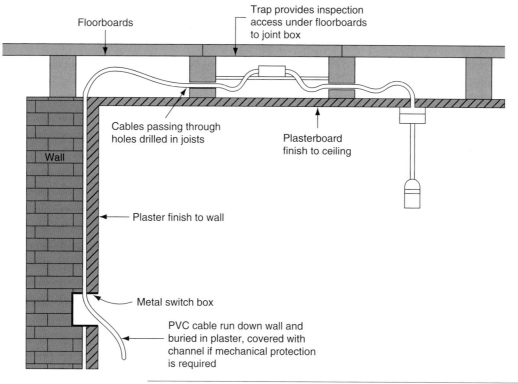

Floorboards

Trap provides inspection access under floorboards to joint box

Cables passing through holes drilled in joists

Plasterboard finish to ceiling

Wall

Plaster finish to wall

Metal switch box

PVC cable run down wall and buried in plaster, covered with channel if mechanical protection is required

FIGURE 14.21
A concealed PVC sheathed wiring system.

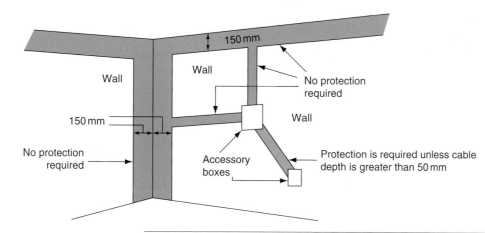

150 mm

Wall

Wall

No protection required

150 mm

Wall

No protection required

Accessory boxes

Protection is required unless cable depth is greater than 50 mm

FIGURE 14.22
Permitted cable routes.

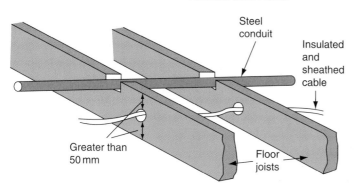

Steel conduit

Insulated and sheathed cable

Greater than 50 mm

Floor joists

Notes:

1. Maximum diameter of hole should be 0.25 × joist depth.
2. Holes on centre line in a zone between 0.25 and 0.4 × span.
3. Maximum depth of notch should be 0.125 × joist depth.
4. Notches on top in a zone between 0.1 and 0.25 × span.
5. Holes in the same joist should be at least 3 diameters apart.

FIGURE 14.23
Correct installation of cables in floor joists.

has been below 0°C for the previous 24h because the insulation becomes brittle at low temperatures and may be damaged during installation.

Conduit installations

A **conduit** is a tube, channel or pipe in which insulated conductors are contained. The conduit, in effect, replaces the PVC outer sheath of a cable, providing mechanical protection for the insulated conductors. A conduit installation can be rewired easily or altered at any time, and this flexibility, coupled with mechanical protection, makes conduit installations popular for commercial and industrial applications. There are three types of conduit used in electrical installation work: steel, PVC and flexible.

STEEL CONDUIT

Steel conduits are made to a specification defined by BS 4568 and are either heavy gauge welded or solid drawn. Heavy gauge is made from a sheet of steel welded along the seam to form a tube and is used for most electrical installation work. Solid drawn conduit is a seamless tube which is much more expensive and only used for special gas-tight, explosion-proof or flameproof installations.

Conduit is supplied in 3.75m lengths and typical sizes are 16, 20, 25 and 32mm. Conduit tubing and fittings are supplied in a black enamel finish for internal use or hot galvanized finish for use on external or damp installations. A wide range of fittings is available and the conduit is fixed using saddles or pipe hooks, as shown in Fig. 14.24.

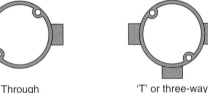

Back outlet box   Terminal box   Through box   'T' or three-way box

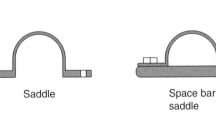

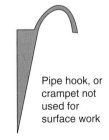

Saddle   Space bar saddle   Distance saddle   Pipe hook, or crampet not used for surface work

FIGURE 14.24
Conduit fittings and saddles.

Metal conduits are threaded with stocks and dies and bent using special bending machines. The metal conduit is also utilized as the CPC and, therefore, all connections must be screwed up tightly and all burrs removed so that cables will not be damaged as they are drawn into the conduit. Metal conduits containing a.c. circuits must contain phase and neutral conductors in the same conduit to prevent eddy currents flowing, which would result in the metal conduit becoming hot (Regulations 521.5.2, 522.8.1 and 522.8.11).

PVC CONDUIT

PVC conduit used on typical electrical installations is heavy gauge standard impact tube manufactured to BS 4607. The conduit size and range of fittings are the same as those available for metal conduit. PVC conduit is most often joined by placing the end of the conduit into the appropriate fitting and fixing with a PVC solvent adhesive. PVC conduit can be bent by hand using a bending spring of the same diameter as the inside of the conduit. The spring is pushed into the conduit to the point of the intended bend and the conduit then bent over the knee. The spring ensures that the conduit keeps its circular shape. In cold weather, a little warmth applied to the point of the intended bend often helps to achieve a more successful bend.

The advantages of a PVC conduit system are that it may be installed much more quickly than steel conduit and is non-corrosive, but it does not have the mechanical strength of steel conduit. Since PVC conduit is an insulator it cannot be used as the CPC and a separate earth conductor must be run to every outlet. It is not suitable for installations subjected to temperatures below 25°C or above 60°C. Where luminaires are suspended from PVC conduit boxes, precautions must be taken to ensure that the lamp does not raise the box temperature or that the mass of the luminaire supported by each box does not exceed the maximum recommended by the manufacturer (IEE Regulations 522.1 and 522.2). PVC conduit also expands much more than metal conduit and so long runs require an expansion coupling to allow for conduit movement and help to prevent distortion during temperature changes.

All conduit installations must be erected first before any wiring is installed (IEE Regulation 522.8.2). The radius of all bends in conduit must not cause the cables to suffer damage, and therefore the minimum radius of bends given in Table 4E of the *On Site Guide* applies (IEE Regulation 522.8.3). All conduits should terminate in a box or fitting and meet the boxes or fittings at right angles, as shown in Fig. 14.25. Any unused conduit-box entries should be blanked off and all boxes covered with a box lid, fitting or accessory to provide complete enclosure of the conduit system. Conduit runs should be separate from other services, unless intentionally bonded, to prevent arcing occurring from a faulty circuit within the conduit, which might cause the pipe of another service to become punctured.

When drawing cables into conduit they must first be *run off* the cable drum. That is, the drum must be rotated as shown in Fig. 14.26 and not allowed to *spiral off*, which will cause the cable to twist.

307

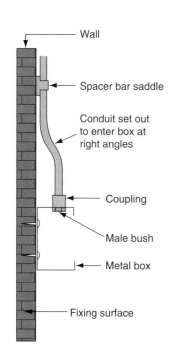

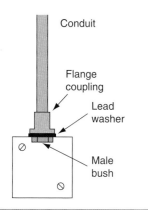

FIGURE 14.25
Terminating conduits.

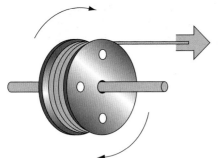

Cables *run off* will not twist, a short length of conduit can be used as an axle for the cable drum

Cables allowed to *spiral off* a drum will become twisted

FIGURE 14.26
Running off cable from a drum.

Definition

Flexible conduit manufactured to BS 731-1:1993 is made of interlinked metal spirals often covered with a PVC sleeving.

Definition

Single-PVC insulated conductors are usually drawn into the installed conduit to complete the installation.

Cables should be fed into the conduit in a manner which prevents any cable crossing over and becoming twisted inside the conduit. The cable insulation must not be damaged on the metal edges of the draw-in box. Cables can be pulled in on a draw wire if the run is a long one. The draw wire itself may be drawn in on a fish tape, which is a thin spring steel or plastic tape.

A limit must be placed on the number of bends between boxes in a conduit run and the number of cables which may be drawn into a conduit to prevent the cables being strained during wiring. Appendix 5 of the *On Site Guide* gives a guide to the cable capacities of conduits and trunking.

FLEXIBLE CONDUIT

Flexible conduit manufactured to BS 731-1: 1993 is made of interlinked metal spirals often covered with a PVC sleeving. The tubing must not be relied upon to provide a continuous earth path and, consequently, a separate CPC must be run either inside or outside the flexible tube (Regulation 543.2.1).

Flexible conduit is used for the final connection to motors so that the vibrations of the motor are not transmitted throughout the electrical installation and to allow for modifications to be made to the final motor position and drive belt adjustments.

CONDUIT CAPACITIES

Single-PVC insulated conductors are usually drawn into the installed conduit to complete the installation. Having decided upon the type, size and number of cables required for a final circuit, it is then necessary to select the appropriate size of conduit to accommodate those cables.

The tables in Appendix 5 of the *On Site Guide* describe a 'factor system' for determining the size of conduit required to enclose a number of conductors. The tables are shown in Tables 14.6 and 14.7. The method is as follows:

- Identify the cable factor for the particular size of conductor, see Table 14.6.

- Multiply the cable factor by the number of conductors, to give the sum of the cable factors.

- Identify the appropriate part of the conduit factor table given by the length of run and number of bends, see Table 14.7.

- The correct size of conduit to accommodate the cables is that conduit which has a factor equal to or greater than the sum of the cable factors.

Table 14.6 Conduit Cable Factors. Adapted from the IEE *On Site Guide* by Kind Permission of the Institution of Electrical Engineers

Cable factors for use in conduit in long straight runs over 3m, or runs of any length incorporating bends

Type of conductor	Conductor cross-sectional area (mm^2)	Cable factor
Solid or stranded	1	16
	1.5	22
	2.5	30
	4	43
	6	58
	10	105
	16	145
	25	217

The inner radius of a conduit bend should be not less than 2.5 times the outside diameter of the conduit.

Example 1

Six 2.5 mm^2 PVC insulated cables are to be run in a conduit containing two bends between boxes 10 m apart. Determine the minimum size of conduit to contain these cables.

From Table 14.6:

$$\text{The factor for one 2.5 mm}^2 \text{ cable} = 30$$
$$\text{The sum of the cable factors} = 6 \times 30$$
$$= 180$$

From Table 14.7, a 25 mm conduit, 10 m long and containing two bends, has a factor of 260. A 20 mm conduit containing two bends only has a factor of 141 which is less than 180, the sum of the cable factors and, therefore, 25 mm conduit is the minimum size to contain these cables.

Example 2

Ten 1.0 mm^2 PVC insulated cables are to be drawn into a plastic conduit which is 6 m long between boxes and contains one bend. A 4.0 mm PVC insulated CPC is also included. Determine the minimum size of conduit to contain these conductors.

From Table 14.6:

$$\text{The factor for one 1.0 mm cable} = 16$$
$$\text{The factor for one 4.0 mm cable} = 43$$
$$\text{The sum of the cable factors} = (10 \times 16) + (1 \times 43)$$
$$= 203$$

From Table 14.7, a 20 mm conduit, 6 m long and containing one bend, has a factor of 233. A 16 mm conduit containing one bend only has a factor of 143 which is less than 203, the sum of the cable factors and, therefore, 20 mm conduit is the minimum size to contain these cables.

Table 14.7 Conduit Cable Factors. Adapted from the IEE On Site Guide by Kind Permission of the Institution of Electrical Engineers

Cable factors for runs incorporating bends and long straight runs

Length of run (m)	Straight				One bend				Two bends				Three bends				Four bends			
Conduit diameter (mm)	16	20	25	32	16	20	25	32	16	20	25	32	16	20	25	32	16	20	25	32
1	Covered by Tables A and B				188	303	543	947	177	286	514	900	158	256	463	818	130	213	388	692
1.5					182	294	528	923	167	270	487	857	143	233	422	750	111	182	333	600
2					177	286	514	900	158	256	463	818	130	213	388	692	97	159	292	529
2.5					171	278	500	878	150	244	442	783	120	196	358	643	86	141	260	474
3	179	290	521	911	167	270	487	857	143	233	422	750	111	182	333	600				
3.5	177	286	514	900	162	263	475	837	136	222	404	720	103	169	311	563				
4	174	282	507	889	158	256	463	818	130	213	388	692	97	159	292	529				
4.5	171	278	500	878	154	250	452	800	125	204	373	667	91	149	275	500				
5	167	270	487	857	150	244	442	783	120	196	358	643	86	141	260	474				
6	162	263	475	837	143	233	422	750	111	182	333	600								
7	158	256	463	818	136	222	404	720	103	169	311	563								
8	154	250	452	800	130	213	388	692	97	159	292	529								
9	150	244	442	783	125	204	373	667	91	149	275	500								
10					120	196	358	643	86	141	260	474								

Additional factors: For 38 mm diameter use 1.4 × (32 mm factor)
For 50 mm diameter use 2.6 × (32 mm factor)
For 63 mm diameter use 4.2 × (32 mm factor)

Trunking installations

A **trunking** is an enclosure provided for the protection of cables which is normally square or rectangular in cross-section, having one removable side. Trunking may be thought of as a more accessible conduit system and for industrial and commercial installations it is replacing the larger conduit sizes. A trunking system can have great flexibility when used in conjunction with conduit; the trunking forms the background or framework for the installation, with conduits running from the trunking to the point controlling the current-using apparatus. When an alteration or extension is required it is easy to drill a hole in the side of the trunking and run a conduit to the new point. The new wiring can then be drawn through the new conduit and the existing trunking to the supply point.

Trunking is supplied in 3 m lengths and various cross-sections measured in millimetres from 50×50 up to 300×150. Most trunking is available in either steel or plastic.

METALLIC TRUNKING

Metallic trunking is formed from mild steel sheet, coated with grey or silver enamel paint for internal use or a hot-dipped galvanized coating where damp conditions might be encountered and made to a specification defined by BS EN 500 85. A wide range of accessories is available, such as 45° bends, 90° bends, tee and four-way junctions, for speedy on-site assembly. Alternatively, bends may be fabricated in lengths of trunking, as shown in Fig. 14.27. This may be necessary or more convenient if a bend or set is non-standard, but it does take more time to fabricate bends than merely to bolt on standard accessories.

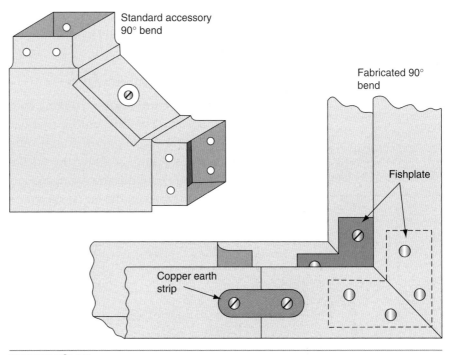

Standard accessory 90° bend

Fabricated 90° bend

Fishplate

Copper earth strip

FIGURE 14.27

Alternative trunking bends.

When fabricating bends the trunking should be supported with wooden blocks for sawing and filing, in order to prevent the sheet-steel vibrating or becoming deformed. Fish plates must be made and riveted or bolted to the trunking to form a solid and secure bend. When manufactured bends are used, the continuity of the earth path must be ensured across the joint by making all fixing screw connections very tight, or fitting a separate copper strap between the trunking and the standard bend. If an earth continuity test on the trunking is found to be unsatisfactory, an insulated CPC must be installed inside the trunking. The size of the protective conductor will be determined by the largest cable contained in the trunking, as described by Table 54.7 of the IEE Regulations. If the circuit conductors are less than $16\,mm^2$, then a $16\,mm^2$ CPC will be required.

NON-METALLIC TRUNKING

Trunking and trunking accessories are also available in high-impact PVC. The accessories are usually secured to the lengths of trunking with a PVC solvent adhesive. PVC trunking, like PVC conduit, is easy to install and is non-corrosive. A separate CPC will need to be installed and non-metallic trunking may require more frequent fixings because it is less rigid than metallic trunking. All trunking fixings should use round-headed screws to prevent damage to cables since the thin sheet construction makes it impossible to countersink screw heads.

MINI-TRUNKING

Mini-trunking is very small PVC trunking, ideal for surface wiring in domestic and commercial installations such as offices. The trunking has a cross-section of $16 \times 16\,mm$, $25 \times 16\,mm$, $38 \times 16\,mm$ or $38 \times 25\,mm$ and is ideal for switch drops or for housing auxiliary circuits such as telephone or audio equipment wiring. The modern square look in switches and sockets is complemented by the mini-trunking which is very easy to install (see Fig. 14.28).

SKIRTING TRUNKING

Skirting trunking is a trunking manufactured from PVC or steel and in the shape of a skirting board is frequently used in commercial buildings such as hospitals, laboratories and offices. The trunking is fitted around the walls of a room at either the skirting board level or at the working surface level and contains the wiring for socket outlets and telephone points which are mounted on the lid, as shown in Fig. 14.28.

> **Definition**
>
> *Mini-trunking* is very small PVC trunking, ideal for surface wiring in domestic and commercial installations such as offices.

> **Definition**
>
> *Skirting trunking* is a trunking manufactured from PVC or steel and in the shape of a skirting board is frequently used in commercial buildings such as hospitals, laboratories and offices.

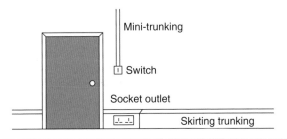

FIGURE 14.28

Typical installation of skirting trunking and mini-trunking.

Where any trunking passes through walls, partitions, ceilings or floors, short lengths of lid should be fitted so that the remainder of the lid may be removed later without difficulty. Any damage to the structure of the buildings must be made good with mortar, plaster or concrete in order to prevent the spread of fire. Fire barriers must be fitted inside the trunking every 5 m, or at every floor level or room dividing wall, if this is a shorter distance, as shown in Fig. 14.29(a).

Where trunking is installed vertically, the installed conductors must be supported so that the maximum unsupported length of non-sheathed cable does not exceed 5 m. Figure 14.29(b) shows cables woven through insulated pin supports, which is one method of supporting vertical cables.

PVC insulated cables are usually drawn into an erected conduit installation or laid into an erected trunking installation. Table 5D of the *On Site Guide* only gives factors for conduits up to 32 mm in diameter, which would indicate that conduits larger than this are not in frequent or common use. Where a cable enclosure greater than 32 mm is required because of the number or size of the conductors, it is generally more economical and convenient to use trunking.

TRUNKING CAPACITIES

The **ratio** of the space occupied by all the cables in a conduit or trunking to the whole space enclosed by the conduit or trunking is known as the **space factor**. Where sizes and types of cable and trunking are not covered

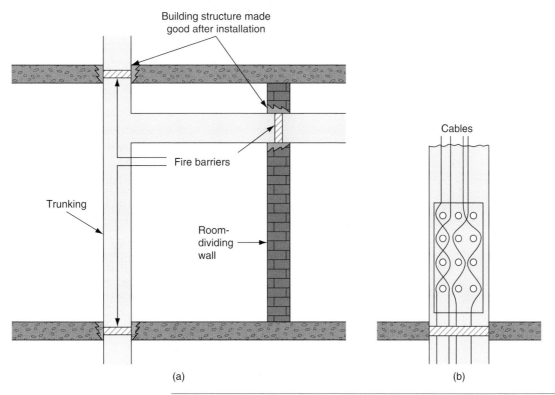

FIGURE 14.29
Installation of trunking: (a) fire barriers in trunking and (b) cable supports in vertical trunking.

by the tables in the *On Site Guide* a space factor of 45% must not be exceeded. This means that the cables must not fill more than 45% of the space enclosed by the trunking. The tables take this factor into account.

To calculate the size of trunking required to enclose a number of cables:

- Identify the cable factor for the particular size of conductor, see Table 14.8.

- Multiply the cable factor by the number of conductors to give the sum of the cable factors.

- Consider the factors for trunking and shown in Table 14.9. The correct size of trunking to accommodate the cables is that trunking which has a factor equal to or greater than the sum of the cable factors.

Example

Calculate the minimum size of trunking required to accommodate the following single-core PVC cables:

20 × 1.5 mm   solid conductors

20 × 2.5 mm   solid conductors

21 × 4.0 mm   stranded conductors

16 × 6.0 mm   stranded conductors

From Table 14.8, the cable factors are:

for 1.5 mm solid cable – 8.0

for 2.5 mm solid cable – 11.9

for 4.0 mm stranded cable – 16.6

for 6.0 mm stranded cable – 21.2

The sum of the cable terms is:

$(20 \times 8.0) + (20 \times 11.9) + (21 \times 16.6) + (16 \times 21.2) = 1085.8$. From table 14.9, 75 × 38 mm trunking has a factor of 1146 and, therefore, the minimum size of trunking to accommodate these cables is 75 × 38 mm, although a larger size, say 75 × 50 mm would be equally acceptable if this was more readily available as a standard stock item.

Segregation of circuits

Where an installation comprises a mixture of low-voltage and very low-voltage circuits such as mains lighting and power, fire alarm and telecommunication circuits, they must be separated or *segregated* to prevent electrical contact (IEE Regulation 528.1).

For the purpose of these regulations various circuits are identified by one of two bands as follows:

Band I telephone, radio, bell, call and intruder alarm circuits, emergency circuits for fire alarm and emergency lighting.

Band II mains voltage circuits.

Table 14.8 Trunking Cable Factors. Adapted from the IEE *On Site Guide* by Kind Permission of the Institution of Electrical Engineers

Cable factors for trunking

Type of conductor	Conductor cross-sectional area (mm²)	PVC BS 6004 Cable factor	Thermosetting BS 7211 Cable factor
Solid	1.5	8.0	8.6
	2.5	11.9	11.9
Stranded	1.5	8.6	9.6
	2.5	12.6	13.9
	4	16.6	18.1
	6	21.2	22.9
	10	35.3	36.3
	16	47.8	50.3
	25	73.9	75.4

Notes: These factors are for metal trunking and may be optimistic for plastic trunking where the cross-sectional area available may be significantly reduced from the nominal by the thickness of the wall material.
The provision of spare space is advisable; however, any circuits added at a later date must take into account grouping. Appendix 4, BS 7671.

Table 14.9 Trunking Cable Factors. Adapted from the IEE *On Site Guide* by Kind Permission of the Institution of Electrical Engineers

Factors for trunking

Dimensions of trunking (mm × mm)	Factor	Dimensions of trunking (mm × mm)	Factor
50 × 38	767	200 × 100	8572
50 × 50	1037	200 × 150	13001
75 × 25	738	200 × 200	17429
75 × 38	1146	225 × 38	3474
75 × 50	1555	225 × 50	4671
75 × 75	2371	225 × 75	7167
100 × 25	993	225 × 100	9662
100 × 38	1542	225 × 150	14652
100 × 50	2091	225 × 200	19643
100 × 75	3189	225 × 225	22138
100 × 100	4252	300 × 38	4648
150 × 38	2999	300 × 50	6251
150 × 50	3091	300 × 75	9590
150 × 75	4743	300 × 100	12929
150 × 100	6394	300 × 150	19607
150 × 150	9697	300 × 200	26285
200 × 38	3082	300 × 225	29624
200 × 50	4145	300 × 300	39428
200 × 75	6359		

Space factor – 45% with trunking thickness taken into account.

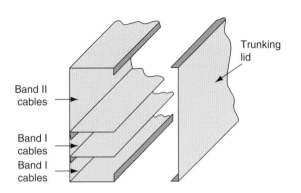

FIGURE 14.30
Segregation of cables in trunking.

When Band I circuits are insulated to the same voltage as Band II circuits, they may be drawn into the same compartment.

When trunking contains rigidly fixed metal barriers along its length, the same trunking may be used to enclose cables of the separate Bands without further precautions, provided that each band is separated by a barrier, as shown in Fig. 14.30.

Multi-compartment PVC trunking cannot provide band segregations since there is no metal screen between the Bands. This can only be provided in PVC trunking if screened cables are drawn into the trunking.

Cable tray installations

Cable tray is a sheet-steel channel with multiple holes. The most common finish is hot-dipped galvanized but PVC-coated tray is also available. It is used extensively on large industrial and commercial installations for supporting MI and SWA cables which are laid on the cable tray and secured with cable ties through the tray holes.

Cable tray should be adequately supported during installation by brackets which are appropriate for the particular installation. The tray should be bolted to the brackets with round-headed bolts and nuts, with the round head inside the tray so that cables drawn along the tray are not damaged.

The tray is supplied in standard widths from 50 to 900 mm, and a wide range of bends, tees and reducers is available. Figure 14.31 shows a factory made 90° bend at B. The tray can also be bent using a cable tray bending machine to create bends such as that shown at A in Fig. 14.31. The installed tray should be securely bolted with round-headed bolts where lengths or accessories are attached, so that there is a continuous earth path which may be bonded to an electrical earth. The whole tray should provide a firm support for the cables and therefore the tray fixings must be capable of supporting the weight of both the tray and cables.

PVC/SWA cable installations

Steel wire armoured PVC insulated cables are now extensively used on industrial installations and often laid on cable tray. This type of installation has the advantage of flexibility, allowing modifications to be made speedily as the need arises. The cable has a steel wire armouring giving mechanical

Definition

Cable tray is a sheet-steel channel with multiple holes. The most common finish is hot-dipped galvanized but PVC-coated tray is also available. It is used extensively on large industrial and commercial installations for supporting MI and SWA cables which are laid on the cable tray and secured with cable ties through the tray holes.

Definition

Steel wire armoured PVC insulated cables are now extensively used on industrial installations and often laid on cable tray.

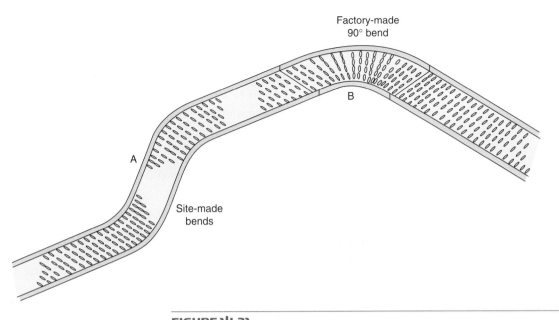

Factory-made
90° bend

B

A

Site-made
bends

FIGURE 14.31
Cable tray with bends.

protection and permitting it to be laid directly in the ground or in ducts, or it may be fixed directly or laid on a cable tray. Figure 6.2 shows a PVC/SWA cable.

It should be remembered that when several cables are grouped together the current rating will be reduced according to the correction factors given in Appendix 4 (Table 4C1) of the IEE Regulations.

The cable is easy to handle during installation, is pliable and may be bent to a radius of eight times the cable diameter. The PVC insulation would be damaged if installed in ambient temperatures over 70°C or below 0°C, but once installed the cable can operate at low temperatures.

The cable is terminated with a simple gland which compresses a compression ring on to the steel wire armouring to provide the earth continuity between the switchgear and the cable.

MI Cable installations

Mineral insulated cables are available for general wiring as:

- light-duty MI cables for voltages up to 600V and sizes from 1.0 to 10 mm²,
- heavy-duty MI cables for voltages up to 1000V and sizes from 1.0 to 150 mm².

Figure 6.3 (p. 126) shows an MI cable and termination.

The cables are available with bare sheaths or with a PVC oversheath. The cable sheath provides sufficient mechanical protection for all but the most severe situations, where it may be necessary to fit a steel sheath or conduit over the cable to give extra protection, particularly near floor level in some industrial situations.

The cable may be laid directly in the ground, in ducts, on cable tray or clipped directly to a structure. It is not affected by water, oil or the cutting fluids used in engineering and can withstand very high temperature or even fire. The cable diameter is small in relation to its current carrying capacity and it should last indefinitely if correctly installed because it is made from inorganic materials. These characteristics make the cable ideal for Band I emergency circuits, boiler-houses, furnaces, petrol stations and chemical plant installations.

The cable is supplied in coils and should be run off during installation and not spiralled off, as described in Fig. 14.26 for conduit. The cable can be work hardened if over-handled or over-manipulated. This makes the copper outer sheath stiff and may result in fracture. The outer sheath of the cable must not be penetrated, otherwise moisture will enter the magnesium oxide insulation and lower its resistance. To reduce the risk of damage to the outer sheath during installation, cables should be straightened and formed by hammering with a hide hammer or a block of wood and a steel hammer. When bending MI cables the radius of the bend should not cause the cable to become damaged and clips should provide adequate support (Regulation 522.8.5), see Table 14.5.

The cable must be prepared for termination by removing the outer copper sheath to reveal the copper conductors. This can be achieved by using a rotary stripper tool or, if only a few cables are to be terminated, the outer sheath can be removed with side cutters, peeling off the cable in a similar way to peeling the skin from a piece of fruit with a knife. When enough conductor has been revealed, the outer sheath must be cut off square to facilitate the fitting of the sealing pot, and this can be done with a ringing tool. All excess magnesium oxide powder must be wiped from the conductors with a clean cloth. This is to prevent moisture from penetrating the seal by capillary action.

Cable ends must be terminated with a special seal to prevent the entry of moisture. Figure 6.3 shows a brass screw-on seal and gland assembly, which allows termination of the MI cables to standard switchgear and conduit fittings. The sealing pot is filled with a sealing compound, which is pressed in from one side only to prevent air pockets forming, and the pot closed by crimping home the sealing disc with an MI crimping tool such as that shown in Fig 7.4. Such an assembly is suitable for working temperatures up to 105°C. Other compounds or powdered glass can increase the working temperature up to 250°C.

The conductors are not identified during the manufacturing process and so it is necessary to identify them after the ends have been sealed. A simple continuity or polarity test, as described later in this chapter, can identify the conductors which are then sleeved or identified with coloured markers.

Connection of MI cables can be made directly to motors, but to absorb the vibrations a 360° loop should be made in the cable just before the termination. If excessive vibration is to be expected the MI cable should be terminated in a conduit through box and the final connection made by flexible conduit.

Copper MI cables may develop a green incrustation or patina on the surface, even when exposed to normal atmospheres. This is not harmful and should not be removed. However, if the cable is exposed to an environment which might encourage corrosion, an MI cable with an overall PVC sheath should be used.

Support and fixing methods for electrical equipment

Individual conductors may be installed in trunking or conduit and individual cables may be clipped directly to a surface or laid on a tray using the wiring system which is most appropriate for the particular installation. The installation method chosen will depend upon the contract specification, the fabric of the building and the type of installation – domestic, commercial or industrial.

It is important that the wiring systems and fixing methods are appropriate for the particular type of installation and compatible with the structural materials used in the building construction. The electrical installation must be compatible with the installed conditions, must not damage the fabric of the building or weaken load-bearing girders or joists.

The installation designer must ask himself the following questions:

- Does this wiring system meet the contract specification?

- Is the wiring system compatible with this particular installation?

- Do I need to consider any special regulations such as those required by agricultural and horticultural installations, swimming pools or flameproof installations?

- Will this type of electrical installation be aesthetically acceptable and compatible with the other structural materials?

The installation electrician must ask himself the following questions:

- Am I using materials and equipment which meet the relevant British Standards and the contract specification?

- Am I using an appropriate fixing method for this wiring system or piece of equipment?

- Will the structural material carry the extra load that my conduits and cables will place upon it?

- Will my fixings and fittings weaken the existing fabric of the building?

- Will the electrical installation interfere with other supplies and services?

- Will all terminations and joints be accessible upon completion of the erection period? (IEE Regulations 513.1 and 526.3.)

- Will the materials being used for the electrical installation be compatible with the intended use of the building?

- Am I working safely and efficiently and in accordance with the IEE Regulations (BS 7671)?

A domestic installation usually calls for a PVC insulated and sheathed wiring system. These cables are generally fixed using plastic clips incorporating a masonry nail which means that the cables can be fixed to wood, plaster or brick with almost equal ease.

Cables must be run straight and neatly between clips fixed at equal distances and providing adequate support for the cable so that it does not become damaged by its own weight (IEE Regulation 522.8.4) and shown in Table 14.5.

A commercial or industrial installation might call for a conduit or trunking wiring system. A conduit is a tube, channel or pipe in which insulated conductors are contained. The conduit, in effect, replaces the PVC outer sheath of a cable, providing mechanical protection for the insulated conductors. A conduit installation can be rewired easily or altered at any time and this flexibility, coupled with mechanical protection, makes conduit installations popular for commercial and industrial applications. Steel conduits and trunking are, however, much heavier than single cables and, therefore, need substantial and firm fixings and supports. A wide range of support brackets is available for fixing conduit, trunking and tray installations to the fabric of a commercial or industrial installation. Some of these are shown in Fig. 14.32.

When a heavier or more robust fixing is required to support cabling or equipment a nut and bolt or screw fixing is called for. Wood screws may be screwed directly into wood but when fixing to stone, brick or concrete it is first necessary to drill a hole in the masonry material which is then plugged with a material (usually plastic) to which a screw can be secured.

For the most robust fixing to masonry materials an expansion bolt such as that made by Rawlbolt should be used.

For lightweight fixings to hollow partitions or plasterboard a spring toggle can be used. Plasterboard cannot support a screw fixing directly into itself but the spring toggle spreads the load over a larger area, making the fixing suitable for light loads.

Let us look in a little more detail at individual joining, support and fixing methods.

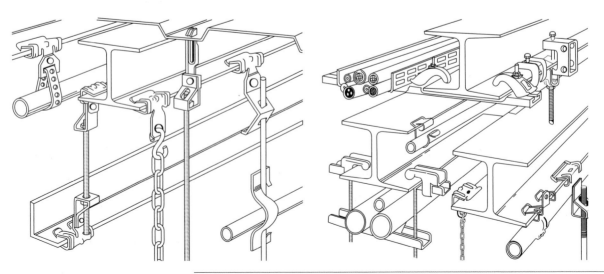

FIGURE 14.32

Same manufactured girder supports for electrical equipment.

Joining materials

Plastic can be joined with an appropriate solvent. Metals may be welded, brazed or soldered, but the most popular method of on-site joining of metals on electrical installations is by nuts and bolts or rivets.

A nut and bolt joint may be considered a temporary fastening since the parts can easily be separated if required by unscrewing the nut and removing the bolt. A rivet is a permanent fastening since the parts riveted together cannot be easily separated.

Two pieces of metal joined by a bolt and nut and by a machine screw and nut are shown in Fig. 14.33. The nut is tightened to secure the joint. When joining trunking or cable trays, a round head machine screw should be used with the head inside to reduce the risk of damage to cables being drawn into the trunking or tray.

Thin sheet material such as trunking is often joined using a pop riveter. Special rivets are used with a hand tool, as shown in Fig. 14.34. Where possible, the parts to be riveted should be clamped and drilled together with a clearance hole for the rivet. The stem of the rivet is pushed into the nose bush of the riveter until the alloy sleeve of the rivet is flush with the nose bush (a). The rivet is then placed in the hole and the handles squeezed together (b). The alloy sleeve is compressed and the rivet stem will break off when the rivet is set and the joint complete (c). To release the broken-off stem piece, the nose bush is turned upwards and the handles opened sharply. The stem will fall out and is discarded (d).

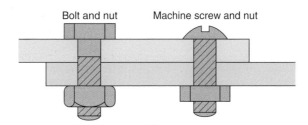

FIGURE 14.33
Joining of metals.

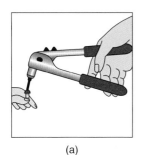 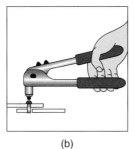 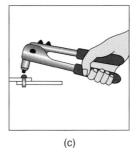 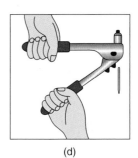

(a)   (b)   (c)   (d)

FIGURE 14.34
Metal joining with pop rivets.

Bracket supports

Conduit and trunking may be fixed directly to a surface such as a brick wall or concrete ceiling, but where cable runs are across girders or other steel framework, spring steel clips may be used but support brackets or clips often require manufacturing.

The brackets are usually made from flat iron, which is painted after manufacturing to prevent corrosion. They may be made on-site by the electrician or, if many brackets are required, the electrical contractor may make a working sketch with dimensions and have the items manufactured by a blacksmith or metal fabricator.

The type of bracket required will be determined by the installation, but Fig. 14.35 gives some examples of brackets which may be modified to suit particular circumstances.

FIXING METHODS

PVC insulated and sheathed wiring systems are usually fixed with PVC clips in order to comply with IEE Regulation 522.8.3 and 4 and shown in Table 14.5. The clips are supplied in various sizes to hold the cable firmly, and the fixing nail is a hardened masonry nail. Figure 14.36 shows a cable clip of this type. The use of a masonry nail means that fixings to wood, plaster, brick or stone can be made with equal ease.

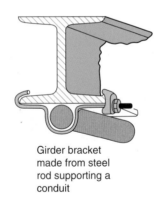

Girder bracket made from steel rod supporting a conduit

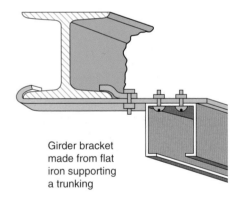

Girder bracket made from flat iron supporting a trunking

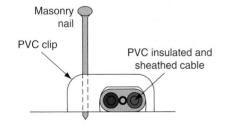

Masonry nail

PVC clip

PVC insulated and sheathed cable

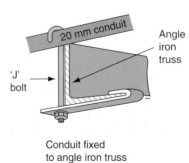

20 mm conduit

'J' bolt

Angle iron truss

Conduit fixed to angle iron truss with flat iron and 'J' bolt

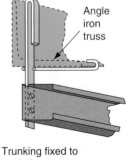

Angle iron truss

Trunking fixed to angle iron truss by a variation of flat iron and 'J' bolt

FIGURE 14.36
PVC insulated and sheathed cable clip.

FIGURE 14.35
Bracket supports for conduits and trunking.

When heavier cables, trunking, conduit or luminaires have to be fixed a screw fixing is often needed. Wood screws may be screwed directly into wood but when fixing to brick, stone, plaster or concrete it is necessary to drill a hole in the masonry material, which is then plugged with a material to which the screw can be secured.

PLASTIC PLUGS

A **plastic plug** is made of a hollow plastic tube split up to half its length to allow for expansion. Each size of plastic plug is colour coded to match a wood screw size.

A hole is drilled into the masonry, using a masonry drill of the same diameter, to the length of the plastic plug (see Fig. 14.37). The plastic plug is inserted into the hole and tapped home until it is level with the surface of the masonry. Finally, the fixing screw is driven into the plastic plug until it becomes tight and the fixture is secure.

EXPANSION BOLTS

The most well known **expansion bolt** is made by Rawlbolt and consists of a split iron shell held together at one end by a steel ferrule and a spring wire clip at the other end. Tightening the bolt draws up an expanding bolt inside the split iron shell, forcing the iron to expand and grip the masonry. Rawlbolts are for heavy-duty masonry fixings (see Fig. 14.38).

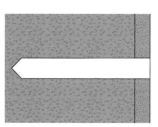
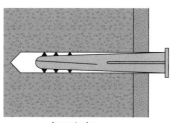
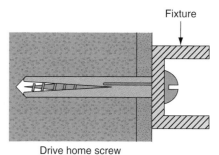

| Drill hole in masonry | Insert plug | Drive home screw |

FIGURE 14.37
Screw fixing to plastic plug.

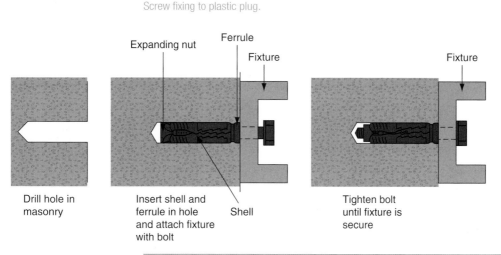

Drill hole in masonry

Insert shell and ferrule in hole and attach fixture with bolt

Tighten bolt until fixture is secure

FIGURE 14.38
Expansion bolt fixing.

Definition

A *spring toggle bolt* provides one method of fixing to hollow partition walls which are usually faced with plasterboard and a plaster skimming.

A hole is drilled in the masonry to take the iron shell and ferrule. The iron shell is inserted with the spring wire clip end first so that the ferrule is at the outer surface. The bolt is passed through the fixture, located in the expanding nut and tightened until the fixing becomes secure.

SPRING TOGGLE BOLTS

A **spring toggle bolt** provides one method of fixing to hollow partition walls which are usually faced with plasterboard and a plaster skimming. Plasterboard and plaster wall or ceiling surfaces are not strong enough to support a load fixed directly into the plasterboard, but the spring toggle spreads the load over a larger area, making the fixing suitable for light loads (see Fig. 14.39).

A hole is drilled through the plasterboard and into the cavity. The toggle bolt is passed through the fixture and the toggle wings screwed into the bolt. The toggle wings are compressed and passed through the hole in the plasterboard and into the cavity where they spring apart and rest on the cavity side of the plasterboard. The bolt is tightened until the fixing becomes firm. The bolt of the spring toggle cannot be removed after fixing without the loss of the toggle wings. If it becomes necessary to remove and refix the fixture a new toggle bolt will have to be used.

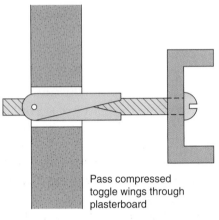

Pass compressed toggle wings through plasterboard

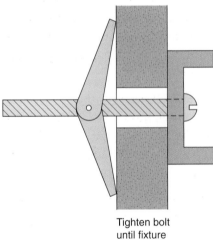

Tighten bolt until fixture is secure

FIGURE 14.39
Spring toggle bolt fixing.

Special installations or locations

All electrical installations and installed equipment must be safe to use and free from the dangers of electric shock, but some installations or locations require special consideration because of the inherent dangers of the installed conditions. The danger may arise because of the corrosive or explosive nature of the atmosphere, because the installation must be used in damp or low-temperature conditions or because there is a need to provide additional mechanical protection for the electrical system Part 7 of the IEE Regulations deals with these special installations or locations. In this section we will consider some of the installations which require special consideration.

Temporary construction site installations

Temporary electrical supplies provided on construction sites can save many man hours of labour by providing the energy required for fixed and portable tools and lighting which speeds up the completion of a project. However, construction sites are dangerous places and the temporary electrical supply which is installed to assist the construction process must comply with all of the relevant wiring regulations for permanent installations (Regulation 110.1). All equipment must be of a robust construction in order to fulfil the on-site electrical requirements while being exposed to rough handling, vehicular nudging, the wind, rain and sun. All equipment socket outlets, plugs and couplers must be of the industrial type to BS EN 60439 and BS EN 60309 and specified by Regulation 704.511.1 as shown in Fig. 14.40.

Where an electrician is not permanently on site, MCBs are preferred so that overcurrent protection devices can be safely reset by an unskilled person.

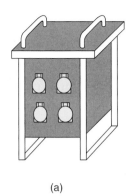

(a)

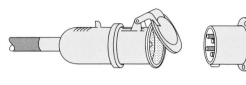

(b)

FIGURE 14.40

110 V distribution unit and cable connector, suitable for construction site electrical supplies:
(a) reduced-voltage distribution unit incorporating industrial sockets to BS EN 60309 and (b) industrial plug and connector.

The British Standards Code of Practice 1017, *The Distribution of Electricity on Construction and Building Sites,* advises that protection against earth faults may be obtained by first providing a low impedance path, so that overcurrent devices can operate quickly as described in Chapter 12, and secondly by fitting an RCD in addition to the overcurrent protection device (IEE Regulation 704.410.3.10). The 17th edition of the IEE Regulations considers construction sites very special locations, devoting the whole of Section 704 to their requirements. A construction site installation should be tested and inspected in accordance with Part 6 of the Wiring Regulations every 3 months throughout the construction period.

The source of supply for the temporary installation may be from a petrol or diesel generating set or from the local supply company. When the local electricity company provides the supply, the incoming cable must be terminated in a waterproof and locked enclosure to prevent unauthorized access and provide metering arrangements.

IEE Regulations 704.313, 704.410.3.10 and 411.8 tells us that reduced low voltage is strongly *preferred* for portable hand lamps and tools used on construction and demolition sites.

The distribution of electrical supplies on a construction site would typically be as follows:

- 400 V three phase for supplies to major items of plant having a rating above 3.75 kW such as cranes and lifts. These supplies must be wired in armoured cables.

- 230 V single phase for supplies to items of equipment which are robustly installed such as floodlighting towers, small hoists and site offices. These supplies must be wired in armoured cable unless run inside the site offices.

- 110 V single phase for supplies to all mobile hand tools and all mobile lighting equipment. The supply is usually provided by a reduced voltage distribution unit which incorporates splashproof sockets fed from a centre-tapped 110 V transformer. This arrangement limits the voltage to earth to 55 V, which is recognized as safe

325

in most locations. A 110V distribution unit is shown in Fig. 14.40. Edison screw lamps are used for 110V lighting supplies so that they are not interchangeable with 230V site office lamps.

There are occasions when even a 110V supply from a centre-tapped transformer is too high, for example, supplies to inspection lamps for use inside damp or confined places. In these circumstances a safety extra-low voltage (SELV) supply would be required.

Industrial plugs have a keyway which prevents a tool from one voltage being connected to the socket outlet of a different voltage. They are also colour coded for easy identification as follows:

400 V – red

230 V – blue

110 V – yellow

50 V – white

25 V – violet.

Agricultural and horticultural installations

Especially adverse installation conditions are to be encountered on agricultural and horticultural installations because of the presence of livestock, vermin, dampness, corrosive substances and mechanical damage. The 17th edition of the IEE Wiring Regulations considers these installations very special locations and has devoted the whole of Section 705 to their requirements. In situations accessible to livestock the electrical equipment should be of a type which is appropriate for the external influences likely to occur, and should have at least protection IP44, that is, protection against solid objects and water splashing from any direction (Regulation 705.512.2, see also Fig. 13.3).

In buildings intended for livestock, all fixed wiring systems must be inaccessible to the livestock and cables liable to be attacked by vermin must be suitably protected (IEE Regulation 705.513.2).

PVC cables enclosed in heavy-duty PVC conduit are suitable for installations in most agricultural buildings. All exposed and extraneous metalwork must be provided with supplementary equipotential bonding in areas where livestock is kept (Regulation 705.415.2.1). In many situations, waterproof socket outlets to BS 196 must be installed. All socket outlet circuits must be protected by an RCD complying with the appropriate British Standard and the operating current must not exceed 30 mA.

Cables buried on agricultural or horticultural land should be buried at a depth not less than 600 mm, or 1000 mm where the ground may be cultivated, and the cable must have an armour sheath and be further protected by cable tiles. Overhead cables must be insulated and installed so that they are clear of farm machinery or placed at a minimum height of 6 m to comply with Regulation 705.522.

Horses and cattle have a very low body resistance, which makes them susceptible to an electric shock at voltages lower than 25V rms. The sensitivity of farm animals to electric shock means that they can be contained by an electric fence. An animal touching the fence receives a short pulse of electricity which passes through the animal to the general mass of earth and back to an earth electrode sunk near the controller, as shown in Fig. 14.41. The pulses are generated by a capacitor–resistor circuit inside the controller which may be mains or battery operated (capacitor–resistor circuits are discussed in *Advanced Electrical Installation Work*). There must be no risk to any human coming into contact with the controller, which should be manufactured to BS 2632. The output voltage of the controller must not exceed 10kV and the energy must not be greater than 5J. The duration of the pulse must not be greater than 1.5ms and the pulse must never have a frequency greater than one pulse per second. This shock level is very similar to that which can be experienced by touching a spark plug lead on a motor car. The energy levels are very low at 5J. There are 3.6 million joules of energy in 1kWh.

Earth electrodes connected to the earth terminal of an electric fence controller must be separate from the earthing system of any other circuit and should be situated outside the resistance area of any electrode used for protective earthing. The electric fence controller and the fence wire must be installed so that they do not come into contact with any power, telephone or radio systems, including poles. Agricultural and horticultural installations should be tested and inspected in accordance with Part 6 of the Wiring Regulations every 3 years.

327

Caravans and caravan sites

The electrical installations on caravan sites, and within caravans, must comply in all respects with the wiring regulations for buildings. All the dangers which exist in buildings are present in and around caravans, including the added dangers associated with repeated connection and disconnection of the supply and the flexing of the caravan installation in a moving vehicle. The 17th edition of the Regulations has devoted Section 721 to the electrical installation in caravans and motor caravans and Section 708 to caravan parks.

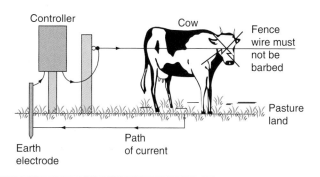

FIGURE 14.41

Farm animal control by electric fence.

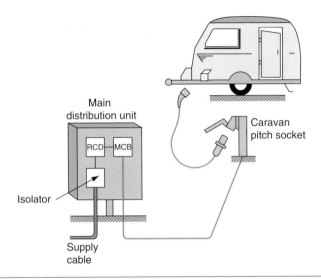

Main
distribution unit

RCD—MCB

Caravan
pitch socket

Isolator

Supply
cable

FIGURE 14.42
Electrical supplies to caravans.

Caravans

- every caravan pitch must have at least one 16 A industrial socket
- each socket must have RCD and overcurrent protection
- IEE Regulations 708.553.

Touring caravans must be supplied from a 16 A industrial type socket outlet adjacent to the caravan park pitch, similar to that shown in Fig. 14.40. Each socket outlet must be provided with individual overcurrent protection and an individual residual current circuit breaker with a rated tripping current of 30 mA (IEE Regulations 708.553.1.12 and 708.553.1.13). The distance between the caravan connector and the site socket outlet must be not more than 20 m (Regulation 708.512.3). These requirements are shown in Fig. 14.42.

The supply cables must be installed outside the pitch area and be buried at a depth of at least 0.6 m (IEE Regulation 708.521.1.1).

The caravan or motor caravan must be provided with a mains isolating switch and an RCD to break all live conductors (Regulations 721.411). An adjacent notice detailing how to connect and disconnect the supply safely must also be provided, as shown in Regulation 721.514. Electrical equipment must not be installed in fuel storage compartments (Regulation 721.528.3.5). Caravans flex when being towed, and therefore the installation must be wired in flexible or stranded conductors of at least 1.5 mm cross-section. The conductors must be supported on horizontal runs at least every 25 cm and the metalwork of the caravan and chassis must be bonded with 4.0 mm² cable.

The wiring of the extra low-voltage battery supply must be run in such a way that it does not come into contact with the 230 V wiring system (Regulation 721.528.1).

The caravan should be connected to the pitch socket outlet by means of a flexible cable, not longer than 25 m, and having a minimum cross-sectional area of 2.5 mm² or as detailed in Table 721.

Because of the mobile nature of caravans it is recommended that the electrical installation be tested and inspected at intervals considered appropriate, between 1 and 3 years but not exceeding 3 years (Regulation 721.514.1).

Static electricity

Definition

Static electricity is a voltage charge which builds up to many thousands of volts between two surfaces when they rub together.

Static electricity is a voltage charge which builds up to many thousands of volts between two surfaces when they rub together. A dangerous situation occurs when the static charge has built up to a potential capable of striking an arc through the airgap separating the two surfaces.

Static charges build up in a thunderstorm. A lightning strike is the discharge of the thunder cloud, which might have built up to a voltage of 100 MV, to the general mass of earth which is at zero volts. Lightning discharge currents are of the order of 20 kA, hence the need for lightning conductors on vulnerable buildings in order to discharge the energy safely. We looked at Lighting Protection Systems in Chapter 13.

Static charge builds up between any two insulating surfaces or between an insulating surface and a conducting surface, but it is not apparent between two conducting surfaces.

A motor car moving through the air builds up a static charge which sometimes gives the occupants a minor shock as they step out and touch the door handle.

A nylon overall and nylon bed sheets build up static charge which is the cause of the 'crackle' when you shake them. Many flammable liquids have the same properties as insulators, and therefore liquids, gases, powders and paints moving through pipes build up a static charge.

Petrol pumps, operating theatre oxygen masks and car spray booths are particularly at risk because a spark in these situations may ignite the flammable liquid, powder or gas.

So how do we protect ourselves against the risks associated with static electrical charges? I said earlier that a build-up of static charge is not apparent between two conducting surfaces, and this gives a clue to the solution. Bonding surfaces together with equipotential bonding conductors prevents a build-up of static electricity between the surfaces. If we use large-diameter pipes, we reduce the flow rates of liquids and powders and, therefore, we reduce the build-up of static charge. Hospitals use cotton sheets and uniforms, and use bonding extensively in operating theatres. Rubber, which contains a proportion of graphite, is used to manufacture antistatic trolley wheels and surgeons' boots. Rubber constructed in this manner enables any build-up of static charge to 'leak' away. Increasing humidity also reduces static charge because the water droplets carry away the static charge, thus removing the hazard.

Computer supplies

Every modern office now contains computers, and many systems are linked together or networked. Most computer systems are sensitive to variations

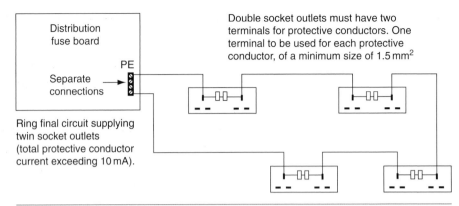

FIGURE 14.43

Recommended method of connecting IT equipment to socket outlets.

or distortions in the mains supply and many computers incorporate filters which produce *high protective conductor currents* of around 2 or 3 mA. This is clearly not a fault current, but is typical of the current which flows in the CPC of IT equipment under normal operating conditions. IEE Regulation 543.7.1 and 4 deals with the earthing requirements for the installation of equipment having high protective conductor currents. IEE Guidance Note 7 recommends that IT equipment should be connected to double sockets as shown in Fig. 14.43.

Clean supplies

Supplies to computer circuits must be 'clean' and 'secure'. Mainframe computers and computer networks are sensitive to mains distortion or interference, which is referred to as 'noise'. Noise is mostly caused by switching an inductive circuit which causes a transient spike, or by brush gear making contact with the commutator segments of an electric motor. These distortions in the mains supply can cause computers to 'crash' or provoke errors and are shown in Fig. 14.44.

To avoid this, a 'clean' supply is required for the computer network. This can be provided by taking the ring or radial circuits for the computer supplies from a point as close as possible to the intake position of the electrical supply to the building. A clean earth can also be taken from this point, which is usually one core of the cable and not the armour of an SWA cable, and distributed around the final wiring circuit. Alternatively, the computer supply can be cleaned by means of a filter such as that shown in Fig. 14.45.

Secure supplies

The mains electrical supply in the United Kingdom is extremely reliable and secure. However, the loss of supply to a mainframe computer or computer network for even a second can cause the system to 'crash', and hours or even days of work can be lost.

One solution to this problem is to protect 'precious' software systems with an uninterruptible power supply (UPS). A UPS is essentially a battery supply electronically modified to provide a clean and secure a.c. supply.

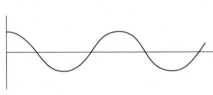

Clean supply

Spikes, caused by an over voltage transient surging through the mains

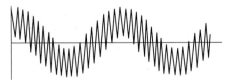

'Noise': unwanted electrical signals picked up by power lines or supply cords

FIGURE 14.44

Distortions in the a.c. mains supply.

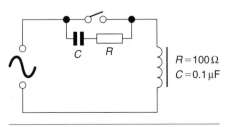

$R = 100\,\Omega$
$C = 0.1\,\mu\text{F}$

FIGURE 14.45

A simple noise suppressor.

331

An *UPS* is essentially a battery supply electronically modified to provide a clean and secure a.c. supply. The UPS is plugged into the mains supply and the computer systems are plugged into the UPS.

The UPS is plugged into the mains supply and the computer systems are plugged into the UPS.

An **UPS** to protect a small network of, say, six PCs is physically about the size of one PC hard drive and is usually placed under or at the side of an operator's desk.

It is best to dedicate a ring or radial circuit to the UPS and either to connect the computer equipment permanently or to use non-standard outlets to discourage the unauthorized use and overloading of these special supplies by, for example, kettles or the cleaner's vaccum.

Finally, remember that most premises these days contain some computer equipment and systems. Electricians intending to isolate supplies for testing or modification should first check and then check again before they finally isolate the supply in order to avoid loss or damage to computer systems.

Optical fibre cables

The introduction of fibre-optic cable systems and digital transmissions will undoubtedly affect future cabling arrangements and the work of the electrician. Networks based on the digital technology currently being used so successfully by the telecommunications industry are very likely to become the long-term standard for computer systems. Fibre-optic systems dramatically reduce the number of cables required for control and communications systems, and this will in turn reduce the physical room required for these systems. Fibre-optic cables are also immune to electrical noise when run parallel to mains cables and, therefore, the present rules of segregation and screening may change in the future. There is no spark risk if the cable is accidentally cut and, therefore, such circuits are intrinsically safe. Intrinsic safety is described in Chapter 13 under the heading Hazardous Area Installations.

Optical fibre cables are communication cables made from optical-quality plastic, the same material from which spectacle lenses are manufactured. The energy is transferred down the cable as digital pulses of laser light as against current flowing down a copper conductor in electrical installation terms. The light pulses stay within the fibre-optic cable because of a scientific principle known as 'total internal refraction' which means that the laser light bounces down the cable and when it strikes the outer wall it is always deflected inwards and, therefore, does not escape out of the cable, as shown in Fig. 14.46.

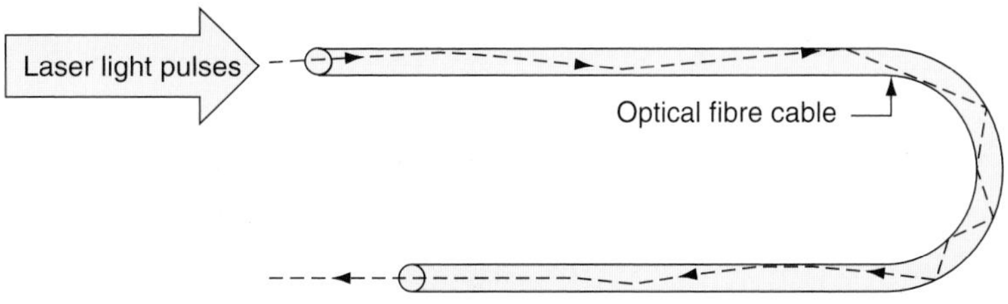

Optical fibre cables are communication cables made from optical-quality plastic, the same material from which spectacle lenses are manufactured. The energy is transferred down the cable as digital pulses of laser light as against current flowing down a copper conductor in electrical installation terms.

FIGURE 14.46
Digital pulses of laser light down an optical fibre cable.

The cables are very small because the optical quality of the conductor is very high and signals can be transmitted over great distances. They are cheap to produce and lightweight because these new cables are made from high-quality plastic and not high-quality copper. Single-sheathed cables are often called 'simplex' cables and twin sheathed cables 'duplex', that is, two simplex cables together in one sheath. Multi-core cables are available containing up to 24 single fibres.

Fibre-optic cables look like steel wire armour cables (but of course are lighter) and should be installed in the same way and given the same level of protection as SWA cables. Avoid tight-radius bends if possible and kinks at all costs. Cables are terminated in special joint boxes which ensure cable ends are cleanly cut and butted together to ensure the continuity of the light pulses. Fibre-optic cables are Band I circuits when used for data transmission and must therefore be segregated from other mains cables to satisfy the IEE Regulations.

The testing of fibre-optic cables requires that special instruments be used to measure the light attenuation (i.e. light loss) down the cable. Finally, when working with fibre-optic cables, electricians should avoid direct eye contact with the low-energy laser light transmitted down the conductors.

Fire alarm circuits (BS 5839 and BS EN 54-2: 1998)

Through one or more of the various statutory Acts, all public buildings are required to provide an effective means of giving a warning of fire so that life and property may be protected. An effective system is one which gives a warning of fire while sufficient time remains for the fire to be put out and any occupants to leave the building.

Fire alarm circuits are wired as either normally open or normally closed. In a *normally open circuit*, the alarm call points are connected in parallel with each other so that when any alarm point is initiated the circuit is completed and the sounder gives a warning of fire. The arrangement is shown in Fig. 14.47. It is essential for some parts of the wiring system to continue

Definition

Fire alarm circuits are wired as either normally open or normally closed. In a *normally open circuit*, the alarm call points are connected in parallel with each other so that when any alarm point is initiated the circuit is completed and the sounder gives a warning of fire.

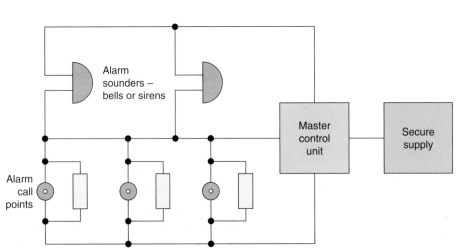

FIGURE 14.47

A simple normally open fire alarm circuit.

operating even when attacked by fire. For this reason the master control and sounders should be wired in MI or FP200 cable. The alarm call points of a normally open system must also be wired in MI or FP200 cable, unless a monitored system is used. In its simplest form this system requires a high-value resistor to be connected across the call-point contacts, which permits a small current to circulate and operate an indicator, declaring the circuit healthy. With a monitored system, PVC insulated cables may be used to wire the alarm call points.

In a **normally closed circuit**, the alarm call points are connected in series to normally closed contacts as shown in Fig. 14.48. When the alarm is initiated, or if a break occurs in the wiring, the alarm is activated. The sounders and master control unit must be wired in MI or FP200 cable, but the call points may be wired in PVC insulated cable since this circuit will always 'fail safe'.

Alarm call points

Manually operated alarm call points should be provided in all parts of a building where people may be present, and should be located so that no one need walk for more than 30 m from any position within the premises in order to give an alarm. A breakglass manual call point is shown in Fig. 14.49. They should be located on exit routes and, in particular, on the floor landings of staircases and exits to the street. They should be fixed at a height of 1.4 m above the floor at easily accessible, well illuminated and conspicuous positions.

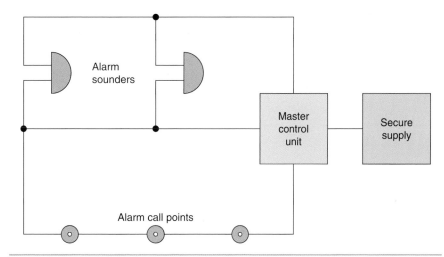

FIGURE 14.48

A simple normally closed fire alarm circuit.

Automatic detection of fire is possible with heat and smoke detectors. These are usually installed on the ceilings and at the top of stair wells of buildings because heat and smoke rise. Smoke detectors tend to give a faster response than heat detectors, but whether manual or automatic call points are used, should be determined by their suitability for the particular installation. They should be able to discriminate between a fire and the normal environment in which they are to be installed.

FIGURE 14.49

Breakglass manual call point.

FIGURE 14.50
Typical fire alarm sounders.

Sounders

The positions and numbers of **sounders** should be such that the alarm can be distinctly heard above the background noise in every part of the premises. The sounders should produce a minimum of 65 dB, or 5 dB above any ambient sound which might persist, for more than 30 s. If the sounders are to arouse sleeping persons then the minimum sound level should be increased to 75 dB at the bedhead. Bells, hooters or sirens may be used but in any one installation they must all be of the same type. Examples of sounders are shown in Fig. 14.50. Normal speech is about 5 dB.

Fire alarm design considerations

Since all fire alarm installations must comply with the relevant statutory regulations, good practice recommends that contact be made with the local fire prevention officer at the design stage in order to identify any particular local regulations and obtain the necessary certification.

Larger buildings must be divided into zones so that the location of the fire can be quickly identified by the emergency services. The zones can be indicated on an indicator board situated in, for example, a supervisor's office or the main reception area.

In selecting the zones, the following rules must be considered:

1. Each zone should not have a floor area in excess of 2000 m^2.

2. Each zone should be confined to one storey, except where the total floor area of the building does not exceed 300 m^2.

3. Staircases and very small buildings should be treated as one zone.

4. Each zone should be a single fire compartment. This means that the walls, ceilings and floors are capable of containing the smoke and fire.

At least one fire alarm sounder will be required in each zone, but all sounders in the building must operate when the alarm is activated.

The main sounders may be silenced by an authorized person, once the general public have been evacuated from the building, but the current must be diverted to a supervisory buzzer which cannot be silenced until the system has been restored to its normal operational state.

A fire alarm installation may be linked to the local fire brigade's control room by the telecommunication network, if the permission of the fire authority and local telecommunication office is obtained.

The electricity supply to the fire alarm installation must be secure in the most serious conditions. In practice the most reliable supply is the mains supply, backed up by a 'standby' battery supply in case of mains failure. The supply should be exclusive to the fire alarm installation, fed from a separate switch fuse, painted red and labelled, 'Fire Alarm – Do Not Switch Off'. Standby battery supplies should be capable of maintaining the system in full normal operation for at least 24 h and, at the end of that time, be capable of sounding the alarm for at least 30 min.

Fire alarm circuits are Band I circuits and consequently cables forming part of a fire alarm installation must be physically segregated from all Band II circuits unless they are insulated for the highest voltage (IEE Regulations 528.1 and 560.7.1).

Intruder alarms

The installation of security alarm systems in the United Kingdom is already a multi-million-pound business and yet it is also a relatively new industry. As society becomes increasingly aware of crime prevention, it is evident that the market for security systems will expand.

Not all homes are equally at risk, but all homes have something of value to a thief. Properties in cities are at highest risk, followed by homes in towns and villages, and at least risk are homes in rural areas. A nearby motorway junction can, however, greatly increase the risk factor. Flats and maison-ettes are the most vulnerable, with other types of property at roughly equal risk. Most intruders are young, fit and foolhardy opportunists. They ideally want to get in and away quickly but, if they can work unseen, they may take a lot of trouble to gain access to a property by, for example, removing the glass from a window.

Most intruders are looking for portable and easily saleable items such as video recorders, television sets, home computers, jewellery, cameras, silverware, money, cheque books or credit cards. The Home Office has stated that only 7% of homes are sufficiently protected against intruders, although 75% of householders believe they are secure. Taking the simplest precautions will reduce the risk, while installing a security system can greatly reduce the risk of a successful burglary.

Security lighting

Security lighting is the first line of defence in the fight against crime. 'Bad men all hate the light and avoid it, for fear their practices should be shown up' (John 3:20). A recent study carried out by Middlesex University has shown that in two London boroughs the crime figures were reduced by improving the lighting levels. Police forces agree that homes and public buildings which are externally well illuminated are a much less attractive target for the thief.

Security lighting installed on the outside of the home may be activated by external detectors. These detectors sense the presence of a person outside the protected property and additional lighting is switched on. This will deter most potential intruders while also acting as courtesy lighting for visitors (Fig. 14.51).

Passive infra-red detectors

Passive infra-red (PIR) detector units allow a householder to switch on lighting units automatically whenever the area covered is approached by a moving body whose thermal radiation differs from the background. This type of detector is ideal for driveways or dark areas around the protected

---

Definition

Security lighting is the first line of defence in the fight against crime.

FIGURE 14.51

Security lighting reduces crime.

Definition

PIR detector units allow a householder to switch on lighting units automatically whenever the area covered is approached by a moving body whose thermal radiation differs from the background.

Definition

An *intruder alarm system* serves as a deterrent to a potential thief and often reduces home insurance premiums.

Definition

A *perimeter protection system* places alarm sensors on all external doors and windows so that an intruder can be detected as he or she attempts to gain access to the protected property.

Definition

A *movement or heat detector* placed in a room will detect the presence of anyone entering or leaving that room.

property. It also saves energy because the lamps are only switched on when someone approaches the protected area. The major contribution to security lighting comes from the 'unexpected' high-level illumination of an area when an intruder least expects it. This surprise factor often encourages the potential intruder to 'try next door'.

PIR detectors are designed to sense heat changes in the field of view dictated by the lens system. The field of view can be as wide as 180°, as shown by the diagram in Fig. 14.52. Many of the 'better' detectors use a split lens system so that a number of beams have to be broken before the detector switches on the security lighting. This capability overcomes the problem of false alarms, and a typical PIR is shown in Fig. 14.53.

PIR detectors are often used to switch tungsten halogen floodlights because, of all available luminaires, tungsten halogen offers instant high-level illumination. Light fittings must be installed out of reach of an intruder in order to prevent sabotage of the security lighting system.

Intruder alarm systems

Alarm systems are now increasingly considered to be an essential feature of home security for all types of homes and not just property in high-risk areas. An **intruder alarm system** serves as a deterrent to a potential thief and often reduces home insurance premiums. In the event of a burglary they alert the occupants, neighbours and officials to a possible criminal act and generate fear and uncertainty in the mind of the intruder which encourages a more rapid departure. Intruder alarm systems can be broadly divided into three categories – those which give perimeter protection, space protection or trap protection. A system can comprise one or a mixture of all three categories.

A **perimeter protection system** places alarm sensors on all external doors and windows so that an intruder can be detected as he or she attempts to gain access to the protected property. This involves fitting proximity switches to all external doors and windows.

A **movement or heat detector** placed in a room will detect the presence of anyone entering or leaving that room. PIR detectors and ultrasonic detectors

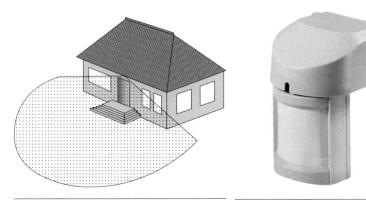

FIGURE 14.52
PIR detector and field of detection.

FIGURE 14.53
A typical PIR detector.

give space protection. Space protection does have the disadvantage of being triggered by domestic pets but it is simpler and, therefore, cheaper to install. Perimeter protection involves a much more extensive and, therefore, expensive installation, but is easier to live with.

Trap protection places alarm sensors on internal doors and pressure pad switches under carpets on through routes between, for example, the main living area and the master bedroom. If an intruder gains access to one room he cannot move from it without triggering the alarm.

PROXIMITY SWITCHES

These are designed for the discreet protection of doors and windows. They are made from moulded plastic and are about the size of a chewing-gum packet, as shown in Fig. 14.54. One moulding contains a reed switch, the other a magnet, and when they are placed close together the magnet maintains the contacts of the reed switch in either an open or closed position. Opening the door or window separates the two mouldings and the switch is activated, triggering the alarm.

PIR DETECTORS

These are activated by a moving body which is warmer than the surroundings. The PIR shown in Fig. 14.55 has a range of 12 m and a detection zone of 110° when mounted between 1.8 and 2 m high.

Intruder alarm sounders

Alarm sounders give an audible warning of a possible criminal act. Bells or sirens enclosed in a waterproof enclosure, such as shown in Fig. 14.56, are suitable. It is usual to connect two sounders on an intruder alarm installation, one inside to make the intruder apprehensive and anxious, hopefully encouraging a rapid departure from the premises, and one outside. The outside sounder should be displayed prominently since the installation of

Definition

Trap protection places alarm sensors on internal doors and pressure pad switches under carpets on through routes between, for example, the main living area and the master bedroom.

FIGURE 14.54
Proximity switches for perimeter protection.

FIGURE 14.55
PIR intruder alarm detector.

FIGURE 14.56
Intruder alarm sounder.

FIGURE 14.57

Intruder alarm control panel with remote keypad.

338

an alarm system is thought to deter the casual intruder and a ringing alarm encourages neighbours and officials to investigate a possible criminal act.

CONTROL PANEL

The control panel, such as that shown in Fig. 14.57, is at the centre of the intruder alarm system. All external sensors and warning devices radiate from the control panel. The system is switched on or off at the control panel using a switch or coded buttons. To avoid triggering the alarm as you enter or leave the premises, there are exit and entry delay times to allow movement between the control panel and the door.

SUPPLY

The supply to the intruder alarm system must be secure and this is usually achieved by an a.c. mains supply and battery back-up. Nickel–cadmium rechargeable cells are usually mounted in the sounder housing box.

Design considerations

It is estimated that there is now a 5% chance of being burgled, but the installation of a security system does deter a potential intruder. Every home in Britain will almost certainly contain electrical goods, money or valuables of value to an intruder. Installing an intruder alarm system tells the potential intruder that you intend to make his job difficult, which in most cases encourages him to look for easier pickings.

The type and extent of the intruder alarm installation, and therefore the cost, will depend upon many factors including the type and position of the building, the contents of the building, the insurance risk involved and the peace of mind offered by an alarm system to the owner or occupier of the building.

The designer must ensure that an intruder cannot sabotage the alarm system by cutting the wires or pulling the alarm box from the wall. Most systems will trigger if the wires are cut and sounders should be mounted in any easy-to-see but difficult-to-reach position.

Intruder alarm circuits are Band I circuits and should, therefore, be segregated from mains supply cables which are designated as Band II circuits or insulated to the highest voltage present if run in a common enclosure with Band II cables (IEE Regulation 528.1).

Closed circuit television

Closed circuit television (CCTV) is now an integral part of many security systems. CCTV systems range from a single monitor with just one camera dedicated to monitoring perhaps a hotel car park, through to systems with many internal and external cameras connected to several locations for monitoring perhaps a shopping precinct.

CCTV cameras are also required to operate in total darkness when flood-lighting is impractical. This is possible by using infra-red lighting which renders the scene under observation visible to the camera while to the human eye it appears to be in total darkness.

FIGURE 14.58
CCTV camera.

Cameras may be fixed or movable under remote control, such as those used for motorway traffic monitoring. Typically an external camera would be enclosed in a weatherproof housing such as those shown in Fig. 14.58. Using remote control, the camera can be panned, tilted or focused and have its viewing screen washed and wiped.

Pictures from several cameras can be multiplexed on to a single co-axial video cable, together with all the signals required for the remote control of the camera.

A permanent record of the CCTV pictures can be stored and replayed by incorporating a video tape recorder into the system as is the practice in most banks and building societies.

Security cameras should be robustly fixed and cable runs designed so that they cannot be sabotaged by a potential intruder.

Emergency lighting (BS 5266 and BS EN 1838)

Emergency lighting should be planned, installed and maintained to the highest standards of reliability and integrity, so that it will operate satisfactorily when called into action, no matter how infrequently this may be.

Emergency lighting is not required in private homes because the occupants are familiar with their surroundings, but in public buildings people are in unfamiliar surroundings. In an emergency people do not always act rationally, but well illuminated and easily identified exit routes can help to reduce panic.

Emergency lighting is provided for two reasons; to illuminate escape routes, called 'escape' lighting; and to enable a process or activity to continue after a normal lights failure, called 'standby' lighting.

Escape lighting is usually required by local and national statutory authorities under legislative powers. The escape lighting scheme should be planned so that identifiable features and obstructions are visible in the lower levels of illumination which may prevail during an emergency. Exit routes should be clearly indicated by signs and illuminated to a uniform level, avoiding bright and dark areas.

Standby lighting is required in hospital operating theatres and in industry, where an operation or process once started must continue, even if the mains lighting fails. Standby lighting may also be required for security reasons. The cash points in public buildings may need to be illuminated at all times to discourage acts of theft occurring during a mains lighting failure.

Emergency supplies

Since an emergency occurring in a building may cause the mains supply to fail, the emergency lighting should be supplied from a source which is independent from the main supply. In most premises the alternative power supply would be from batteries, but generators may also be used. Generators can have a large capacity and duration, but a major disadvantage is the delay of time while the generator runs up to speed and takes over the load. In some premises a delay of more than 5 s is considered

> **Definition**
>
> *Emergency lighting* is not required in private homes because the occupants are familiar with their surroundings, but in public buildings people are in unfamiliar surroundings. In an emergency people do not always act rationally, but well illuminated and easily identified exit routes can help to reduce panic.

> **Definition**
>
> Emergency lighting is provided for two reasons; to illuminate escape routes, called 'escape' lighting; and to enable a process or activity to continue after a normal lights failure, called 'standby' lighting.

339

unacceptable, and in these cases a battery supply is required to supply the load until the generator can take over.

The emergency lighting supply must have an adequate capacity and rating for the specified duration of time (IEE Regulation 313.2). BS 5266 and BS EN 1838 states that after a battery is discharged by being called into operation for its specified duration of time, it should be capable of once again operating for the specified duration of time following a recharge period of not longer than 24 h. The duration of time for which the emergency lighting should operate will be specified by a statutory authority but is normally 1–3 h. The British Standard states that escape lighting should operate for a minimum of 1 hour. Standby lighting operation time will depend upon financial considerations and the importance of continuing the process or activity.

There are two possible modes of operation for emergency lighting installations: maintained and non-maintained.

MAINTAINED EMERGENCY LIGHTING

In a maintained system the **emergency lamps** are continuously lit using the normal supply when this is available, and change over to an alternative supply when the mains supply fails. The advantage of this system is that the lamps are continuously proven healthy and any failure is immediately obvious. It is a wise precaution to fit a supervisory buzzer in the emergency supply to prevent accidental discharge of the batteries, since it is not otherwise obvious which supply is being used.

Maintained emergency lighting is normally installed in theatres, cinemas, discotheques and places of entertainment where the normal lighting may be dimmed or extinguished while the building is occupied. The emergency supply for this type of installation is often supplied from a central battery, the emergency lamps being wired in parallel from the low-voltage supply as shown in Fig. 14.59. Escape sign lighting units used in commercial facilities should be wired in the maintained mode.

Definition

In a maintained system the *emergency lamps* are continuously lit using the normal supply when this is available, and change over to an alternative supply when the mains supply fails.

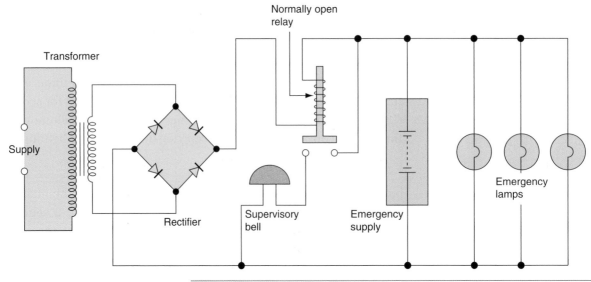

FIGURE 14.59

Maintained emergency lighting.

Definition

In a *non-maintained system* the emergency lamps are only illuminated if the normal mains supply fails.

NON-MAINTAINED EMERGENCY LIGHTING

In a **non-maintained system** the emergency lamps are only illuminated if the normal mains supply fails. Failure of the main supply de-energizes a solenoid and a relay connects the emergency lamps to a battery supply, which is maintained in a state of readiness by a trickle charge from the normal mains supply. When the normal supply is restored, the relay solenoid is energized, breaking the relay contacts, which disconnects the emergency lamps, and the charger recharges the battery. Figure 14.60 illustrates this arrangement.

The disadvantage with this type of installation is that broken lamps are not detected until they are called into operation in an emergency, unless regularly maintained. The emergency supply is usually provided by a battery contained within the luminaire, together with the charger and relay, making the unit self-contained. Self-contained units are cheaper and easier to install than a central battery system, but the central battery can have a greater capacity and duration, and permit a range of emergency lighting luminaires to be installed.

Maintenance

The contractor installing the emergency lighting should provide a test facility which is simple to operate and secure against unauthorized interference. The emergency lighting installation must be segregated completely from any other wiring, so that a fault on the main electrical installation cannot damage the emergency lighting installation (IEE Regulation 528.1). Figure 14.30 shows a trunking which provides for segregation of circuits.

The batteries used for the emergency supply should be suitable for this purpose. Motor vehicle batteries are not suitable for emergency lighting applications, except in the starter system of motor-driven generators. The fuel supply to a motor-driven generator should be checked. The battery room of a central battery system must be well ventilated and, in the case of a motor-driven generator, adequately heated to ensure rapid starting in cold weather.

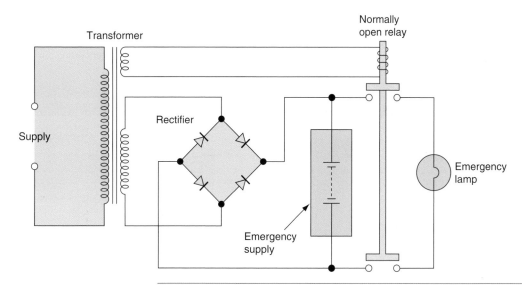

FIGURE 14.60

Non-maintained emergency lighting.

The British Standard recommends that the full load should be carried by the emergency supply for at least 1 h in every 6 months. After testing, the emergency system must be carefully restored to its normal operative state. A record should be kept of each item of equipment and the date of each test by a qualified or responsible person. It may be necessary to produce the record as evidence of satisfactory compliance with statutory legislation to a duly authorized person.

Self-contained units are suitable for small installations of up to about 12 units. The batteries contained within these units should be replaced about every 5 years, or as recommended by the manufacturer.

Primary cells

A **primary cell** cannot be recharged. Once the active chemicals are exhausted, the cell must be discarded.

Primary cells, in the form of Leclanche cells, are used extensively as portable power sources for radios and torches and have an emf of 1.5V. Larger voltages are achieved by connecting cells in series. Thus, a 6V supply can be provided by connecting four cells in series.

Mercury primary cells have an emf of 1.35V, and can have a very large capacity in a small physical size. They have a long shelf life and leakproof construction, and are used in watches and hearing aids.

Secondary cells

A **secondary cell** has the advantage of being rechargeable. If the cell is connected to a suitable electrical supply, electrical energy is stored on the plates of the cell as chemical energy. When the cell is connected to a load, the chemical energy is converted to electrical energy.

A lead-acid cell is a secondary cell. Each cell delivers about 2V, and when six cells are connected in series a 12V battery is formed of the type used on motor vehicles. Figure 14.61 shows the construction of a lead-acid battery.

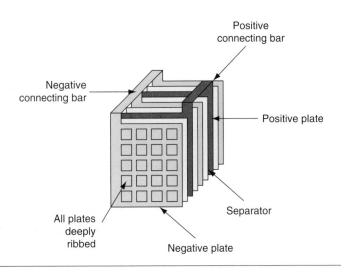

FIGURE 14.61

The construction of a lead-acid battery.

> **Definition**
>
> A *primary cell* cannot be recharged. Once the active chemicals are exhausted, the cell must be discarded.

> **Definition**
>
> A *secondary cell* has the advantage of being rechargeable. If the cell is connected to a suitable electrical supply, electrical energy is stored on the plates of the cell as chemical energy.

342

A lead-acid battery is constructed of lead plates which are deeply ribbed to give maximum surface area for a given weight of plate. The plates are assembled in groups, with insulating separators between them. The separators are made of a porous insulating material, such as wood or ebonite, and the whole assembly is immersed in a dilute sulphuric acid solution in a plastic container.

Battery rating

The capacity of a cell to store charge is a measure of the total quantity of electricity which it can cause to be displaced around a circuit after being fully charged. It is stated in ampere hours, abbreviation Ah, and calculated at the 10 h rate which is the steady load current which would completely discharge the battery in 10 h. Therefore, a 50 Ah battery will provide a steady current of 5 A for 10 h.

Maintenance of lead-acid batteries

- The plates of the battery must always be covered by the dilute sulphuric acid. If the level falls, it must be topped up with distilled water.

- Battery connections must always be tight and should be covered with a thin coat of petroleum jelly.

- The specific gravity or relative density of the battery gives the best indication of its state of charge. A discharged cell will have a specific gravity of 1.150, which will rise to 1.280 when fully charged. The specific gravity of a cell can be tested with a hydrometer.

- To maintain a battery in good condition it should be regularly trickle charged. A rapid charge or discharge encourages the plates to buckle, and may cause permanent damage.

- The room used to charge a battery must be well ventilated because the charged cell gives off hydrogen and oxygen, which are explosive in the correct proportions (IEE Regulation 560.6.3).

Telephone socket outlets

The installation of telecommunications equipment could, for many years, only be undertaken by British Telecom engineers, but today an electrical contractor may now supply and install telecommunications equipment.

On new premises the electrical contractor may install sockets and the associated wiring to the point of intended line entry, but the connection of the incoming line to the installed master socket must only be made by the telephone company's engineer.

On existing installations, additional secondary sockets may be installed to provide an extended plug-in facility as shown in Fig. 14.62. Any number of secondary sockets may be connected in parallel, but the number of telephones which may be connected at any one time is restricted.

Each telephone or extension bell is marked with a ringing equivalence number (REN) on the underside. Each exchange line has a maximum

343

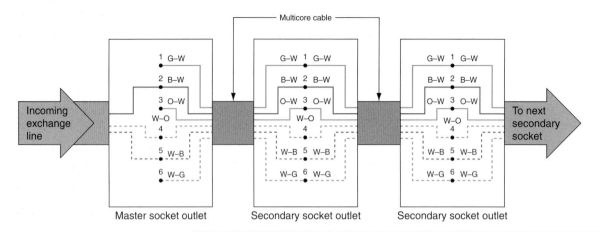

FIGURE 14.62

Telephone circuit outlet connection diagram.

Table 14.10	Telephone Cable Identification	
Code	Base colour	Stripe
G–W	Green	White
B–W	Blue	White
O–W	Orange	White
W–O	White	Orange
W–B	White	Blue
W–G	White	Green

Table 14.11 Telephone Socket Terminal Identification. Terminals 1 and 6 are Frequently Unused, and Therefore Four-Core Cable May Normally Be Installed. Terminal 4, on the Incoming Exchange Line, Is Only Used on a PBX Line for Earth Recall

Socket terminal	Circuit
1	Spare
2	Speech circuit
3	Bell circuit
4	Earth recall
5	Speech circuit
6	Spare

capacity of REN 4 and therefore, the total REN values of all the connected telephones must not exceed four if they are to work correctly.

An extension bell may be connected to the installation by connecting the two bell wires to terminals 3 and 5 of a telephone socket. The extension bell must be of the high impedance type having an REN rating. All equipment connected to a BT exchange line must display the green circle of approval.

The multi-core cable used for wiring extension socket outlets should be of a type intended for use with telephone circuits, which will normally be between 0.4 and 0.68 mm in cross-section. Telephone cable conductors are identified in Table 14.10 and the individual terminals in Table 14.11. The conductors should be connected as shown in Fig. 14.62.

Telecommunications cables are Band I circuits and must be segregated from Band II circuits containing mains cables (IEE Regulations 528.1).

Bathroom installations

Rooms containing a fixed bath tub or shower basin are considered an area of increased shock risk and, therefore, additional regulations are specified in Section 701 of the IEE Regulations. This is to reduce the risk of electric shock to people in circumstances where body resistance is lowered because of contact with water. The regulations can be summarized as follows:

- Socket outlets must not be installed and no provision is made for connection of portable appliances unless the socket outlet can be fixed 3 metres horizontally beyond the zone 1 boundary within the bath or shower room (IEE Regulation 701.512.4).

- Only shaver sockets which comply with BS EN 60742, that is, those which contain an isolating transformer, may be installed in zone 2 or outside the zones in the bath or shower room (IEE Regulation 701.512.4).

- All circuits in a bath or shower room, that is both power and lighting, must be additionally protected by an RCD having a rated maximum operating current of 30 mA (IEE Regulation 701.411.3.3).

- There are restrictions as to where appliances, switchgear and wiring accessories may be installed. See Zones for bath and shower rooms below.

- Local supplementary equipotential bonding (IEE Regulation 701.415.2) must be provided to all gas, water and central heating pipes in addition, to metallic baths, *unless the following two requirements are both met*:

  (i) all bathroom circuits, both lighting and power, are protected by a 30 mA RCD and

  (ii) the bath or shower is located in a building with protective equipotential bonding in place as described in Fig. 6.7 (IEE Regulation 411.3.1.2).

Note: Local supplementary equipotential bonding may be an additional requirement of the 'Local Authority' regulations in, for example, licensed premises, student accommodation and rented property.

Zones for bath and shower rooms

Locations that contain a bath or shower are divided in zones or separate areas as shown in Fig. 14.63.

Zone 0 – the bath tub or shower basin itself, which can contain water and is, therefore, the most dangerous zone

Zone 1 – the next most dangerous zone in which people stand in water

Zone 2 – the next most dangerous zone in which people might be in contact with water

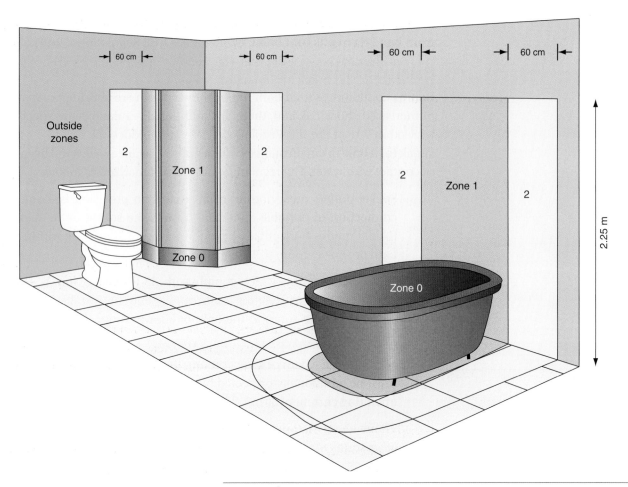

FIGURE 14.63
Bathroom Zone Dimensions. (Any window recess is zone 2.)

Outside Zones – people are least likely to be in contact with water but are still in a potentially dangerous environment and the general Regulations of BS 7671 apply.

- spaces under the bath which are accessible '*only with the use of a tool*' are outside the zones

- spaces under the bath which are accessible '*without the use of a tool*' are zone 1.

Electrical equipment and accessories are restricted within the zones.

Zone 0 – being the most potentially dangerous zone for all practical purposes, no electrical equipment can be installed in this zone. However, the Regulations permit that where SELV fixed equipment with a rated voltage not exceeding 12V a.c. cannot be located elsewhere, it may be installed in this zone (IEE Regulation 701.55). The electrical equipment must have at least IP $\times$ 7, protection against total immersion in water (IEE Regulation 701.512.2).

Zone 1 – water heaters, showers and shower pumps and SELV fixed equipment may be installed in zone 1. The electrical equipment must have at least IPX4 protection against water splashing from any direction. If the electrical equipment may be exposed to water jets from, for example, commercial cleaning equipment, then the electrical equipment must have IPX5 protection. (The Index of Protection codes were discussed in Chapter 13 and shown at Fig. 13.3.)

Zone 2 – luminaries and fans, and equipment from zone 1 plus shaver units to BS EN 60742 may be installed in zone 2. The electrical equipment must be suitable for installation in that zone according to the manufacturer's instructions and have at least IPX4 protection against splashing or IPX5 protection if commercial cleaning is anticipated.

Outside Zones – appliances are allowed plus accessories except socket outlets unless the location containing the bath or shower is very big and the socket outlet can be installed at least 3 m horizontally beyond the zone 1 boundary (IEE Regulation 701.512.3) and has additional RCD protection (IEE Regulation 701.411.3.3).

If underfloor heating is installed in these areas it must have an overall earthed metallic grid or the heating cable must have an earthed metallic sheath, which is connected to the protective conductor of the supply circuit (IEE Regulation 701.753).

347

Supplementary equipotential bonding

Modern plumbing methods make considerable use of non-metals (PTFE tape on joints, for example). Therefore, the metalwork of water and gas installations cannot be relied upon to be continuous throughout.

The IEE Regulations describe the need to consider additional protection by supplementary equipotential bonding in situations where there is a high risk of electric shock (e.g. in kitchens and bathrooms) (IEE Regulation 415.2).

In kitchens, supplementary bonding of hot and cold taps, sink tops and exposed water and gas pipes *is only required* if an earth continuity test proves that they are not already effectively and reliably connected to the protective equipotential bonding, having negligible impedance, by the soldered pipe fittings of the installation. If the test proves unsatisfactory, the metalwork must be bonded using a single-core copper conductor with PVC green/yellow insulation, which will normally be 4 mm^2 for domestic installations but must comply with Regulation 543.1.1.

In rooms containing a fixed bath or shower, supplementary equipotential bonding conductors *must* be installed to reduce to a minimum the risk of an electric shock unless the following *two* conditions are met:

  (i) all bathroom circuits are protected by a fuse or MCB plus a 30 mA RCD and

  (ii) the bathroom is located in a building with a protective equipotential bonding system in place (IEE Regulation 701.415.2). Such a system is shown in Fig. 6.7.

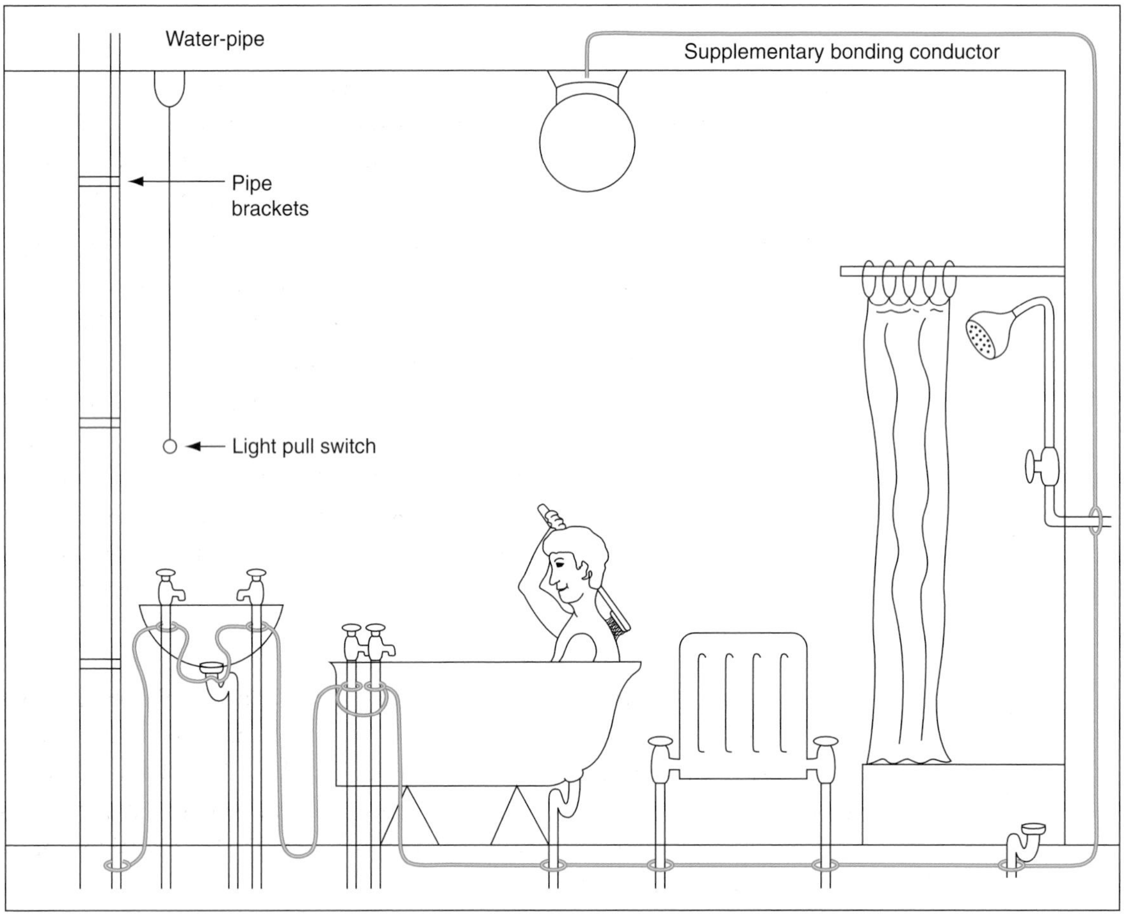

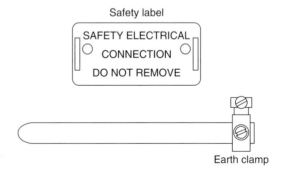

FIGURE 14.64

Supplementary bonding in bathrooms to metal pipework.

FIGURE 14.65

Typical earth bonding clamp.

Bonding conductors in domestic premises will normally be of $4\,mm^2$ copper with PVC insulation to comply with Regulation 543.1.1 and must be connected between all exposed metalwork (e.g. between metal baths, bath and sink taps, shower fittings, metal waste pipes and radiators, as shown in Fig. 14.64.

The bonding connection must be made to a cleaned pipe, using a suitable bonding clip. Fixed at or near the connection must be a permanent label saying 'Safety electrical connection – do not remove' (Regulation 514.3) as shown in Fig. 14.65.

Inspection and testing techniques

The **testing of an installation** implies the use of instruments to obtain readings. However, a test is unlikely to identify a cracked socket outlet, a chipped or loose switch plate, a missing conduit-box lid or saddle, so it is also necessary to make a visual inspection of the installation.

All new installations must be inspected and tested during erection and upon completion before being put into service. All **existing installations** should be periodically inspected and tested to ensure that they are safe and meet the regulations (IEE Regulations 610 to 634).

The method used to test an installation may inject a current into the system. This current must not cause danger to any person or equipment in contact with the installation, even if the circuit being tested is faulty. The test procedures must be followed carefully and in the correct sequence, as indicated by Regulation 612.1. This ensures that the protective conductors are correctly connected and secure before the circuit is energized.

Visual inspection

The installation must be visually inspected before testing begins. The aim of the **visual inspection** is to confirm that all equipment and accessories are undamaged and comply with the relevant British and European Standards, and also that the installation has been securely and correctly erected. Regulation 611.3 gives a checklist for the initial visual inspection of an installation, including:

- connection of conductors
- identification of conductors
- routing of cables in safe zones
- selection of conductors for current carrying capacity and volt drop
- connection of single-pole devices for protection or switching in phase conductors only
- correct connection of socket outlets, lampholders, accessories and equipment
- presence of fire barriers, suitable seals and protection against thermal effects
- methods of 'basic protection' against electric shock, including the insulation of live parts and placement of live parts out of reach by fitting appropriate barriers and enclosures
- methods of 'fault protection' against electric shock including the presence of earthing conductors for both protective bonding and supplementary bonding.
- prevention of detrimental influences (e.g. corrosion)
- presence of appropriate devices for isolation and switching
- presence of undervoltage protection devices

Definition

The *testing of an installation* implies the use of instruments to obtain readings. However, a test is unlikely to identify a cracked socket outlet, a chipped or loose switch plate, a missing conduit-box lid or saddle, so it is also necessary to make a visual inspection of the installation.

Definition

All *existing installations* should be periodically inspected and tested to ensure that they are safe and meet the regulations of the IEE (Regulations 610 to 634).

Definition

The installation must be visually inspected before testing begins. The aim of the *visual inspection* is to confirm that all equipment and accessories are undamaged and comply with the relevant British and European Standards, and also that the installation has been securely and correctly erected.

Key Fact

Testing

- All new installations must be inspected and tested during erection and upon completion
- All existing installations must be inspected and tested periodically
- IEE Regulations 610 to 634.

- choice and setting of protective devices
- labelling of circuits, fuses, switches and terminals
- selection of equipment and protective measures appropriate to external influences
- adequate access to switchgear and equipment
- presence of danger notices and other warning notices
- presence of diagrams, instruction and similar information
- appropriate erection method.

The checklist is a guide, it is not exhaustive or detailed, and should be used to identify relevant items for inspection, which can then be expanded upon. For example, the first item on the checklist, connection of conductors, might be further expanded to include the following:

- Are connections secure?
- Are connections correct? (conductor identification)
- Is the cable adequately supported so that no strain is placed on the connections?
- Does the outer sheath enter the accessory?
- Is the insulation undamaged?
- Does the insulation proceed up to but not *into* the connection?

This is repeated for each appropriate item on the checklist.

Those tests which are relevant to the installation must then be carried out in the sequence given in Regulation 612.1 and the *On Site Guide* for reasons of safety and accuracy. These tests are as follows:

Before the supply is connected

1. Test for continuity of protective conductors, including protective equipotential and supplementary bonding.
2. Test the continuity of all ring final circuit conductors.
3. Test for insulation resistance.
4. Test for polarity using the continuity method.
5. Test the earth electrode resistance.

With the supply connected

6. Recheck polarity using a voltmeter or approved test lamp.
7. Test the earth fault loop impedance.
8. Carry out functional testing (e.g. operation of RCDs).

If any test fails to comply with the Regulations, then *all* the preceding tests must be repeated after the fault has been rectified. This is because the earlier test results may have been influenced by the fault (Regulation 612.1).

There is an increased use of electronic devices in electrical installation work, for example, in dimmer switches and ignitor circuits of discharge lamps. These devices should temporarily be disconnected so that they are not damaged by the test voltage of, for example, the insulation resistance test (Regulation 612.3).

Approved test instruments

Definition

The *test instruments* and *test leads* used by the electrician for testing an electrical installation must meet all the requirements of the relevant regulations.

The **test instruments** and **test leads** used by the electrician for testing an electrical installation must meet all the requirements of the relevant regulations. The Health and Safety Executive has published Guidance Notes GS 38 for test equipment used by electricians. The IEE Regulations (BS 7671) also specify the test voltage or current required to carry out particular tests satisfactorily. All test equipment must be chosen in accordance with the relevant parts of BS EN 61557. All testing must, therefore, be carried out using an 'approved' test instrument if the test results are to be valid. The test instrument must also carry a **calibration certificate**, otherwise the recorded results may be void. *Calibration certificates usually last for a year. Test instruments must, therefore, be tested and recalibrated each year by an approved supplier.* This will maintain the accuracy of the instrument to an acceptable level, usually within 2% of the true value.

Definition

All testing must, therefore, be carried out using an 'approved' test instrument if the test results are to be valid. The test instrument must also carry a *calibration certificate*, otherwise the recorded results may be void.

Modern digital test instruments are reasonably robust, but to maintain them in good working order they must be treated with care. An approved test instrument costs equally as much as a good-quality camera; it should, therefore, receive the same care and consideration. Let us look at the requirements of four often used test meters.

CONTINUITY TESTER

To measure accurately the resistance of the conductors in an electrical installation we must use an instrument which is capable of producing an open circuit voltage of between 4 and 24V a.c. or d.c., and deliver a short-circuit current of not less than 200mA (Regulation 612.2.1). The functions of continuity testing and insulation resistance testing are usually combined in one test instrument.

INSULATION RESISTANCE TESTER

The test instrument must be capable of detecting insulation leakage between live conductors and between live conductors and earth. To do this and comply with Regulation 612.3 the test instrument must be capable of producing a test voltage of 250V, 500V or 1000V and deliver an output current of not less than 1mA at its normal voltage.

EARTH FAULT LOOP IMPEDANCE TESTER

The test instrument must be capable of delivering fault currents as high as 25A for up to 40ms using the supply voltage. During the test, the instrument does an Ohm's law calculation and displays the test result as a resistance reading.

RCD TESTER

Where circuits are protected by an RCD we must carry out a test to ensure that the device will operate very quickly under fault conditions and within the time limits set by the IEE Regulations. The instrument must, therefore, simulate a fault and measure the time taken for the RCD to operate. The instrument is, therefore, calibrated to give a reading measured in milliseconds to an in-service accuracy of 10%.

If you purchase good-quality 'approved' test instruments and leads from specialist manufacturers they will meet all the Regulations and Standards and therefore give valid test results. However, to carry out all the tests required by the IEE Regulations will require a number of test instruments and this will represent a major capital investment in the region of £1000.

Let us now consider the individual tests.

1 Testing the continuity of protective conductors (612.2)

The object of the test is to ensure that the CPC is correctly connected, is electrically sound and has a total resistance which is low enough to permit the overcurrent protective device to operate within the disconnection time requirements of Regulation 411.4.6, should an earth fault occur. Every protective conductor must be separately tested from the consumer's main protective earthing terminal to verify that it is electrically sound and correctly connected, including the protective equipotential bonding conductors and supplementary bonding conductors.

A d.c. test using an ohmmeter continuity tester is suitable where the protective conductors are of copper or aluminium up to 35 mm². The test is made with the supply disconnected, measuring from the consumer's main protective earthing terminal to the far end of each CPC, as shown in Fig. 14.66.

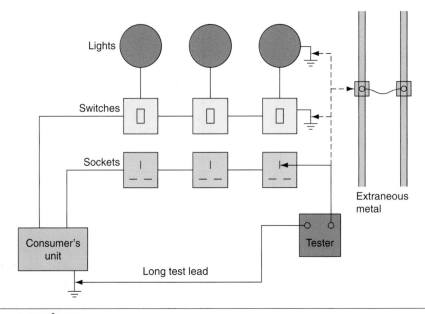

FIGURE 14.66

Testing continuity of protective conductors.

Table 14.12	Resistance Values of Some Metallic Containers	
Metallic sheath	Size (mm)	Resistance at 20°C (mΩ/m)
Conduit	20	1.25
	25	1.14
	32	0.85
Trunking	50 × 50	0.949
	75 × 75	0.526
	100 × 100	0.337

The resistance of the long test lead is subtracted from these readings to give the resistance value of the CPC. The result is recorded on an installation schedule such as that given in Appendix 6 of the IEE Regulations.

A satisfactory test result for the bonding conductors will be in the order of 0.05 Ω or less (IEE Guidance Note 3).

Where steel conduit or trunking forms the protective conductor, the standard test described above may be used, but additionally the enclosure must be visually checked along its length to verify the integrity of all the joints.

If the inspecting engineer has grounds to question the soundness and quality of these joints then the phase–earth loop impedance test described later in this chapter should be carried out.

If, after carrying out this further test, the inspecting engineer still questions the quality and soundness of the protective conductor formed by the metallic conduit or trunking then a further test can be done using an a.c. voltage not greater than 50V at the frequency of the installation and a current approaching 1.5 times the design current of the circuit, but not greater than 25 A.

This test can be done using a low-voltage transformer and suitably connected ammeters and voltmeters, but a number of commercial instruments are available such as the Clare tester, which give a direct reading in ohms.

Because fault currents will flow around the earth fault loop path, the measured resistance values must be low enough to allow the overcurrent protective device to operate quickly. For a satisfactory test result, the resistance of the protective conductor should be consistent with those values calculated for a line conductor of similar length and cross-sectional area. Values of resistance per metre for copper and aluminium conductors are shown in Table 12.1. The resistances of some other metallic containers are given in Table 14.12.

Example

The CPC for a ring final circuit is formed by a 1.5 mm² copper conductor of 50 m approximate length. Determine a satisfactory continuity test value for the CPC using the value given in Table 9A of the *On Site Guide*. From Table 9A (shown in Table 12.1 of Chapter 12).

Resistance/metre for a
1.5 mm² Copper conductor = 12.10mΩ/m
Therefore,

$$\text{the resistance of 50 mm} = 50 \times 12.10 \times 10^{-3}$$
$$= 0.605\,\Omega$$

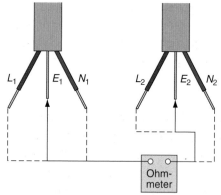

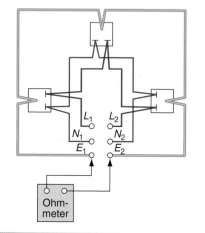

FIGURE 14.67

Step 1 test: measuring the resistance of phase, neutral and protective conductors.

The protective conductor resistance values calculated by this method can only be an approximation since the length of the CPC can only be estimated. Therefore, in this case, a satisfactory test result would be obtained if the resistance of the protective conductor was about 0.6 Ω. A more precise result is indicated by the earth fault loop impedance test which is carried out later in the sequence of tests.

2 Testing for continuity of ring final circuit conductors 612.2.2

The object of the test is to ensure that all ring circuit cables are continuous around the ring, that is, that there are no breaks and no interconnections in the ring and that all connections are electrically and mechanically sound. This test also verifies the polarity of each socket outlet.

The test is made with the supply disconnected, using an ohmmeter as follows:

Disconnect and separate the conductors of both legs of the ring at the main fuse. There are three steps to this test.

Step 1

Measure the resistance of the phase conductors (L_1 and L_2), the neutral conductors (N_1 and N_2) and the protective conductors (E_1 and E_2) at the mains position as shown in Fig. 14.67. End-to-end live and neutral conductor readings should be approximately the same (i.e. within 0.05 Ω) if the ring is continuous. The protective conductor reading will be 1.67 times as great as these readings if 2.5/1.5 mm cable is used. Record the results on a table such as that shown in Table 14.13.

Step 2

The live and neutral conductors should now be temporarily joined together as shown in Fig. 14.68. An ohmmeter reading should then be taken between live and neutral at *every* socket outlet on the ring circuit. The readings obtained should be substantially the same, provided that there are no breaks or multiple loops in the ring. Each reading should have a value of approximately

Table 14.13 Table Which May Be Used to Record the Readings Taken When Carrying Out the Continuity of Ring Final Circuit Conductors Tests According to IEE Regulation 612.2.2

Test	Ohmmeter connected to	Ohmmeter readings	This gives a value for
Step 1	L_1 and L_2 N_1 and N_2 E_1 and E_2		r_1 r_2
Step 2	Live and neutral at each socket		
Step 3	Live and earth at each socket		$R_1 + R_2$
As a check ($R_1 + R_2$) value should equal			$r_1 + r_2/4$

half the live and neutral ohmmeter readings measured in step 1 of this test. Sockets connected as a spur will have a slightly higher value of resistance because they are fed by only one cable, while each socket on the ring is fed by two cables. Record the results on a table such as that shown in Table 14.13.

Step 3

Where the CPC is wired as a ring, for example where twin and earth cables or plastic conduit is used to wire the ring, temporarily join the live and CPCs together as shown in Fig. 14.69. An ohmmeter reading should then be taken between live and earth at *every* socket outlet on the ring. The readings obtained should be substantially the same provided that there are no breaks or multiple loops in the ring. This value is equal to $R_1 + R_2$ for the circuit. Record the results on an installation schedule such as that given in Appendix 6 of the IEE Regulations or a table such as that shown in Table 14.13. The step 3 value of $R_1 + R_2$ should be equal to $(r_1 + r_2)/4$, where r_1 and r_2 are the ohmmeter readings from step 1 of this test (see Table 14.13).

3 Testing insulation resistance (612.3)

The object of the test is to verify that the quality of the insulation is satisfactory and has not deteriorated or short circuited. The test should be made at the consumer's unit with the mains switch off, all fuses in place and all switches closed. Neon lamps, capacitors and electronic circuits

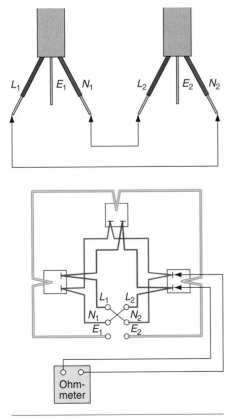

FIGURE 14.68

Step 2 test: connection of mains conductors and test circuit conditions.

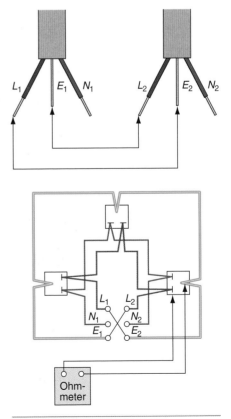

FIGURE 14.69

Step 3 test: connection of mains conductors and test circuit conditions.

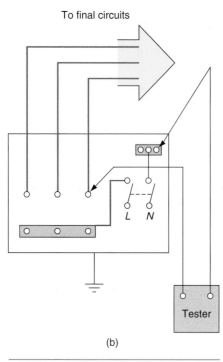

To final circuits

(a)

Tester

To final circuits

(b)

Tester

FIGURE 14.70
Insulation resistance test.

should be disconnected, since they will respectively glow, charge up or be damaged by the test.

There are two tests to be carried out using an insulation resistance tester which must have a test voltage of 500 V d.c. for 230 V and 400 V installations. These are line and neutral conductors to earth and between line conductors. The procedures are:

Phase and neutral conductors to earth

1. Remove all lamps.

2. Close all switches and circuit breakers.

3. Disconnect appliances.

4. Test separately between the line conductor and earth, *and* between the neutral conductor and earth, for *every* distribution circuit at the consumer's unit as shown in Fig. 14.70(a). Record the results on a schedule of test results such as that given in Appendix 6 of the IEE Regulations.

Between line conductors

1. Remove all lamps.

2. Close all switches and circuit breakers.

3. Disconnect appliances.

4. Test between line and neutral conductors of *every* distribution circuit at the consumer's unit as shown in Fig. 14.70(b) and record the result.

The insulation resistance readings for each test must be not less than 1.0 MΩ for a satisfactory result (IEE Regulation 612.3.2).

Where the circuit includes electronic equipment which might be damaged by the insulation resistance test, a measurement between all live conductors (i.e. live and neutral conductors connected together) and the earthing arrangements may be made. The insulation resistance of these tests should be not less than 1.0 MΩ (IEE Regulation 612.3.3).

Although an insulation resistance reading of 1.0 MΩ complies with the Regulations, the IEE Guidance Notes tell us that much higher values than this can be expected and that a reading of less than 2 MΩ might indicate a latent but not yet visible fault in the installation. In these cases each circuit should be separately tested to obtain a reading greater than 2 MΩ.

4 Testing polarity (612.6)

The object of this test is to verify that all fuses, circuit breakers and switches are connected in the line or live conductor only, and that all socket outlets are correctly wired and Edison screw-type lampholders have the centre contact connected to the live conductor. It is important to make a polarity test on the installation since a visual inspection will only indicate conductor identification.

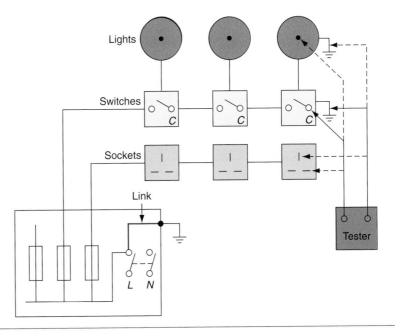

FIGURE 14.71
Polarity test.

The test is done with the supply disconnected using an ohmmeter or continuity tester as follows:

1. Switch off the supply at the main switch.

2. Remove all lamps and appliances.

3. *Fix a temporary link* between the line and earth connections on the consumer's side of the main switch.

4. Test between the 'common' terminal and earth at each switch position.

5. Test between the centre pin of any Edison screw lampholders and any convenient earth connection.

6. Test between the live pin (i.e. the pin to the right of earth) and earth at each socket outlet as shown in Fig. 14.71.

For a satisfactory test result the ohmmeter or continuity meter should read very close to zero for each test.

Remove the test link and record the results on a schedule of test results such as that given in Appendix 6 of the IEE Regulations.

5 Testing earth electrode resistance (612.7)

Low-voltage supplies having earthing arrangements which are independent of the supply cable are classified as TT systems. For this type of supply it is necessary to sink an earth electrode into the general mass of earth, which then forms a part of the earth return in conjunction with an RCD.

To verify the resistance of an electrode used in this way, the following test method may be applied:

1. Disconnect the installation protective equipotential bonding from the earth electrode to ensure that the test current passes only through the earth electrode.

2. Switch off the consumer's unit to isolate the installation.

3. Using a line earth loop impedance tester, test between the incoming line conductor and the earth electrode.

4. Reconnect the protective bonding conductors when the test is completed.

Record the result on a schedule of test results such as that given in Appendix 6 of the IEE Regulations.

The IEE Guidance Note 3 tells us that an acceptable value for the measurement of the earth electrode resistance will be about 200 Ω or less.

Providing the first five tests were satisfactory, the supply may now be switched on and the final tests completed with the supply connected.

6 Testing polarity: Supply connected

Using an approved voltmeter or test lamp and probes which comply with the HSE Guidance Note GS38, again carry out a polarity test to verify that all fuses, circuit breakers and switches are connected in the live conductor. Test from the common terminal of switches to earth, the live pin of each socket outlet to earth and the centre pin of any Edison screw lampholders to earth. In each case the voltmeter or test lamp should indicate the supply voltage for a satisfactory result.

7 Testing earth fault loop impedance (supply connected) (612.9)

The object of this test is to verify that the impedance of the whole earth fault current loop line to earth is low enough to allow the overcurrent protective device to operate within the disconnection time requirements of Regulations 411.3.2.2 and 411.4.6 and 7, should a fault occur.

The whole earth fault current loop examined by this test is comprised of all the installation protective conductors, the protective earthing terminal and protective conductors, the earthed neutral point and the secondary winding of the supply transformer and the line conductor from the transformer to the point of the fault in the installation.

The test will, in most cases, be done with a purpose-made line earth loop impedance tester which circulates a current in excess of 10 A around the loop for a very short time, so reducing the danger of a faulty circuit. The test is made with the supply switched on, and carried out from the furthest point of *every* final circuit, including lighting, socket outlets and any fixed appliances. Record the results on a schedule of test results.

Purpose-built testers give a readout in ohms and a satisfactory result is obtained when the loop impedance does not exceed the appropriate values given in Tables 41.2 and 41.3 of the IEE Regulations, see also Table 12.2. in Chapter 12.

8 Additional Protection: testing of RCD: Supply connected (612.13)

The object of the test is to verify the effectiveness of the RCD, that it is operating with the correct sensitivity and proving the integrity of the electrical and mechanical elements. The test must simulate an appropriate fault condition and be independent of any test facility incorporated in the device.

When carrying out the test, all loads normally supplied through the device are disconnected.

The testing of a ring circuit protected by a general-purpose RCD to BS EN 61008 in a split-board consumer unit is carried out as follows:

1. Using the standard lead supplied with the test instrument, disconnect all other loads and plug in the test lead to the socket at the centre of the ring (i.e. the socket at the furthest point from the source of supply).

2. Set the test instrument to the tripping current of the device and at a phase angle of 0°.

3. Press the test button – the RCD should trip and disconnect the supply within 200 ms.

4. Change the phase angle from 0° to 180° and press the test button once again. The RCD should again trip within 200 ms. Record the highest value of these two results on a schedule of test results such as that given in Appendix 6 of the IEE Regulations.

5. Now set the test instrument to 50% of the rated tripping current of the RCD and press the test button. The RCD should *not trip* within 2 s. This test is testing the RCD for inconvenience *or* nuisance tripping.

6. Finally, the effective operation of the test button incorporated within the RCD should be tested to prove the integrity of the mechanical elements in the tripping device. This test should be repeated every 3 months.

If the RCD fails any of the above tests it should be changed for a new one.

Where the RCD has a rated tripping current not exceeding 30 mA and has been installed to reduce the risk associated with 'basic' and or 'fault' protection as indicated in Regulation 411.1, a residual current of 150 mA should cause the circuit breaker to open within 40 ms.

Certification and reporting

Following the inspection and testing of an installation, a certificate should be given by the electrical contractor or responsible person to the person ordering the work (Chapters 62 and 63 of the IEE Regulations).

The certificate should be based upon the model form set out in Appendix 6 of the IEE Regulations. It should include the test values which verify that the installation complies with the regulations for electrical installations at the time of testing.

An 'Electrical Installation certificate' should be used for the initial certification of a new electrical installation or for an alteration or addition to an existing installation.

All installations should be tested and inspected periodically and a 'periodic inspection' certificate issued. Suggested frequency of periodic inspections are given below:

- domestic installations – 10 years
- commercial installations – 5 years
- industrial installations – 3 years
- agricultural installations – 3 years
- caravan installations – 3 years
- caravan site installations – 1 year
- temporary installations on construction sites – 3 months

See Table 3.2 of IEE Guidance Note 3 for a full list of the maximum period between inspections for all types of installation.

Construction site safety

In this chapter we have looked at many different types of electrical installations and wiring systems. Most of this type of work is carried out alongside other trades as a part of the construction process. Electricians have an important part to play in any construction project. However, a construction site has the potential to be hazardous because of the temporary and changing nature of the building environment. We have looked at the common causes of accidents at work and how to control the risks associated with various hazards in earlier chapters of this book. To make your work environment safe always:

- behave responsibly and sensibly
- assess the hazards before starting work
- wear appropriate personal protective equipment (PPE) and take all necessary precautions
- keep you work area clean and tidy and use barriers to isolate potential hazards
- carry out a safe isolation procedure and prove the supply 'dead' before beginning work on 'live' circuits or equipment. While some 'live testing' is permitted by 'competent' persons, 'live working' is not permissible under the Electricity at Work Regulations. Secure electrical

Key Facts

Waste

- clean up before you leave the job
- put waste material in the correct skip
- recycle used tubes and lamps
- get rid of waste responsibly.

isolation is discussed in Chapter 8 and a voltage indicator, which is suitable for proving the supply 'dead', is shown at Fig. 8.10.

- when the job is finished, have a good clean up and dispose of all waste materials responsibly. What goes into the skip for normal disposal is usually a matter of common sense, but under the new Hazardous Waste Regulations fluorescent tubes and lamps are classified as hazardous and the responsible way to dispose of these is to recycle them. In Chapter 8 under the sub-heading 'Disposing of Waste' we discussed this topic in detail.

Check your Understanding

When you have completed these questions, check out the answers at the back of the book.
Note: more than one multiple choice answer may be correct.

1. An electricity supply which uses the cable sheath to provide an earth is called a:
   a. PME supply
   b. TN-S system
   c. TN-C-S system
   d. TT system.

2. An electricity supply in which the protective and neutral functions are combined is called:
   a. PME system
   b. TN-S system
   c. TN-C-S system
   d. TT system.

3. An electricity supply in which the earthing arrangements must be provided by the consumer is called a:
   a. PME system
   b. TN-S system
   c. TN-C-S system
   d. TT system.

4. The main lighting circuit in a room having only one entrance would probably be:
   a. pull switch control
   b. intermediate switch control
   c. one-way switch control
   d. two-way switch control.

5. The main lighting circuit in a room having two entrances would probably be:
   a. pull switch control
   b. intermediate switch control
   c. one-way switch control
   d. two-way switch control.

6. The main lighting in a long corridor with many switches would probably be:
   a. pull switch control
   b. intermediate switch control
   c. one-way switch control
   d. two-way switch control.

7. Identify the energy-efficient lamps on the following list:

    a. GLS lamp

    b. CFLs

    c. fluorescent tubes

    d. mini-spots.

8. A PVC insulated and sheathed wiring system would be suitable for the following type of installation:

    a. commercial

    b. domestic

    c. horticultural

    d. industrial.

9. A PVC conduit installation would be suitable for the following type of installation:

    a. commercial

    b. domestic

    c. horticultural

    d. industrial.

10. A steel conduit installation would be suitable for the following type of installation:

    a. commercial

    b. domestic

    c. horticultural

    d. industrial.

11. A steel trunking installation would be suitable for the following type of installation:

    a. commercial

    b. domestic

    c. horticultural

    d. industrial.

12. Which of the following fixing methods would be suitable for holding a lightweight load on a plasterboard partition:

    a. cable clip

    b. Rawlbolt

    c. screw fixing to plastic plug

    d. spring toggle.

13. Which of the following fixing methods would be suitable for holding a medium weight load on to a brick wall:

   a. cable clip

   b. Rawlbolt

   c. screw fixing to plastic plug

   d. spring toggle.

14. Which of the following fixing methods would be suitable for holding a PVC insulated and sheathed cable on to a wood surface such as a ceiling joist:

   a. cable clip

   b. Rawlbolt

   c. screw fixing to plastic plug

   d. spring toggle.

15. Which of the following fixing methods would be suitable for securing a heavy electric motor to a concrete bed:

   a. cable clip

   b. Rawlbolt

   c. screw fixing to plastic plug

   d. spring toggle.

16. Electrical supplies on construction sites for portable tools should be at:

   a. 25 V

   b. 110 V

   c. 230 V

   d. 400 V.

17. Electric supplies on construction sites for major items such as cranes should be at:

   a. 25 V

   b. 110 V

   c. 230 V

   d. 400 V.

18. Electrical supplies on construction sites for site offices should be at:

   a. 25 V

   b. 110 V

   c. 230 V

   d. 400 V.

19. Every caravan pitch must have at least one 16 A socket outlet for connection to a caravan. The socket outlet must be:

    a. of the industrial type

    b. protected from overcurrent

    c. protected additionally by an RCD

    d. buried at a depth of 600 mm.

20. All final circuits to a bathroom, both power and lighting, must have:

    a. overcurrent protection

    b. RCD protection

    c. IPX7 protection

    d. Zone 1 protection.

21. Identify from the list below the test which is used to verify that all fuses, circuit breakers and switches are connected in the line conductor:

    a. continuity test

    b. earth fault loop impedance

    c. functional test

    d. polarity test.

22. The Electricity at Work Regulations state that:

    a. everyone may work 'live'

    b. no one may work 'live'

    c. only competent people may work 'live'

    d. only instructed and skilled people may work 'live'.

23. Use a sketch with notes of explanation to show the earthing and protective bonding arrangements at the mains intake position of a 230 V cable sheath supply.

24. Use a sketch with notes of explanation to show the earthing and protective bonding arrangements at the mains intake position of a 230 V TT system.

25. State the advantages of a conduit trunking and tray cable enclosure system for a commercial installation such as a shopping centre.

26. Compare PVC/SWA cables with MI cables and give their advantages, disadvantages and typical applications.

27. Use a sketch with notes of explanation to show how mini-trunking and skirting trunking could be used to contain all the electrical supplies in a school's computing classroom.

28. Explain the meaning of 'segregation' of circuits.

29. Use a sketch with notes of explanation to show how trunking and tray may be suspended from the girders of a building structure by appropriate brackets.

30. State five regulations which are specifically relevant to special installations or locations on agricultural premises.

31. State five regulations which are specifically relevant to special installations or locations on horticultural premises.

32. State five regulations which are specifically relevant to special installations or locations on caravan parks and caravans.

33. Use a sketch to show how the computer power supplies of question 27 above should be connected.

34. Very briefly state the meaning of:
    - noise
    - clean supplies and
    - secure supplies
    - with regard to computer power supplies.

38. In five brief statements summarize the main requirements of emergency lighting.

39. In five brief statements summarize the main requirements of bathroom electrical installations.

40. Use bullet points to state why the IEE Regulations require all new installations to be both inspected and tested.

41. Use bullet points to describe a safe isolation procedure.

42. Who can carry out live testing?
    Why is live working not allowed?

43. Why do we carry out a continuity of protective conductors test?
    State the values of a satisfactory test result.

44. Why do we carry out an insulation resistance test?
    State the values of a satisfactory test result.

45. Why do we carry out a polarity test?
    State the values or indications of a satisfactory result.

Answers to Check your understanding

Chapter 1

1.	d	**2.**	a	**3.**	a, b
4.	b, c	**5.**	d	**6.**	d
7.	d	**8.**	c	**9.**	a, b
10.	b, c, d	**11.**	a, b	**12.**	c, d
13.	b, c, d	**14.**	c, d	**15.**	c
16.	A				

17 to 25 – answers in text of Chapter 1.

Chapter 2

1.	d	**2.**	a	**3.**	c
4.	b	**5.**	c	**6.**	d
7.	b	**8.**	a	**9.**	b
10.	c	**11.**	a	**12.**	d
13.	d	**14.**	a, b, c		

15 to 24 – answers in text of Chapter 2.

Chapter 3

1.	c	**2.**	d	**3.**	b
4.	c	**5.**	d	**6.**	b

7 to 12 – answers in text of Chapter 3.

Chapter 4

1. a		**2.** c		**3.** b	
4. d		**5.** a		**6.** d	
7. b		**8.** a		**9.** c	
10. d		**11.** c		**12.** b	
13. b		**14.** c		**15.** d	

16 to 21 – answers in text of Chapter 4.

Chapter 5

1. a, c	**2.** b, d	**3.** a, c
4. b, d	**5.** a, c	**6.** b, d
7. c	**8.** d	**9.** c
10. d	**11.** d	**12.** a
13. d	**14.** a	**15.** c
16. d	**17.** c	**18.** d

19 to 30 – answers in text of Chapter 5.

Chapter 6

1. b	**2.** c	**3.** d
4. c, d	**5.** c	**6.** d
7. c	**8.** d	**9.** a
10. b	**11.** c	**12.** d
13. c	**14.** d	

15 to 24 – answers in text of Chapter 6.

Chapter 7

1. a, b	**2.** a	**3.** d
4. c	**5.** c, d	**6.** b, d

7 to 13 – answers in text of Chapter 7.

Chapter 8

1. a, c	**2.** b, d	**3.** d
4. b	**5.** c	**6.** b, c, d
7. c	**8.** d	**9.** c
10. a	**11.** c	**12.** c
13. a, b, d		

14 to 23 – answers in text of Chapter 8.

Chapter 9

1. b   **2.** c   **3.** a

4. d

5 to 12 – answers in text of Chapter 9.

Chapter 10

1. a   **2.** c   **3.** d

4. c   **5.** a   **6.** b

7. c   **8.** c   **9.** c

10. a, d   **11.** c   **12.** b

13. b   **14.** c

15 to 21 – answers in text of Chapter 10.

Chapter 11

1. c   **2.** d   **3.** a, c

4. a, b   **5.** b   **6.** a

7. c   **8.** b   **9.** b

10. c

11 to 17 – answers in text of Chapter 11.

Chapter 12

1. a   **2.** d   **3.** c

4. b   **5.** b   **6.** b

7. c   **8.** a   **9.** d

10. d   **11.** c   **12.** c

13. b   **14.** b

15 to 28 – answers in text of Chapter 12

Chapter 13

1. a, c   **2.** c   **3.** a, b, d

4. d   **5.** b   **6.** c

7. c   **8.** d   **9.** d

10. c

11 to 20 – answers in text of Chapter 13.

Chapter 14

1.	b	**2.**	a, c	**3.**	d
4.	c	**5.**	d	**6.**	b
7.	b, c	**8.**	b	**9.**	c
10.	a, d	**11.**	a, d	**12.**	d
13.	c	**14.**	a	**15.**	b
16.	b	**17.**	d	**18.**	c
19.	a, b, c	**20.**	a, b	**21.**	d
22.	b				

23 to 45 – answers in text of Chapter 14.

Appendix A: Abbreviations, symbols and codes

Abbreviations used in electronics for multiples and submultiples

T	Tera	10^{12}
G	Giga	10^9
M	mega or meg	10^6
K	Kilo	10^3
D	Deci	10^{-1}
C	Centi	10^{-2}
M	Milli	10^{-3}
M	Micro	10^{-6}
N	Nano	10^{-9}
P	Pico	10^{-12}

Terms and symbols used in electronics

Term	Symbol
Approximately equal to	$\simeq$
Proportional to	$\propto$
Infinity	∞
Sum of	Σ
Greater than	$>$
Less than	$<$
Much greater than	$\gg$
Much less than	$\ll$
Base of natural logarithms	e
Common logarithms of x	$\log x$
Temperature	θ
Time constant	T
Efficiency	η
Per unit	p.u.

Appendix B: Health and Safety Executive (HSE) publications and information

HSE Books, Information Leaflets and Guides may be obtained from HSE Books, P.O. Box 1999, Sudbury, Suffolk CO10 6FS

HSE Infoline – Telephone No. 01541 545500 or write to HSE Information Centre, Broad Lane, Sheffield S3 7HO

HSE home page on the World Wide Web http:/www.open.gov.uk/hse/hsehome.htm

The Health and Safety Poster (Figure 1.2) and other HSE publications are available from www.hsebooks.com

Environmental Health Department of the Local Authority Look in the local telephone directory under the name of the authority

HSE AREA OFFICES

01 **South West**
Inter City House, Mitchell Lane, Victoria Street, Bristol BS1 6AN
Telephone: 01171 290681

02 **South**
Priestley House, Priestley Road,
Basingstoke RG24 9NW Telephone: 01256 473181

03 **South East**
3 East Grinstead House, London Road,
East Grinstead, West Sussex RH19 1RR
Telephone: 01342 326922

05 **London North**
Maritime House, 1 Linton Road, Barking,
Essex IG11 8HF Telephone: 0208 594 5522

06 **London South**
1 Long Lane, London SE1 4PG
Telephone: 0207 407 8911

07 **East Anglia**
39 Baddow Road, Chelmsford, Essex CM2 OHL
Telephone: 0207 407 8911

08 **Northern Home Counties**
14 Cardiff Road, Luton, Beds LU1 1PP
Telephone: 01 582 34121

09 **East Midlands**
Belgrave House, 1 Greyfriars, Northampton NN1 2BS
Telephone: 01604 21233

10 **West Midlands**
McLaren Building, 2 Masshouse Circus, Queensway
Birmingham B4 7NP
Telephone: 0121 200 2299

11 **Wales**
Brunel House, Nizalan Road, Cardiff CF2 1SH
Telephone: 02920 473777

12 **Marches**
The Marches House, Midway, Newcastle-under-Lyme,
Staffs ST5 1DT Telephone: 01782 717181

13 **North Midlands**
Brikbeck House, Trinity Square, Nottingham NG1 4AU
Telephone: 0115 470712

14 **South Yorkshire**
Sovereign House, 40 Silver Street, Sheffield S1 2ES
Telephone: 0114 739081

15 **West and North Yorkshire**
8 St Paul's Street, Leeds LS1 2LE
Telephone: 0113 446191

16 **Greater Manchester**
Quay House, Quay Street, Manchester M3 3JB
Telephone: 0161 831 7111

17 **Merseyside**
The Triad, Stanley Road, Bootle L20 3PG
Telephone: 01229 922 7211

18 **North West**
Victoria House, Ormskirk Road, Preston PR1 1HH
Telephone: 01772 59321

19 **North East**
Arden House, Regent Centre, Gosforth, Newcastle upon Tyne NE3 3JN
Telephone: 0191 284 8448

20 **Scotland East**
Belford House, 59 Belford Road, Edinburgh EH4 3UE
Telephone: 0181 225 1313

21 **Scotland West**
314 St Vincent Street, Glasgow G3 8XG
Telephone: 0141 204 2646

Glossary of terms

Acceleration

Acceleration is the rate of change in velocity with time.

$$\text{Acceleration} = \frac{\text{Velocity}}{\text{Time}} = (\text{m/s}^2)$$

Accident

An accident may be defined as an uncontrolled event causing injury or damage to an individual or property.

Alarm call points

Manually operated alarm call points should be provided in all parts of a building where people may be present, and should be located so that no one need to walk for more than 30 m from any position within the premises in order to give an alarm.

Alloy

An alloy is a mixture of two or more metals.

Appointed person

An appointed person is someone who is nominated to take charge when someone is injured or becomes ill, including calling an ambulance if required. The appointed person will also look after the first aid equipment, including re-stocking the first aid box.

Approved test instruments

The test instruments and test leads used by the electrician for testing an electrical installation must meet all the requirements of the relevant regulations. All testing must, therefore, be carried out using an 'approved' test instrument if the test results are to be valid. The test instrument must also carry a calibration certificate, otherwise the recorded results may be void.

Basic protection

Basic protection is provided by the insulation of live parts in accordance with Section 416 of the IEE Regulations.

Bonding conductor

A protective conductor providing equipotential bonding.

Bonding

The linking together of the exposed or extraneous metal parts of an electrical installation.

Cable tray

Cable tray is a sheet-steel channel with multiple holes. The most common finish is hot-dipped galvanized but PVC-coated tray is also available. It is used extensively on large industrial and commercial installations for supporting MI and SWA cables which are laid on the cable tray and secured with cable ties through the tray holes.

Capacitive reactance

Capacitive reactance (X_C) is the opposition to an a.c. current in a capacitive circuit. It causes the current in the circuit to lead ahead of the voltage.

Centrifugal force	Centrifugal force is the force acting away from the centre, the opposite to centripetal force.
Centripetal force	Centripetal force is the force acting towards the centre when a mass attached to a string is rotated in a circular path.
Circuit protective conductor (CPC)	A protective conductor connecting exposed conductive parts of equipment to the main earthing terminal.
Cohesive or adhesive force	Cohesive or adhesive force is the force required to hold things together.
Compact fluorescent lamps (CFLs)	CFLs are miniature fluorescent lamps designed to replace ordinary GLS lamps.
Competent person	A competent person is anyone who has the necessary technical skills, training and expertise to safely carry out the particular activity.
Compressive force	Compressive force is the force pushing things together.
Conductor	A conductor is a material, usually a metal, in which the electrons are loosely bound to the central nucleus. These electrons can easily become 'free electrons' which allows heat and electricity to pass easily through the material.
Conduit	A conduit is a tube, channel or pipe in which insulated conductors are contained.
Corrosion	The destruction of a metal by chemical action.
Delivery notes	Delivery note is used to confirm that goods have been delivered by the supplier, who will then send out an invoice requesting payment.
Duty holder	Duty holder, this phrase recognizes the level of responsibility which electricians are expected to take on as a part of their job in order to control electrical safety in the work environment. Everyone has a duty of care, but not everyone is a duty holder. The person who exercises 'control over the whole systems, equipment and conductors' and is the Electrical Company's representative on-site is a duty holder.
Earth	The conductive mass of the earth. Whose electrical potential is taken as zero.
Earthing	The act of connecting the exposed conductive parts of an installation to the main protective earthing terminal of the installation.
Efficiency of any machine	The ratio of the output power to the input power is known as the *efficiency* of the machine. The symbol for efficiency is the Greek letter 'eta' (η). In general, $$\eta = \frac{\text{Power output}}{\text{Power input}}$$
Electric current	The drift of electrons within a conductor is known as an electric current, measured in amperes and given the symbol I.
Electric shock	Electric shock occurs when a person becomes part of the electrical circuit.
Electrical force	Electrical force is the force created by an electrical field.
Electrotechnical industry	The electrotechnical industry is made up of a variety of individual companies, all providing a service within their own specialism to a customer, client or user.

Emergency lighting

Emergency lighting is not required in private homes because the occupants are familiar with their surroundings, but in public buildings people are in unfamiliar surroundings. In an emergency people do not always act rationally, but well-illuminated and easily identified exit routes can help to reduce panic.

Emergency switching

Emergency switching involves the rapid disconnection of the electrical supply by a single action to remove or prevent danger.

Escape/standby lighting

Emergency lighting is provided for two reasons; to illuminate escape routes, called 'escape' lighting; and to enable a process or activity to continue after a normal lights failure, called 'standby' lighting.

Expansion bolts

The most well-known expansion bolt is made by Rawlbolt and consists of a split iron shell held together at one end by a steel ferrule and a spring wire clip at the other end. Tightening the bolt draws up an expanding bolt inside the split iron shell, forcing the iron to expand and grip the masonry. Rawlbolts are for heavy-duty masonry fixings.

Exposed conductive parts

The metalwork of an electrical appliance or the trunking and conduit of an electrical system which can be touched because they are not normally live, but which may become live under fault conditions.

Extraneous conductive parts

The structural steelwork of a building and other service pipes such as gas, water, radiators and sinks.

Faraday's law

Faraday's law which states that *when a conductor cuts or is cut by a magnetic field, an emf is induced in that conductor.*

Fault protection

Fault protection is provided by protective equipotential bonding and automatic disconnection of the supply (by a fuse or miniature circuit breaker, MCB) in accordance with IEE Regulations 411.3 to 6.

Ferrous

A word used to describe all metals in which the main constituent is iron.

Fire alarm circuits

Fire alarm circuits are wired as either normally open or normally closed. In a *normally open circuit*, the alarm call points are connected in parallel with each other so that when any alarm point is initiated the circuit is completed and the sounder gives a warning of fire. In a *normally closed circuit*, the alarm call points are connected in series to normally closed contacts. When the alarm is initiated, or if a break occurs in the wiring, the alarm is activated.

Fire

Fire is a chemical reaction which will continue if fuel, oxygen and heat are present.

First aid

First aid is the initial assistance or treatment given to a casualty for any injury or sudden illness before the arrival of an ambulance, doctor or other medically qualified person.

First aider

A first aider is someone who has undergone a training course to administer first aid at work and holds a current first aid certificate.

Flashpoint

The lowest temperature at which sufficient vapour is given off from a flammable substance to form an explosive gas–air mixture is called the flashpoint.

377

Flexible conduit

Flexible conduit manufactured to BS 731-1: 1993 is made of interlinked metal spirals often covered with a PVC sleeving.

Fluorescent lamp

A fluorescent lamp is a linear arc tube, internally coated with a fluorescent powder, containing a low-pressure mercury vapour discharge.

Force

The presence of a force can only be detected by its effect on a body. A force may cause a stationary object to move or bring a moving body to rest.

Friction force

Friction force is the force which resists or prevents the movement of two surfaces in contact.

Functional switching

Functional switching involves the switching on or off, or varying the supply, of electrically operated equipment in normal service.

Fuse

A fuse is the weakest link in the circuit. Under fault conditions it will melt when an overcurrent flows, protecting the circuit conductors from damage.

Gravitational force

Gravitational force is the force acting towards the centre of the earth due to the effect of gravity.

Hazard risk assessment

Employers of more than five people must document the risks at work and the process is known as hazard risk assessment.

Hazard

A hazard is something with the 'potential' to cause harm, for example, chemicals, electricity or working above ground.

Hazardous area

An area in which an explosive gas–air mixture is present is called a *hazardous area*, and any electrical apparatus or equipment within a hazardous area must be classified as flameproof to protect the safety of workers.

Heating, magnetic or chemical

The three effects of an electric current: When an electric current flows in a circuit it can have one or more of the following three effects: *heating, magnetic* or *chemical*.

Impedance

The total opposition to current flow in an a.c. circuit is called impedance and given the symbol Z.

Inductive reactance

Inductive reactance (X_L) is the opposition to an a.c. current in an inductive circuit. It causes the current in the circuit to lag behind the applied voltage.

Inertial force

Inertial force is the force required to get things moving, to change direction or stop.

Inspection and testing techniques

The testing of an installation implies the use of instruments to obtain readings. However, a test is unlikely to identify a cracked socket outlet, a chipped or loose switch plate, a missing conduit-box lid or saddle, so it is also necessary to make a visual inspection of the installation. All existing installations should be periodically inspected and tested to ensure that they are safe and meet the regulations of the IEE (Regulations 610 to 634).

Instructed person

An instructed person is a person adequately advised or supervised by skilled persons to be able to avoid the dangers which electricity may create.

Insulator

An insulator is a material, usually a non-metal, in which the electrons are very firmly bound to the nucleus and, therefore, will not allow heat or electricity to pass through it. Good insulating materials are PVC, rubber, glass and wood.

Intrinsically safe circuit

An intrinsically safe circuit is one in which no spark or thermal effect is capable of causing ignition of a given explosive atmosphere.

Intruder alarm systems

An intruder alarm system serves as a deterrent to a potential thief and often reduces home insurance premiums.

Isolation

Isolation is defined as cutting off the electrical supply to a circuit or item of equipment in order to ensure the safety of those working on the equipment by making dead those parts which are live in normal service.

Job sheets

A job sheet or job card carries information about a job which needs to be done, usually a small job.

Lamp

A lamp is a device for converting electrical energy into light energy.

Lever

A lever is any rigid body which pivots or rotates about a fixed axis or fulcrum. Load force × Distance from fulcrum = Effort force × Distance from fulcrum

Levers and turning force

A lever allows a heavy load to be lifted or moved by a small effort.

Luminaire

A luminaire is equipment which supports an electric lamp and distributes or filters the light created by the lamp.

Magnesium oxide

The conductors of mineral insulated metal sheathed (MICC) cables are insulated with compressed magnesium oxide

Magnetic field

The region of space through which the influence of a magnet can be detected is called the magnetic field of that magnet.

Magnetic force

Magnetic force is the force created by a magnetic field.

Magnetic hysteresis

Magnetic hysteresis loops describe the way in which different materials respond to being magnetized.

Magnetic poles

The places on a magnetic material where the lines of flux are concentrated are called the magnetic poles.

Maintained emergency lighting

In a maintained system the emergency lamps are continuously lit using the normal supply when this is available, and change over to an alternative supply when the mains supply fails.

Manual handling

Manual handling is lifting, transporting or supporting loads by hand or by bodily force.

Mass

Mass is a measure of the amount of material in a substance, such as metal, plastic, wood, brick or tissue, which is collectively known as a body. The mass of a body remains constant and can easily be found by comparing it on a set of balance scales with a set of standard masses. The SI unit of mass is the kilogram (kg).

Mechanics

Mechanics is the scientific study of 'machines', where a machine is defined as a device which transmits motion or force from one place to another.

379

Metallic trunking	Metallic trunking is formed from mild steel sheet, coated with grey or silver enamel paint for internal use or a hot-dipped galvanized coating where damp conditions might be encountered.
Mini-trunking	Mini-trunking is very small PVC trunking, ideal for surface wiring in domestic and commercial installations such as offices.
Movement or heat detector	A movement or heat detector placed in a room will detect the presence of anyone entering or leaving that room.
Mutual inductance	A mutual inductance of 1 henry exists between two coils when a uniformly varying current of 1 ampere per second in one coil produces an emf of 1 volt in the other coil.
Non-ferrous	Metals which *do not* contain iron are called non-ferrous. They are non-magnetic and resist rusting. Copper, aluminium, tin, lead, zinc and brass are examples of non-ferrous metals.
Non-maintained emergency lighting	In a non-maintained system the emergency lamps are only illuminated if the normal mains supply fails.
Non-statutory regulations and codes of practice	Non-statutory regulations and codes of practice interpret the statutory regulations telling us how we can comply with the law.
Ohm's law	Ohm's law says that the current passing through a conductor under constant temperature conditions is proportional to the potential difference across the conductor.
Optical fibre cables	Optical fibre cables are communication cables made from optical-quality plastic, the same material from which spectacle lenses are manufactured. The energy is transferred down the cable as digital pulses of laser light as against current flowing down a copper conductor in electrical installation terms.
Ordinary person	An ordinary person is a person who is neither a skilled person nor an instructed person.
Overload current	An overload current can be defined as a current which exceeds the rated value in an otherwise healthy circuit.
Passive infra-red (PIR) detectors	PIR detector units allow a householder to switch on lighting units automatically whenever the area covered is approached by a moving body whose thermal radiation differs from the background.
People	People may be described as an ordinary person, a skilled person, an instructed person or a competent person.
Perimeter protection system	A perimeter protection system places alarm sensors on all external doors and windows so that an intruder can be detected as he or she attempts to gain access to the protected property.
Person	A person can be described as ordinary, competent, instructed or skilled depending upon that person's skill or ability.

Personal protective equipment (PPE)	PPE is defined as all equipment designed to be worn, or held, to protect against a risk to health and safety.
Phasor	A phasor is a straight line, having definite length and direction, which represents to scale the magnitude and direction of a quantity such as a current, voltage or impedance.
Plastic plugs	A plastic plug is made of a hollow plastic tube split up to half its length to allow for expansion. Each size of plastic plug is colour-coded to match a wood screw size.
Polyvinylchloride (PVC)	PVC used for cable insulation is a thermoplastic polymer.
Potential difference	The potential difference (p.d.) is the change in energy levels measured across the load terminals. This is also called the volt drop or terminal voltage, since emf and p.d. are both measured in volts.
Power factor	Power factor (p.f.) is defined as the cosine of the phase angle between the current and voltage.
Power	Power is the rate of doing work.

$$\text{Power} = \frac{\text{Work done}}{\text{Time taken}} \text{(W)}$$

Pressure or stress	Pressure or stress is a measure of the force per unit area.

$$\text{Pressure or stress} = \frac{\text{Force}}{\text{Area}} \text{(N/m}^2\text{)}$$

Primary cell	A primary cell cannot be recharged. Once the active chemicals are exhausted, the cell must be discarded.
Protective equipotential bonding	This is equipotential bonding for the purpose of safety.
PVC/SWA cable installations	Steel wire armoured PVC insulated cables are now extensively used on industrial installations and often laid on cable tray.
Reasonably practicable or absolute	If the requirement of the regulation is absolute, then that regulation must be met regardless of cost or any other consideration. If the regulation is to be met 'so far as is reasonably practicable', then risks, cost, time, trouble and difficulty can be considered.
Relay	A relay is an electromagnetic switch operated by a solenoid.
Resistance	In any circuit, resistance is defined as opposition to current flow.
Resistivity	The resistivity (symbol ρ – the Greek letter 'rho') of a material is defined as the resistance of a sample of unit length and unit cross-section.
Risk assessments	Risk assessments need to be *suitable* and *sufficient*, not perfect.
Risk	A risk is the 'likelihood' of harm actually being done.
Rubber	Rubber is a tough elastic substance made from the sap of tropical plants.

Safety first – isolation	We must ensure the disconnection and separation of electrical equipment from every source of supply and that this disconnection and separation is secure.
Secondary cells	A secondary cell has the advantage of being rechargeable. If the cell is connected to a suitable electrical supply, electrical energy is stored on the plates of the cell as chemical energy.
Secure supplies	A UPS (uninterruptible power supply) is essentially a battery supply electronically modified to provide a clean and secure a.c. supply. The UPS is plugged into the mains supply and the computer systems are plugged into the UPS.
Security lighting	Security lighting is the first line of defence in the fight against crime.
Shearing force	Shearing force is the force which moves one face of a material over another.
Shock protection	Protection from electric shock is provided by basic protection and fault protection.
Short circuit	A short circuit is an overcurrent resulting from a fault of negligible impedance connected between conductors.
SI units	SI units are based upon a small number of fundamental units from which all other units may be derived.
Silicon rubber	Introducing organic compounds into synthetic rubber produces a good insulating material such as FP200 cables.
Simple machines	A machine is an assembly of parts, some fixed, others movable, by which motion and force are transmitted. With the aid of a machine we are able to magnify the effort exerted at the input and lift or move large loads at the output.
Single PVC insulated conductors	Single PVC insulated conductors are usually drawn into the installed conduit to complete the installation.
Skilled person	A skilled person is a person with technical knowledge or sufficient experience to be able to avoid the dangers which electricity may create.
Skirting trunking	Skirting trunking is a trunking manufactured from PVC or steel and in the shape of a skirting board which is frequently used in commercial buildings such as hospitals, laboratories and offices.
Socket outlets	Socket outlets provide an easy and convenient method of connecting portable electrical appliances to a source of supply.
Sounders	The positions and numbers of sounders should be such that the alarm can be distinctly heard above the background noise in every part of the premises.
Space factor	The ratio of the space occupied by all the cables in a conduit or trunking to the whole space enclosed by the conduit or trunking is known as the *space factor*.
Speed	Speed is concerned with distance travelled and time taken.

Spring toggle bolts

A spring toggle bolt provides one method of fixing to hollow partition walls which are usually faced with plasterboard and a plaster skimming.

Static electricity

Static electricity is a voltage charge which builds up to many thousands of volts between two surfaces when they rub together.

Statutory Regulation

Statutory Regulations have been passed by Parliament and have, therefore, become laws.

Step down transformers

Step down transformers are used to reduce the output voltage, often for safety reasons.

Step up transformers

Step up transformers are used to increase the output voltage. The electricity generated in a power station is stepped up for distribution on the National Grid network.

Switching for mechanical maintenance

The switching for mechanical maintenance requirements is similar to those for isolation except that the control switch must be capable of switching the full load current of the circuit or piece of equipment.

Synthetic rubber

Synthetic rubber is manufactured, as opposed to being produced naturally.

Tensile force

Tensile force is the force pulling things apart.

Thermoplastic polymers

These may be repeatedly warmed and cooled without appreciable changes occurring in the properties of the material.

Thermosetting polymers

Once heated and formed, products made from thermosetting polymers are fixed rigidly. Plug tops, socket outlets and switch plates are made from this material.

Time sheets

A time sheet is a standard form completed by each employee to inform the employer of the actual time spent working on a particular contract or site.

Transformer

A transformer is an electrical machine which is used to change the value of an alternating voltage.

Trap protection

Trap protection places alarm sensors on internal doors and pressure pad switches under carpets on through routes between, for example, the main living area and the master bedroom.

Trunking

A trunking is an enclosure provided for the protection of cables which is normally square or rectangular in cross-section, having one removable side. Trunking may be thought of as a more accessible conduit system.

Velocity

In everyday conversation we often use the word velocity to mean the same as speed, and indeed the units are the same. However, for scientific purposes this is not acceptable since velocity is also concerned with direction.

Visual inspection

The installation must be visually inspected before testing begins. The aim of the visual inspection is to confirm that all equipment and accessories are undamaged and comply with the relevant British and European Standards, and also that the installation has been securely and correctly erected.

Weight

Weight is a measure of the force which a body exerts on anything which supports it. Normally it exerts this force because it is being attracted towards the earth by the force of gravity.

Work done

Work done is dependent upon the force applied times the distance moved in the direction of the force. Work done = Force × Distance moved in the direction of the force (J). The SI unit of work done is the newton metre or joule (symbol J).

Index